Nuclear Instrumentation and Control

Umapathy Mangalanathan ·
Akhilanand Pati Tiwari ·
Uma Gandhi
Editors

Nuclear Instrumentation and Control

Springer

Editors
Umapathy Mangalanathan
Department of Instrumentation and Control Engineering
National Institute of Technology Tiruchirappalli
Tiruchirappalli, India

Uma Gandhi
Department of Instrumentation and Control Engineering
National Institute of Technology Tiruchirappalli
Tiruchirappalli, India

Akhilanand Pati Tiwari
Indian Institute of Technology Mandi
Kamand, Himachal Pradesh, India

ISBN 978-981-97-1282-3 ISBN 978-981-97-1283-0 (eBook)
https://doi.org/10.1007/978-981-97-1283-0

This Springer imprint is published by the registered company Springer Nature Singapore Pte Ltd.
The registered company address is: 152 Beach Road, #21-01/04 Gateway East, Singapore 189721, Singapore

Preface

Nuclear reactors harness the energy released in nuclear fission of heavy nuclei such as the uranium and plutonium nuclei. A number of nuclear reactors of different designs, physical sizes, and capacities are operational in the world, generating nuclear energy for electrical power production in a nuclear power plant (NPP), while many others are used for nuclear research and production of isotopes for numerous applications, e.g., for medical diagnosis and therapy, agriculture, and food preservation, to name a few. The operation of the first nuclear reactor was demonstrated in the early 1940s, and the first commercial nuclear power station started operating in 1950s. The knowledge in the field of nuclear science and engineering has grown ever since. This monograph is a compilation of the advanced knowledge contributed through recent research and experience in mathematical modeling, instrumentation and control design, and operation of nuclear reactors in India. However, some fundamentals have also been included with the intent to make the book self-contained. In it, there are 11 chapters covering radiation monitoring, neutron detectors, thermal and fast reactor instrumentation, gamma detectors for material characterization and imaging, nuclear reactor kinetics and control.

Radiation monitoring systems play a critical role in protecting human health, safeguarding the environment, ensuring regulatory compliance, enhancing emergency preparedness, supporting medical applications, strengthening nuclear security, facilitating research, and advancing public awareness and education about radiation risks and safety measures. By continuously monitoring radiation levels and assessing potential risks, these systems contribute to the safe and responsible use of ionizing radiation in various applications and industries. A comprehensive exploration of radiation monitoring systems in NPPs which includes monitoring locations, types of monitoring, and post-accident monitoring systems, is presented in the monograph.

Measurements in nuclear facilities are of utmost importance for ensuring safe, efficient, and reliable operation as they provide essential information for monitoring reactor conditions, controlling power generation, and maintaining safety systems. Number of parameters that are important for the control and safety of a reactor, include neutron flux, primary coolant inlet and outlet temperatures, coolant channel temperature, moderator temperature, fuel surface temperature, primary system pressure, steam pressure, primary coolant flow rate, steam flow, feed water flow, control, absorber rod position, etc. Among these parameters, the neutron flux is the most important signal in operating the reactor. The need of measuring neutron flux, principles, and types of neutron detectors and its associated signal processing and instrumentation are exclusively brought out in this monograph.

From the inception of nuclear energy, role of the sodium cooled fast reactor (SFR) and its fuel cycle has been recognized for long-term sustainability of nuclear power. SFR technology provides large measures of reactor safety, large thermal efficiency, and high fuel breeding ratio. This leads to better fuel utilization and therefore better economics in power production. A chapter in this monograph deals with various measurement techniques specific to SFR. It also presents the guidelines for designing and deploying safety instrumentation and control systems for SFR.

Radiations are extensively used for advanced materials characterization, medical diagnosis, and specialized medical treatments. Two chapters in this book, provide an overall view of the developments and applications of gamma rays for different advanced materials' characterization studies and also for the studies of biological problems. Scintillation detectors work on the principle of production of visible light due to the interaction of radiation within the active volume of the detector followed by the detection of the light by a photosensor. Inorganic scintillation detectors, in particular, are well known to detect gamma rays with better efficiency and reasonably good resolution. A chapter in this book deliberates upon a few measurement techniques to study the energy resolution and detection efficiency of inorganic scintillation detectors for gamma rays.

Mathematical modeling is an important step in study of physical systems. With this motivation, the point kinetics model of nuclear reactors and its representation in the standard state-space form which is suitable for control analysis and design is presented in this monograph. Further, the reactor behavior has been studied with the nonlinear and linearized models. Interval techniques have been used in designing robust controller for the reactors under uncertainties in modeling and measurement parameters. Interval-based controller design for both the pressurized heavy water and the pressurized water reactors are presented. The recent developments in robotic systems have attracted significant interest in both new and legacy builds in nuclear applications. The control of robots in complex hazardous nuclear environments needs to be developed to overcome its many challenges. A chapter in this book focusses on safety-related control systems involving robots in the nuclear sector.

While writing the chapters, the language has been kept simple and disposition sufficiently detailed so as to be able to disseminate the knowledge effectively to a student not having exposure to specialized knowledge in nuclear science and engineering. As a result, it is expected to be useful to a wide class of learners—undergraduate and graduate students as well as the practicing nuclear engineers and scientists.

Tiruchirappalli, India
March 2024

Umapathy Mangalanathan
Akhilanand Pati Tiwari
Uma Gandhi

Acknowledgements

This monograph is an outcome of the project "Self-powered sensor system for condition monitoring of nuclear waste packages" under the Scheme for Promotion of Academic and Research Collaboration (SPARC), Ministry of Education, Government of India. We would like to thank the Ministry of Education, Government of India, for granting the project, which gave the opportunity to work on this monograph. We wish to place our deep sense of gratitude to the National Institute of Technology (NIT), Tiruchirappalli, India; Homi Bhabha National Institute, Mumbai, India; Indian Institute of Technology Mandi, India; the University of Bristol, Bristol, United Kingdom; the University of Manchester, Manchester, United Kingdom; and Thiagarajar College of Engineering, Madurai, India, for their collaboration in this project. We further wish to register our hearty thanks to Dr. Suresh Kaluvan, postdoctoral researcher, University of Bristol, Bristol; Dr. Robbie Mackenzie, senior research fellow, University of Bristol, Bristol; Dr. Ramya A and Dr. Amit Pal, former research scholars of the Department of Instrumentation and Control Engineering, NIT, Tiruchirappalli, for their research contribution.

We thank all the contributors for spending their valuable time in writing and for their timely response. We wish to thank all the reviewers who read the chapters and made invaluable suggestions. Our sincere thanks are due to Ms. D. Anbu Karthika, academic-administrative assistant, Department of Instrumentation and Control Engineering, NIT, Tiruchirappalli, who copyedited and formatted all the chapters received from the individual contributors. We also express our gratitude to Dr. B. Vasuki, professor, Department of Instrumentation and Control Engineering, NIT Tiruchirappalli; and Mr. M. Ganaadhiban and Ms. Elsa Sharu Johnson for extending their help in formatting process. Finally, we thank the entire team of Springer publications for their cooperation and encouragement in bringing out this monograph.

Tiruchirappalli, India
March 2024

Umapathy Mangalanathan
Akhilanand Pati Tiwari
Uma Gandhi

Contents

Editors and Contributors

About the Editors

Dr. Umapathy Mangalanathan was born in Ramanathapuram, Tamil Nadu, India, in 1967. He received the bachelor's degree in instrumentation and control engineering from the Government College of Technology, Coimbatore, in 1988; master's degree in precision engineering and instrumentation from the Indian Institute of Technology Madras, Chennai, in 1990; and the Ph.D. degree in systems and control engineering from the Indian Institute of Technology Bombay, Mumbai, in 2001. He was Graduate Engineer Trainee in Neyveli Lignite Corporation (NLC) India Limited, Neyveli, for a year and served as Scientist with Defence Research and Development Organization (DRDO), Government of India, Pune, for six years. He joined the National Institute of Technology, Tiruchirappalli, India, as Assistant Professor, in 1996, where he is currently Professor Higher Administrative Grade (HAG) in the Department of Instrumentation and Control Engineering. His research interests include Sensor, Instrumentation, Applications of Smart Materials in Sensing and Control, and Energy harvesting. He received Eminent Engineer Award from IEI (India) Madurai local center in 2013, Best Teacher Award from NIT, Trichy, in 2007, Young Scientist Fellowship Award from TNSCST, Chennai, in 1999, and Commendation Certificate from DRDO, Ministry of Defence, in 1996. He has 220 publications that include book, book chapters, journal articles, and conference papers, also has three Indian patents. He also carried out 14 funded research and development projects from various funding agencies, guided 19 Ph.D. dissertations at NIT Trichy. He served as Dean, Research and Consultancy, Deputy Director of NIT, Trichy, and Head of the department of instrumentation and control engineering, NIT, Trichy.

Dr. Akhilanand Pati Tiwari, (B.S. Engineering, Ph.D. (Systems and Control Engineering), Fellow, Indian National Academy of Engineering), had a distinguished career of more than 37 years at Bhabha Atomic Research Centre, Trombay, Mumbai, India, and at the Homi Bhabha National Institute, Anushaktinagar, Mumbai, India.

He is currently serving as Visiting Professor, School of Computing and Electrical Engineering, Indian Institute of Technology Mandi, India. He has carried out research activities and projects in the areas of Control and Estimation with application to Small and Large Research and Power Nuclear Reactors. He has worked on Research Reactors, namely Dhruva, Purnima-III, and Kamini, and on power reactors, namely 500 MWe Pressurized Heavy Water Reactor (PHWR) units TAPS-3 and 4 and Advanced Heavy Water Reactor (AHWR). His work spans Mathematical Modeling, Simulation, Control Analysis and Design, and Process Instrumentation and Control Design. Among his significant contributions are development and deployment of Liquid Zone Control System and Reactor Regulating System for TAPS-3 and 4 PHWR units and Reactivity Meter for Small PWR-Type Reactors and Prototype Fast Breeder Reactor (PFBR), Kalpakkam. He teaches postgraduate courses 'Modern Control System Design and Simulation' and 'State-space Approach to Reactor Control,' offered at BARC Training School, Mumbai, under auspices of Homi Bhabha National Institute. Akhilanand has contributed several international journal publications, mostly in IEEE Transactions on Nuclear Science, Annals of Nuclear Energy and Nuclear Engineering and Design. He has delivered invited talks in IAEA Technical Meetings. His talks have featured the advanced work carried out by him on spatial control, reactivity measurement, and development of nuclear reactor simulators.

Dr. Uma Gandhi was born in Madurai, India, in 1967. She received the bachelor's degree in instrumentation and control engineering from the Government College of Technology, Coimbatore, in 1989; master's degree in instrumentation engineering from the Madras Institute of Technology, Anna University, Chennai, in 1992; and Ph.D. degree in instrumentation and control engineering from the National Institute of Technology, Tiruchirappalli, in 2009. She was Teaching Research Fellow with the Madras Institute of Technology from 1990 to 1993 and served as Senior Technical Officer in the Department of Science and Technology (DST) funded R&D project in Madras Institute of Technology, Anna University, from 1993 to 1994. She was Head of the Department of Instrumentation and Control Engineering, Sethu Institute of Technology, Madurai, India, from 1995 to 1999. She joined the National Institute of Technology, Tiruchirappalli, India, as Lecturer in 1999, where she is currently Professor in the Department of Instrumentation and Control Engineering. Her research interests include Design and Development of Instrumentation Systems, Design of Sensors and Actuators in Meso- and Micro-scale, MEMS, Energy Harvesting, Bio-medical Instrumentation, and AI for Bio-medical Applications. In 2001, she received Young Scientist Fellowship from Tamil Nadu State Council for Science and Technology under which she worked in the Thin Film Laboratory, Department of Instrumentation, Indian Institute of Science, Bengaluru. She has 150 publications that include book chapters, journal articles, and conference papers and also has three Indian patents. She also carried out 10 funded research and development projects from various funding agencies, guided 13 Ph.D. dissertations at NIT Trichy. She served as Head of the department of instrumentation and control engineering, NIT, Trichy, and currently serving as Dean, Institutional Development and Alumni Relations of NIT, Trichy.

Contributors

G. Anil Kumar Radiation Detectors and Spectroscopy Laboratory, Department of Physics, Indian Institute of Technology Roorkee, Roorkee, Uttrakhand, India

V. Balagi Electronics Division, ESC Building, Bhabha Atomic Research Centre, Mumbai, India

Shohan Banerjee School of Built Environment, Engineering, and Computing, Leeds Beckett University, Leeds, UK

Debayan Bose Department of Computer Science Engineering, IIT Kharagpur, Kharagpur, India

R. Govindaraj Materials Science Group, Indira Gandhi Centre for Atomic Research, Kalpakkam, Tamil Nadu, India

Amitava Gupta Department of Power Engineering and School of Nuclear Studies and Applications, Jadavpur University, Kolkata, India

Guido Herrmann Department of Electrical and Electronic Engineering (EEE), University of Manchester, Manchester, UK

Suresh Kaluvan University of Bristol, Bristol, UK

Umasankari Kannan Retired, Reactor Physics Design Division, Bhabha Atomic Research Centre, Mumbai, India;
Former Senior Professor, Homi Bhabha National Institute, Mumbai, Maharashtra, India

Erwin Jose Lopez Pulgarin Department of Electrical and Electronic Engineering (EEE), University of Manchester, Manchester, UK

G. R. Mackenzie University of Bristol, Bristol, UK

Suriya Murthy Nagamani Indira Gandhi Centre for Atomic Research, Kalpakkam, India

S. Panwar Radiation Detectors and Spectroscopy Laboratory, Department of Physics, Indian Institute of Technology Roorkee, Roorkee, Uttrakhand, India

V. Ranga Radiation Detectors and Spectroscopy Laboratory, Department of Physics, Indian Institute of Technology Roorkee, Roorkee, Uttrakhand, India

Sudipta Samanta Reactor Physics Design Division, Bhabha Atomic Research Centre, Mumbai, India

Tom Scott University of Bristol, Bristol, UK

Robert Skilton Robotics, Repurposing & Decommissioning Directorate (RRDD), United Kingdom Atomic Energy Authority, Abingdon, UK

Sethumadhavan Sridhar Bharatiya Nabikia Electricity Limited (BHAVINI), Kalpakkam, India

V. D. Srivastava Electronics Division, ESC Building, Bhabha Atomic Research Centre, Mumbai, India

A. P. Tiwari School of Computing and Electrical Engineering, Indian Institute of Technology Mandi (IIT Mandi), Kamand, Mandi, HP, India

Kaiqiang Zhang Remote Applications in Challenging Environments (RACE), United Kingdom Atomic Energy Authority, Abingdon, UK

Abbreviations

ABT	Availability-Based Tariff
AHWR	Advanced Heavy Water Reactor
ALARA	As Low As Reasonably Achievable
AMAD	Activity Median Aerodynamic Diameter
AMS	Aeroball Monitoring System
ANNs	Artificial Neural Networks
ANSI	American National Standards Institute
APR	Advanced Power Reactor
BCR	Backup Control Room
BDBA	Beyond Design Basis Accidents
BIBO	Bounded-Input, Bounded-Output
BLE	Bluetooth Low Energy
BNP	Breakdown Noise Pulses
BOEC	Beginning Of Equilibrium Core
Bq	Becquerel
BWRs	Boiling Water Reactors
CAM	Continuous Air Monitoring
CANDU	Canada Deuterium Uranium
CB	Control Building
CCT	Compton Coincidence Technique
CDA	Clad Disruptive Accident
CFDD	Constant Fraction Differential Discriminator
CIC	Compensated Ion Chamber
COTS	Commercial Off-The-Shelf
CPLD	Complex Programmable Logic Devices
CPM	Counts Per Minute
CPS	Counts Per Second
CR	Control Rod
CTBTO	Comprehensive Test Ban Treaty Organization
DAC	Derived Air Concentration
DBS	Design Basis Accidents

NPPAs	Nuclear Power Plant Analyzers
OBE	Operation Basis Earthquake
OBT	Organically Bound Tritium
PAL	Programmable Array Logic
PAMS	Post-accident Monitoring Systems
PAS	Positron Annihilation Spectroscopy
PCM	Per Cent Mille
PET	Positron Emission Tomography
PHT	Primary Heat Transport
PHWR	Pressurized Heavy Water Reactor
PLA	Programmable Logic Arrays
PMFM	Permanent Magnet Flowmeter
PMT	Photomultiplier Tube
PPE	Personal Protection Equipment
PSAR	Preliminary Safety Analysis Report
PWR	Pressurized Water Reactor
QA	Quality Assurance
RAIN	Robotics and Artificial Intelligence for Nuclear
RBMK	Reaktor Bolshoy Moshchnosty Kanalny
RCB	Reactor Containment Building
RFI	Radio-Frequency Interference
ROS	Robotic Operating System
RRS	Reactor Regulating System
SAMG	Severe Accident Management Guidelines
SANS	Small-Angle Neutron Scattering
SAXS	Small-Angle X-ray Scattering
SCRAM	Safety Control Rod Axe Man
SFAC	State Feedback-Assisted Control
SFR	Sodium-Cooled Fast Reactors
SID	Sodium Ionization Detector
SIP	Sputter Ion Pumps
SMC	Sliding Mode Control
SMR	Small Modular Reactor
SOPs	Standard Operating Procedures
SOR	Shut-off Rod
SPND	Self-Powered Neutron Detectors
SRF	Self-Resonant Frequency
SS	Stainless Steel
SSC	Systems Structures and Components
SSE	Safe Shutdown Earthquake
TAC	Time-to-Amplitude Convertor
TARM	Telescopic Articulated Remote Mast
TCH	Time-slotted Channel Hopping
TDPAC	Time Differential Perturbed Angular Correlation
TEDA	Triethylene Diamine

TES	Transition Edge Sensor
TFT	Thin-Film Transistor
TIPs	Traveling In-core Probes
TLD	Thermoluminescence Dosimetry
TPMS	Thermal Power Measurement System
TTL	Transistor–Transistor Logic
UIC	Uncompensated Ion Chamber
UKAEA	UK Atomic Energy Authority
VVERs	Voda-Voda Energo Reactor

Chapter 1
Radiation Monitoring Systems-Current Trends

Suriya Murthy Nagamani

1.1 Introduction

Radiation is a fact of life; it is all pervasive and remains all the time. Cosmic rays and terrestrial radiation are major sources of natural radioactivity coupled with the presence of K-40 radioisotopes inside the human body and a few radioisotopes in food stuff. Conversely, the anthropogenic radiations are mainly from the operation of nuclear power plants, reprocessing of irradiated spent fuel and from associated radiological facilities. The quest for energy for human development has necessitated deriving electricity from non-conventional as well as conventional sources comprising nuclear energy depending on the social, economic, and political considerations of the country. Electricity generating stations that are fuelled with fissile isotopes are designed to efficiently use fissile materials like (U-235, Pu-239, U-233) as nuclear fuel to generate thermal energy which is eventually converted to electrical energy with some defined efficiency. Owing to the risk associated with fission products and activation products that are resulting from nuclear fission, the nuclear safety assumes the utmost importance by the operator and regulator to ensure safeguarding people and the environment.

The progress of nuclear energy for any application depends on consistent and demonstrated expertise in safe operation of NPP by adhering to ten fundamental principles of nuclear safety. Nuclear safety is defined by the International Atomic Energy Agency (IAEA) as "The achievement of proper operating conditions, prevention of accidents or mitigation of accident consequences, resulting in protection of workers, the public and the environment from undue radiation hazards" [1]. Ensuring nuclear safety requires the availability of suitably qualified staff, the establishment of an effective safety culture, and the operation of appropriate safety systems. Radiation monitoring systems play a vital role in monitoring the various parameters that con-

S. M. Nagamani (✉)
Indira Gandhi Centre for Atomic Research, Kalpakkam, India
e-mail: suriya@igcar.gov.in

U. Mangalanathan et al. (eds.), *Nuclear Instrumentation and Control*,
https://doi.org/10.1007/978-981-97-1283-0_1

trol the safe operation of a plant, for example, core monitoring for determining the neutron flux and ensuring the sustained chain reaction. Radiation control at nuclear power plants is performed at various plant states including operating and shutdown. Radiation monitoring is imperative not only during operation of the facility but also during shutdown conditions of Nuclear Power Plant (NPP) owing to the safe operation of the reactor and for protecting people and the environment. The concept of 'Defence in depth' [2] is paramount to the nuclear safety which provides five levels of protection to meet the objectives of nuclear safety, and all levels have defined acceptance criteria with an aim of reducing the probability of harm. Controlling the operation of a nuclear facility is achieved in accordance with a set of operational limits and conditions which are defined in safety analysis to identify safe boundaries of operation. Nuclear instrumentation systems in a NPP are analogous to nerves in the human body for safe, reliable, and cost-effective operation of the NPP. The plant operation for various monitoring including the parameters of core monitoring, neutron flux and distribution, radioactivity levels in primary coolant, moderator, steam, and presence of airborne activity and surface contamination inside the reactor containment building during various plant states to ensure safe operation.

With the advent of modern electronics, novel detector materials, miniaturization of detectors, information and data processing technology, the current trend favors the development and deployment of cheaper, faster, integrated, and smart radiation monitoring systems that can be relied upon for safe operation and maintenance of nuclear power plants.

1.2 Role of Radiation Monitoring in Nuclear Safety

Nuclear safety regulatory requirements demand that a licensee (NPP) is responsible for arrangement of adequate and appropriate monitoring systems to ensure that the environmental discharges are within the stipulated limits. This criterion requires that the licensee shall monitor, control, and record the discharges into the environment. Radiation monitoring in various forms is widely implemented in all controlled areas of the radioactive facility. For a Nuclear Power Plant, it requires installation and operation of complex systems, whereas for a laboratory a few numbers of rudimentary monitoring systems are sufficient [3]. The following are the purposes for which the radiation monitoring instruments are deployed,

- Dose rate measurement inside the controlled areas of various nuclear facilities, including in spent fuel and fresh fuel storage areas.
- Sampling of primary coolant, secondary coolant, and gaseous specimens for determining the specific activity levels.
- Sampling and assay of radionuclides present in the air and liquid stream of the reactor containment building and determination of their energy and activity concentrations.

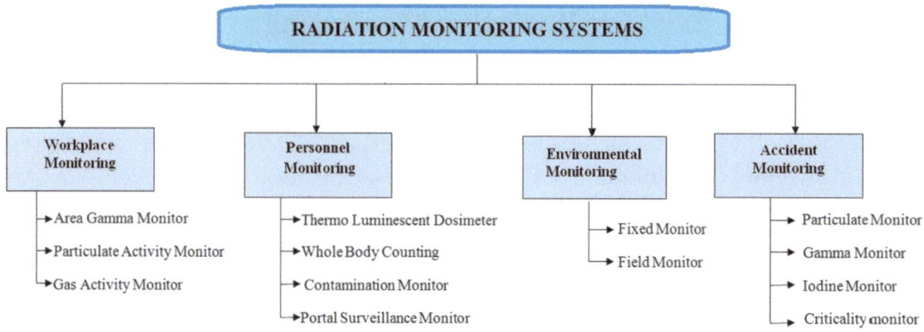

Fig. 1.1 Types of radiation monitoring systems

- Online monitoring of radioactive discharges and collection of samples from the stack, and the quantification of radioactive concentrations using instruments in a laboratory.
- Workplace monitoring involving the dose rate that may give rise to external exposure.
- Air contamination and surface contamination measurements at the workplace.
- Ventilation and duct monitoring.
- Post-accident monitoring.
- Process monitoring.

The radiation monitoring systems for different monitoring strategy as mentioned Fig. 1.1 shall remain operational until decommissioning in commensurate with monitoring requirements.

1.3 Locations for Radiation Monitors in NPP

- Reactor Containment Building
- Reactor Vault area
- Secured storage area meant for fresh fuel with nuclear materials accountancy
- Irradiated fuel storage area and its associated handling machine
- Radioactive waste storage and treatment
- Steam Generator
- Inter-zonal area and exit
- Decontamination unit
- Ventilation ducts
- Stack
- Environment.

1.4 Technical Basis for Radiation Monitoring

Prior to establishing a monitoring programme, it is necessary to identify the appropriate quantities and equipment for monitoring external radiation, surface contamination, and airborne contamination based on the determination of radionuclide inventory and its relative concentrations (nuclide vectors). The monitoring programme should make a clear distinction between the monitoring performed for the 'purpose of controlling operations' and monitoring performed for the 'formal assessment of exposure' to meet regulatory requirements [4].

To select an appropriate monitoring program and techniques, the nuclide vectors and their relative concentration in a nuclear facility should be determined. For this purpose, the following sources of information may be useful

- Operational history of the facility
- Modeling using computer codes
- Alpha, beta, and gamma spectrometry of representative samples
- Radiochemical analysis of representative samples of the radioactive material
- Any combination of the above.

1.5 Equipment Selection

Selection of equipment for different kinds of radiation monitoring plays a vital role in compliance with the monitoring programme [5] in tune with regulatory requirements. The following factors are to be considered which affects the choice of monitoring instruments. These are:

- Quantities to be measured are dose rate, surface or airborne contamination, environmental radioactivity, and respective indications whether in counts per second, Bq/cm^2 or $\mu Sv/h$ or Bq/m^3
- Type of radiation and expected energy range of the relevant radionuclides
- Type testing by the supplier or an authorized agency to demonstrate its suitability to perform satisfactory measurements in the workplace
- Periodic calibration of the equipment to test the precision and accuracy of results
- Detection efficiency for the radionuclides(s) of concern, P-factor and surface emission rate
- Adequate range of dose rates or surface or airborne contamination levels to detect what could be practically encountered. This should cover the required minimum detectable dose rate or surface contamination and airborne contamination level up to the maximum values
- Sensitivity of measurement required, in terms of the Minimum Detectable Activity (MDA) for contamination, the lowest rate for dose rate measurement, and the fraction of a DAC required to be monitored for airborne monitoring

- Whether an instrument can perform more than one function adequately, or whether more than one instrument is required to measure different types of radiation or quantities
- If the monitor is exposed to high radiation intensity, the indication should remain high and the reading should go 'off scale'
- Sensitivity to interfering radiations present (e.g., radio frequency)
- For airborne contamination, the dust levels in the area to be measured
- Speed with which an instrument responds
- Convenience of access for monitoring
- Potential for using installed, rather than hand held equipment
- Possibility of continuous monitoring operation
- For installed equipment particularly, ease of installation and maintenance
- Factors such as handle design, switch accessibility
- Logarithmic/linear/analog scales or digital displays, and ease of use
- Backlit display and/or sufficiently audible output
- Susceptibility to room temperatures, moisture, radiofrequency, magnetic fields
- Inherent safety while operating in a harsh atmosphere like in explosive and combustible locations
- Ability to perform easy decontamination
- Gas supply requirements
- Availability and life expectancy of battery
- Availability of spare parts and serviceability
- Size, weight, and portability of equipment
- Ruggedness, dependability, and ease of servicing
- Initial investment on-going maintenance cost and retiring the instruments.

1.6 Types of Required Monitoring Systems

Radiation monitoring systems are broadly classified into two types, namely: stationary monitoring systems and portable monitoring systems. They should be sensitive, rugged, and reliable to be able to perform in a radiation environment. The desired types and functions are outlined in Table 1.1 [6].

1.6.1 Stationary (Fixed) Monitoring Systems

The pathways of radioactivity release in the work environment require constant monitoring in view of potential risk to occupational workers who are undertaking the radiological work. The extent of such exposure varies with occupancy time and prevailing dose rates. The continuous workplace monitoring for identification of hazards is recommended by regulators and implemented by operators to keep the radiation

Table 1.1 Important types of radiation monitoring instruments

Type	Instrument name	Applications	Radiation measured	Measurement range
Exposure rate and count rate monitoring equipment	Ionization chamber type survey meter	Instantaneous dose rate measurement	X-rays, γ rays, β rays	1 μSv/h–30 mSv/h
	Scintillation type survey meter	Low level radiation monitoring and searching	α rays, β rays, γ rays	1–10^4 counts/s
	GM type survey meter	Measurement of streaming gamma dose rates and surface contamination of beta	β rays, γ rays	1–10^5 counts/m
	Neutron REM counter	Measurement of leaking neutron doses	Neutrons	0.1 μSv/h–10 mSv/h
Personal dosimeter and portable monitoring equipment	Personal dosimeter	Control of personal exposure at nuclear power facilities	β rays, γ rays, neutrons	0.01–1000 mSv/h
	Portable monitoring devices	Outdoor environmental dosimetry	γ rays	10–10^8 nGy/ h
	Low dose Environmental dosemeter	Environmental dosimetry	γ rays	0.01–999 mSv/ h
Detectors	Semiconductor area monitor detector	Monitoring and measurement of air gamma doses at radiation facilities	γ rays	0.1 μSv/h–10 mSv/h
	Air activity monitor semiconductor α-ray and β-ray detector	Assay of alpha and beta rays contained in airborne dust by air filters	α rays, β rays	1–10^5 counts/m
	Scintillation detector	Detector housed in a process monitor	γ rays	Depending on application
	γ-ray ionization chamber detector	Measurement of air gamma doses	γ rays	Depending on application
	High temperature fission chamber	Monitoring the core neutron flux at various power levels	Neutron	0.2 cps/nv
	Boron coated proportional counter	Monitoring neutron for start-up, intermediate and high power levels	Neutron	12 cps/nv

exposures ALARA (As Low As Reasonably Achievable). This monitoring system gives an early warning of possible system malfunction in the plant, warns the operating personnel in case of increasing radioactivity, and prevents the inadvertent release of radioactivity to the environment. The quantities like ambient dose equivalent rate due to gamma and neutron radiation fields, airborne concentration of alpha and beta particulates, and contamination levels on the surface are measured by stationary monitors. Process radiation monitors, ventilation duct monitors, and area radiation monitors are examples of stationary monitoring systems.

1.6.2 Portable Monitoring Systems

Portable instruments are immensely useful where stationary monitors cannot be located in intricate locations in a nuclear facility. This is required to measure the ambient gamma radiation levels in the working areas and plant rooms at specific locations and for specific works. Some of the instruments do have a telescopic arm extendable up to 4 m in length to enable remote measurements, thereby reducing the exposure of health physicists who are performing the radiation survey. Any facility should have an adequate number of handheld measuring instruments to measure external dose equivalent rates having a maximum range of not less than 10 Gy/h (Sv/h) for measurements carried out for post-accident monitoring purposes. Handheld and portable gamma, neutron survey meters, and contamination monitors are examples of portable monitoring systems.

1.7 Characteristics of Radiation Monitoring System

There are several state-of-the-art equipment available for the purpose of detecting and measuring the type and quantity of radiation, data logging, actuations of alarm, and triggering control actions. Radiation monitoring entails characterizing the radiation fields in terms of type and intensity of radiation and energy information in the vicinity of a radiation source and also in the environment. They should have the following properties:

Measurement range: Radiation monitoring instrument intended for use in radiation environment shall have the capability to cover the measurement range required for plant states like operation, fuel handling, shutdown, and accident situations at the nuclear facility.

- A monitoring instrument shall have the capacity of indicating the maximum value of the measuring range, even though the maximum value of the measuring range is surpassed. The monitor shall not fall back to zero on overload.

Energy response: The monitoring instruments shall respond appropriately to the energy of radiation encountered in the workplace. They should fulfill the requirements specified in normative documents such as ISO/IEC standards generally cover the energy range between 80 keV and 1.5 MeV. For reactors having the sources of N-16 radionuclide which emits high-energy gamma of 7.1 MeV, an appropriate monitoring system sensitive to high-energy gamma should be deployed.

Display: The monitoring equipment shall have the proper display indication which guide us to track the variation in the magnitude of measured response in tune with the requirements of the design standards.

- In case of the presence of mixed radiation fields like beta, gamma, and neutron at the point of measurement, the response and effect on the measurement shall be inferred.
- If the instrument is designed to measure the air activity concentration and surface contamination, the appropriate compensation for contributing background shall be considered. The measuring ranges shall exceed the contamination limit values set for the highest class in the controlled area by at least a factor of ten.
- For personal monitoring, the energy response and the measuring accuracy of the radiation dosimeter for occupational radiation exposure in real time shall fulfill the requirements imposed on instruments for monitoring external gamma radiation.
- Radiation monitoring data provided by installed radiation monitors shall be displayed in the main plant control room, and the data should be grouped depending on the purpose of use.
- Data logging shall be done in an uninterrupted manner in such a way that any operational events and accidents at the plant can be analyzed afterward.
- Nuclear data pertaining to the radioactive samples analyzed in the laboratory need not be fed to the control room. However, the data logging shall be archived at a locally available server that may help radiation protection professionals for future analysis.

1.8 Workplace Monitoring

Workplace monitoring refers to comprehensive radiological characterization of the workplace. The objectives of workplace monitoring are to control the radiation exposures to occupational workers and comply with regulatory requirements.

Monitoring of the workplace may be used for different purposes, which include:

(a) Set up and evaluation of the classification of controlled and supervised areas.
(b) Determination of radiological conditions in the workplace, such as expected radiation fields, surface contamination levels, airborne contamination levels, and their spatial and temporal variation.
(c) Supporting the prior radiological evaluation.

(d) Workers understand how, when, and where they are exposed, which can motivate them to take action and/or follow guidance/procedures to reduce their exposure.
(e) Confirmation of the effectiveness of working practices and applying the measures for preventing and controlling radiological hazards (e.g., measures like, adequacy of practices, procedures, supervision, housekeeping, training, and efficacy of engineering controls).
(f) Detection of failures of enclosures containing radioactive sources or departures from operating procedures.
(g) Provision of early warning of elevated gamma dose rate levels or release of airborne contamination.
(h) Provision of information for arranging individual monitoring programme, both for external and internal exposure.
(i) Contribution to the assessment of the exposure of individuals or groups of workers.
(j) Selection and recommendation of appropriate personal protective equipment.
(k) Verification of engineering and administrative controls to maintain the occupational doses ALARA.
(l) Provision of information for the evaluation of doses in the event of unexpected exposures.

For a sound radiation monitoring programme, it is necessary to use the appropriate quantities and equipment for external radiation, surface contamination, and airborne contamination based on the determination of radionuclide inventory and its relative concentrations, in addition to identify, install, and commission the suitable radiation monitoring systems in the workplace. Table 1.2 depicts the general characteristics of radiation detectors deployed in nuclear facilities.

During the routine regulatory inspections, the regulatory body verifies the operational performance of radiation monitoring systems by analyzing the key performance indicator (KPI) of the instruments. In this way, the regulatory body confirms the capability of the licensee to fulfill the safety requirements to assure the reliable and correct operation of the systems and equipment. The regulatory inspection is arranged in a defined periodicity.

1.8.1 Low Range Area Radiation Monitor-LRARM

Background: LRARM is designed and installed to determine the ambient dose equivalent rate due to the gamma radiation field in the controlled area of NPP. This type of monitor is provided to measure the radiological intensity of the gamma radiation field, which is prevalent in accessible areas of the reactor building, spent fuel storage building, radioactive waste building, and, in some cases, in the steam generator building. The LRARM measures the ambient dose equivalent rates and is designed to provide both audio and visual alarms locally in conjunction with a Flasher Hooter for warning the occupational workers whenever the pre-set limit is exceeded. These monitors also provide signal output so that a control room operator can take swift action in case of an alarm.

Description: The device may have an appropriate radiation sensor like energy-compensated GM detector having the energy response in the region of 100 keV–3 MeV. Being energy-compensated GM detector, this detector is capable of measuring only gamma dose rate. Both the detector and the Electronics Unit form an integral unit called the monitor. The measurement range of monitors covers from 1 μSv/h (0.1 mR/h) to 1000 μSv/h (100 mR/h). Two types of area gamma monitors are usually installed in an NPP, depending on the purpose of monitoring. The dose rate level is displayed on the front panel, and its value is compared against the pre-set limit, and a relay contact output is generated for alarm conditions. Relay contact output from this monitor is fed to a Flasher and hooter unit mounted adjacent to this monitor on a common stand. Digital data output is transmitted to the main control room through a local area network (LAN) to register the monitor readings continuously. Such types of radiation monitors are not only for routine area monitoring but also used to isolate the reactor containment building in the event of an elevated radiation field due to accident scenarios, to prevent the release of radioactivity into the outside environment.

Input power supply: It power supply specifications stipulate to have an input power supply of 240 V AC ± 10%, 50 Hz ± 5%. The monitor shall perform satisfactorily at the specified upper and lower limits of 240 V AC ± 10% power supply variation. The monitor as a whole or individual module shall withstand an interruption of input power supply of 240 V AC ± 10% for a duration of 120 ms. The monitor shall also operate satisfactorily for 30% dip in power supply voltage for a duration of 200 ms. Environmental conditions are also defined for area radiation monitors, and they

Table 1.2 Types of radiation detectors used for direct measurements of dose rate

Instruments	Applications	Minimum detectable dose rate (nGy/h)	Relevant comments
Geiger Muller survey meter	Locating contamination, dose rate estimates	100	Durability, reliability, energy dependent & linearity of instrument
Scintillation survey meter	Locating contamination, dose rate estimates	10	Good sensitivity; energy dependent
Scintillation dose monitor	Accurate dose rate measurements	10	AC power may be needed, portable but not a survey instrument
Pressurized ion chamber	Accurate dose rate measurements	10	Accuracy ±20% at 100 nGy/h, AC power may be needed, not a survey instrument

should operate in a temperature range from 8 to 45 °C and relative humidity of 55–85%. The monitor is expected to meet the requirements and perform satisfactorily up to 10^4 Rad of cumulative dose over the prolonged use. Radiation monitors are powered by the most reliable, uninterrupted Class-III power supply.

Monitors that are employed for reactor building isolation logic are protected with an IP65 enclosure for functioning under an adverse accident environment, having a high temperature, presence of a mixture of sodium aerosol, moisture, and sodium hydroxide in case of a fast reactor nuclear facility. This enclosure also protects the sensor and electronics from dust and oil. It is also ensured that the monitors responsible for the reactor safety function are meeting the functional requirements as per the seismic qualification test criteria.

The recommended electronics unit receives the signal in the form of a pulse train from the selected GM detector and gives readout in μSv/h. The unit shall be suitable for wall mounting along with a Flasher Hooter Unit on a common bracket. Suitable mounting brackets shall also be provided. The electronics unit shall consist of a low voltage Power Supply (LVPS), EHT supply, and Signal processing circuits. The required Direct Current voltage for the monitor is generated locally. Short circuit protection for the Direct Current power supply shall be provided. The manufacturer shall provide the complete specification for the power supply based on the requirements of the monitor.

Extra High Tension (EHT) is recommended in commensurate with the requirement of the detectors. Normally, it ranges from 500 to 1500 V for GM tube-based radiation detectors. Signal processing circuit receives signal pulses from the selected GM detector and shall be processed to provide dose rate display and signal output to paper-less recorder for reactor building isolation monitors and dose rate data is logged through DDCS for other monitors and relay contact output of the alarm for flasher hooter unit. A self-check circuit to ensure the healthiness of the detector shall also be provided. The units for RCB isolation shall incorporate an overload circuit (adjustable control) which will change the relay output at a activating point between the full scale reading of the counting instrument and the point where the meter drive circuit no longer increases when the field intensity is increased to ensure that monitor will respond even for count saturation condition. The signal output of isolation monitors of 0–5 V DC corresponding to 1.0 μSv/h to 10^3 μSv/h. The desirable response time for the detector is <5 s. One monitor of the type used in RCB isolation shall undergo a seismic test by performing the shake table test (reference) (1–33 Hz). After the seismic test, the monitor shall be tested for meeting all the functional requirements.

The isolation monitors that have a safety function to protect the nuclear reactor should have been tested for overload due to exposure to radiation. It is recommended to test the 10 times of maximum range of the monitor for 1 h. During this test, the monitor shall not fall back to zero. Functional testing is performed by testing the monitor in response to ISO certified [7] radioactive sources with known radiation field strength.

1.8.2 Gas Activity Monitors

National regulatory requirements and international guidelines demand that the radioactive release from a nuclear power plant through the stack shall be monitored and controlled by adhering to dose constraints and dose limits in order to ensure that both people and the environment remain protected. Therefore, sensitive and reliable monitoring systems are required to detect and quantify the gaseous activity of radionuclides that are released during operational plant states and accident conditions. These monitors are equipped to have real-time data logging, alarm systems, and initiation of control actions. Different measurement systems are used for different types of radionuclides. They are monitored continuously or intermittently, depending on the plant and process requirements. Generally, radioactive effluents are released either in airborne or in liquid form. Neutron interaction with oxygen in water, with nitrogen and argon in air, and with impurity corrosion elements results in the formation of activation products that are radioactive in nature. Radiological impact of short-lived fission products and short-lived activation products is insignificant in reactor effluents [8].

Radioactive gaseous originate from several sources within a nuclear plant are:

- Fission of tramp uranium.
- Leaks from failed fuel rods.
- Diffusion of radioactive gases through intact fuel rods.
- Activation of materials in reactor cooling water.

Nuclear fission of U-235 produces fission product noble gases (FPNG) like Kr-85, Kr-85m, Kr-87, Kr-88 and Xe-133, Xe-133m, Xe-135 and Xe-138. They are chemically inert and hence do not cause any internal exposure. The objectives of the noble gas monitor are

- To measure the activity concentration of radioactive noble gases and their variation with time and space in the workplace
- To perform the measurements during routine operational conditions as well as under emergency situations
- To actuate an alarm when a predetermined activity or concentration is exceeded
- To determine the total gaseous activity discharged
- To sample and retrospective analysis of air or gas containing noble gas.

Gaseous activity monitoring equipment monitors radioactive gas leakage from the block pile areas of the reactor core. The signal output from this monitor is compared with a pre-set level, and alarms are generated within this monitor when the signal level exceeds the set limits.

Equipment: Noble gas monitoring devices are designed to continuously monitor, quantify, and alert for the presence of beta, gamma radioactive noble gases in the working atmosphere or in ducts or in stacks. These monitors should have the capacity to compensate for ambient background samples and should be employed continuously and non-continuously, depending on the requirement. The sample is usually extracted from 1/3 of the stack height using a vacuum pump or from ventilation, or from the work atmosphere, and is allowed to pass through to the detection assembly. The monitor displays a measured volumetric activity and triggers alarms whenever pre-set threshold levels are exceeded. A representative gas activity monitor primarily consists of:

- an air sampling unit
- a flow meter and measurement cell
- a gamma compensation detector
- relevant electronics need
- data processing software, where appropriate
- hooter.

A pre-filter and a desiccant are placed upstream of the measurement cell in order to trap particulate and humidity from the air sampled. In addition, if iodine is expected to be present, the pre-filter shall have an iodine trap, like charcoal. It should be noted that iodine isotopes are decay precursors of noble gases, so the iodine trapped prevents a false indication of the noble gas measurement. In order to maintain the performance of the monitor, such a pre-filter shall not trap or temporarily retain noble gas or decrease the flow rate or the pressure inside the measurement cell over the limits specified by the manufacturer.

Since this measurement is of a gas, piping considerations are not as significant as for those described for alpha-beta airborne contamination. The use of a barometer device will verify that the pressure is in the normal range of operation. The choice of the instrument shall depend on its detection limit and the response time required.

If the monitor is not directly installed in the controlled area, the gas may be sampled and drawn from the controlled area to the measurement cell. In this case, the delay time between the sampling point and the measurement cell shall be taken into account due to the short half-lives of some of the isotopes. Noble gases do not have a tendency to get trapped on the wall of the sampling line.

Description: It consists of Scintillation Detector Assembly (Monitor range: 1–1000 μSv/h) with Photo Multiplier Tube, Pre-amplifier, Interconnecting Cables, and Electronic Processing Unit. Detector assembly mounted in a gas chamber with a volume of 2 L. Air is taken through a sampling pipe from the block pile area to this gas chamber and returns back to the block pile area. Electronic Unit is a Standard Electronic Bin, and this bin is housed in an enclosure as per IP 56 with a Hinged Door. IP 56 standard is used to ensure the safety of electrical enclosures by preventing the penetration of foreign bodies like tools, dirt, and moisture.

The monitor covers a 3-decade log range from 1 to 1000 μSv/h. Pre-amplifier output is transmitted to the Electronic Processing Unit, where the signal is processed and gives two high alarms on the front panel when the signal level exceeds the set level adjusted on the front panel. Digital Data Communication System (DDCS) connectivity is provided for soft transmission of data.

The monitor is intended for continuous use in round-the-clock operation. Similar to the LRARM monitor as mentioned in the previous paragraph, the power supply requirements are 240 V AC ± 10%, 50 Hz + 5% maintained. The monitor shall perform satisfactorily at the specified upper and lower limits of 240 V AC Power Supply variation.

Scintillation detector assembly: This consists of Scintillation Detector and Photomultiplier tube (PMT) (Integral assembly) and shall be suitable for detecting gamma radiation from 1 to 1000 μSv/h. The detector shall be of Integral Assembly consisting of a Scintillator optically coupled to a PMT. Detector assembly mounted in a gas chamber with a volume of 2 L. Suitable Pre-Amplifier shall be provided, and this shall be powered from the DC power supply unit of the Electronic Processing Unit. Detector along with the PM tube and Pre-Amplifier shall be suitably housed in an enclosure.

A chamber designed to collect the gas samples has a volume of 2 L which is adequate for measuring gas activity present in exhaust samples.

The electronic processing unit is built using a standard electronic bin and housed in an enclosure as per IP 56. Suitable cable glands are also provided for both 240 V AC power cable and signal cables. This unit consists of the following modules like: LVPS Module, EHT supply module, pulse amplifier & discriminator module, signal processing unit, and display module.

Pulse amplifier: Pulse amplifier is also equipped in this system, and this unit receives signal input from the pre-amplifier and shall have the provision for pulse shaping and amplification. Pulse amplifier input is compatible with pulse amplifier output. The nominal value of Gain is selected based on the detector output to obtain the required voltage level without saturation. Input of the discriminator shall be compatible with the pulse amplifier output.

The signal processing unit is mounted along with a scintillation detector assembly, LVPS module, EHT supply, linear pulse amplifier, and discriminator. The monitor covers 3-decade log range from 1 to 1000 μSv/h. The output of the discriminator is fed to the processor unit to count and display the dose rate.

Insulation Resistance Test: This test shall be carried out by shorting all the input and output connections, including the power supply line, neutral, and measuring the insulation resistance between the shorted terminals and chassis ground at 500 V DC. The insulation resistance shall be more than 100 MΩ.

Dielectric Strength Test: This test shall be carried out by shorting all the input connections, including power supply line and neutral terminals and measuring the insulation resistance between shorted terminals and chassis ground at 1500 V AC for 60 s. There shall be no flash over.

This test shall be carried out by shorting all the output connections and the insulation resistance between shorted terminals and chassis ground at 500 V AC for 60 s. There shall be no flash over. The monitor shall be capable of withstanding the above without any damage, temporary or permanent.

Functional Tests: All the monitors shall be Tested/Calibrated in known fields which shall be established by a certified source and calculations or by comparison with a standard instrument. All monitors shall be tested for the functional requirements given in this specification.

Burn-in Test: All the monitors shall undergo a burn-in test of 100 h. Duration at the end of which the monitors shall be tested for their conformity to functional requirements. One of the monitors of this type shall be tested as per EMI (Electromagnetic Interference)/EMC (Electromagnetic Compatibility) Test Requirements for Instrumentation & Control Systems.

Testing, Calibration and Maintenance

The fission product noble gas (FPNG) monitoring instrument is required to have a radiological and electronic calibration performed in conformance with national guidance. The obvious way to calibrate noble gas monitors is to use a known activity concentration of the relevant noble gas or gases and to circulate these through the monitor. This requires specialist equipment and is expensive. It is generally performed only during type testing. Type testing should be conducted in accordance with IEC 60761-1.

Functional testing requirements adopted for each instrument should be recommended by the manufacturer or supplier and usually include testing with a radioactive source. Routine (test before first use and periodic) testing generally uses sealed beta contamination sources of a defined nuclide and construction.

1.8.3 Iodine Monitoring

Radioiodine is generated in nuclear reactors as a fission product, and the radioiodine, particularly I-131 is a dosimetrically significant isotope. Coupled with its appreciable fission yield ($\sim$3%), its volatility increases the probability of leak of iodine isotopes into the working environment. Clad rupture occurs due to increased burn up of nuclear fuels inside the reactor due to irradiation, that may result in the release of radioiodine and other fission products into the coolant boundary, eventually to the workplace if the system is not adequately leak proof. Leakage of primary coolant to the containment and the auxiliary buildings will generate airborne iodine owing to vaporization. Spent fuel storage areas and a few areas of the reprocessing plants are the possible sources of radioiodine. Hazards and monitoring of iodine isotopes vary with the nature of the operation of the facility and the physical & chemical properties of iodine. Airborne iodine can occur in different chemical forms in the gaseous effluents from NPPs. Elemental iodine (I_2), organic iodine with methyl iodide (CH_3I) as the simplest organic compound, and hypo-iodous acid (HOI) may be present in significant amounts. In sodium-cooled fast reactors [SFR], the probability of iodine release is insignificant during normal conditions due to effective trapping of iodine by sodium with the formation of sodium iodide. Internal contamination monitoring of occupational exposure to radioiodine is of great concern in radiation protection, and therefore, effective workplace monitoring is required.

In workplace monitoring, the commonly monitored radio isotopes of iodine are I-125, I-131, I-133, and I-135. Besides workplace monitoring, iodine is also monitored at the stack to control the release. NPPs do have a certain apportionment for iodine release to the environment. Iodine reaches out to the public through the air-pasture-cow-human pathway. The thyroid gland has a strong affinity for Iodine molecules, and hence, in case of inhalation of radioiodine, it gets concentrated and delivers a thyroid dose to the exposed individual. Iodine is radiologically significant from the perspective of the workplace and environment and therefore requires continuous monitoring and compliance monitoring. The objectives of iodine monitoring are

- To measure the iodine activity and limit the release of radioiodine in the workplace.
- To recommend suitable PPE to control the internal exposure.
- To demonstrate compliance with authorized discharge limits for airborne radioactive contaminants and with self-imposed operating limits.
- To confirm and conform to regulatory requirements.

Techniques: In the nuclear engineering practices, gaseous form of iodine is the most predominant one and iodine exists as elemental form (I_2), Hydrogen iodide (HIO) , Organic (Methyl iodide) and/or Inorganic form (NaI/CsI). The workplace monitoring of radioactive iodine is achieved either by real-time sampling and monitoring or via sequential sampling and a delayed analysis in the laboratory. The monitoring device has a sampling assembly and a measurement assembly to measure the sampled activity. If the iodine is expected in aerosol form, the techniques for aerosol measurement are applicable. The iodine is generally trapped in a charcoal filter paper or charcoal impregnated with triethylene diamine (TEDA) or silver zeolite in a cartridge.

The most common and notable radioisotope I-131 is measured with a single-channel analyzer or a multichannel analyzer by gamma spectrometry. The latter gives the opportunity to discriminate between iodine isotopes and other radionuclides potentially present, e.g., Cs-137 and Cs-134. Single channel analysis normally uses a NaI(Tl) scintillator, whereas effective correction for any potentially interfering nuclides will require an HP-Ge detector to spectroscopically identify multiple isotopes of iodine. Equipment used for sampling should comprise:

- A pre-filter and a desiccant shall be placed upstream of the collection cartridge to trap particulate and humidity, respectively.
- A sampling assembly.
- A filtering/collection medium.
- Gamma ray detector and associated electronics.
- Alarm device.

The sampling collection is being done with a medium generally made of a cartridge filled with TEDA impregnated charcoal or silver zeolite. In this way, the collection efficiency is close to 100%. Silver zeolite should be used when measurement is required in the presence of noble gases since charcoal can absorb noble gases and impact the measurement result.

Equipment used for real-time measurement

The detection assembly provided for iodine monitoring is installed in the controlled area and stack room to measure and monitor the release of iodine to the workplace and environment, respectively. Adequate shielding has been provided to compensate for the ambient gamma background. The monitor provides the display, signal output, and data output for LAN connectivity so that the monitor's readings are observed in the control room.

This monitor consists of a charcoal-loaded filter cartridge which adsorbs the iodine gas, and a NaI(Tl) based scintillation detector coupled with a photomultiplier tube to estimate the activity on the charcoal sample. To ensure background reduction, a lead shielding assembly (100 mm thickness) has been provided, and the monitor readings are made available in the display unit. The detector operates in single-channel mode and is set for diagnosing the 364 keV gamma. Sample air for monitoring is drawn from the workplace/stack exhaust by means of vacuum pumps and connected to the monitor through SS tubing. The analyzed air samples are exhausted out to the effluent systems.
A typical iodine monitor consists of:

- a suction pump approximately 50 lpm
- charcoal filter in a plastic cartridge
- NaI(Tl) scintillator ($3'' \times 3''$) as gamma detector
- LV power supply module
- linear pulse amplifier module
- electronics and data processing software
- alarm devices with a display in the control room.

Provision shall be made to correct the decay of the adsorbed iodine in the filter. This provision shall be optional and shall be programmable through the keyboard while computing the iodine releases.

The integral release from the stack can be estimated by using

$$I = \frac{C}{\eta} \times \frac{F}{f}$$

I – integral release in Bq
C – counts per second
η – efficiency of counter given by cps/dps
F – Stack flow rate (m^3/h)
f – sample flow rate (m^3/h).

Testing, calibration and maintenance: Tests in IEC 60761-4 [9] are applicable to the measuring assembly. The efficiency of the detection assembly should be determined using simulated sources like Ba-133, which is traceable to national standards applicable for specific geometry. Functional testing of the measurement assembly should be carried out as per the instructions provided by the manufacturer in the operating manual. Due to the short half-life of I-131 (half-life 8 days), functional testing is typically performed with a standard radioactive source having gamma energies close to this; for example, Ba-133 (half-life 10.7 years) and Cs-137 (half-life 30 years). The sampling flow rate of the sampling assembly shall also be verified periodically. Maintenance checks will require periodic leak testing of transfer pipework, checks on filter seals, and integrity checks of the air pump supply. Maintenance checks are defined for sampling. Scintillation detectors will require periodic functional checks to identify loss of sensitivity due to ageing/accumulated dose or cracking due to temperature changes.

1.8.4 Particulate Airborne Activity Monitors

Among the several routes of exposure, inhalation of radioactive airborne dust particles is one of the most important routes of entry of radionuclides into the human body. Entry through inhalation is a relatively complicated process that depends on particle size distribution of the airborne particles, their dynamic behavior, Activity Median Aerodynamic Diameter (AMAD) in air, and the physical and chemical characteristics of the particles after deposition in the respiratory tract. In order to assess the workplace airborne concentration, thereby controlling the internal exposures to occupational workers, the air monitoring is performed.

Methods and Procedures

Regulatory requirements demand that handling, storage, and transport of radioactive materials/sources/fissile materials require a prior risk assessment performed preferably by a Radiation Protection Officer to identify all reasonably foreseeable inhalation and other exposure pathways. It is recommended that the conclusions from the risk assessment be formally recorded, as these will form the technical basis for monitoring use. Migration routes from the potential sources listed above should also be identified. Examples of potential sources (these may be routine or unexpected occurrences) which would result in reasonably foreseeable inhalation exposure pathways and the need for monitoring include:

- Planned activities leading to aerosol generation
- Standard leaks of systems containing radioactive contaminated substances
- Filter change operations
- Failure of pressurized containment
- Gloves in glovebox operations
- Outside of fume hoods
- Transfer of air through airlocks
- Ventilation failures
- Decontamination or maintenance activities

Table 1.3 Air sampling recommendations

Potential annual intake as a fraction of I_{inh}, L	Sampling recommendations
<0.02	Since the potential for intake is too low from this infinitesimal fraction, the air sampling is generally not necessary. However, a defined routine survey is recommended to confirm that contamination levels remain low. A static sampler may be used to verify conditions
≥0.02 and <1.0	Sampling is required. Intermittent samples are suitable near the lower end of the range, depending on the nature of operations. Standards specify that continuous sampling is recommended if activity concentrations exceed 0.1 Derived Air Concentration (DAC) averaged over 40 h or longer. Continuous monitoring with alarm capability is considered where activity concentrations are likely to exceed 1 DAC averaged over 40 h
≥1.0	Continuous monitoring with alarm capability is recommended. Continuous air monitors should be located at each exposure pathway, which has been identified from the risk assessment that, as a result of a breakdown in contamination control or containment, a worker would have to carry out an immediate action to restrict his personal exposure or an action to limit the consequences to the working environment

- Drilling operations
- Use of vacuum cleaners
- Resuspension of spills and leaks (Table 1.3).

Locations

Once the requirements for continuous air monitoring have been determined, it is necessary to identify the position and number of monitors required. Typical locations are close to a person and at representative locations in a room or enclosure. The number, location, airflow pattern, and sensitivity of the monitors should be chosen to provide an alert with the least delay from the onset of the release. Air sampling may also be conducted:

- prior to the initial use of the radioactive materials/sources, or immediately after radiation sources are deployed into the area.
- whenever procedures are revised, new equipment is installed, confinement, or additional sources are handled;
- periodically, to detect the effect of long-term changes in equipment and environment.

- when an incident or accident is suspected or after it has occurred.

Equipment used for continuous air monitoring

In practice, it is recommended to employ a monitor that can make simultaneous and real-time alpha and beta activity measurements. In a special case, where the risk of contamination of a work area is well known and only due to radionuclides emitting alpha particles, it is acceptable to select a monitor that only measures alpha radioactivity. In all other cases, as previously described, it is essential to use a monitor that compensates for the natural radioactivity (hence, alpha measurement is required). Moreover, in case of beta monitoring in an area where there is a very significant variable gamma dose rate, active gamma compensation is recommended.

Modern continuous air monitoring systems have a dual scintillator-based continuous air monitoring system, designed with advanced electronics and display devices. They have the capability of simultaneously measuring specific activity of both alpha & beta activity present dispersed as particulates in the air medium. It is employed in nuclear power plants, radiochemical laboratories, plutonium handling laboratories, and other strategic locations for real-time monitoring. The purpose of real-time monitoring:

i. To provide a prompt alert to breakdowns in control and containment
ii. To initiate immediate action to mitigate harm to the worker or work area
iii. To confirm that operating conditions are satisfactory.

The dual phosphor has zinc sulfide on a much thinner (0.25 mm) plastic scintillator. It essentially has a vacuum pump for sucking the air, a rotameter to measure air flow rate, a chamber with arrangement for trapping of suspended dust particulate onto the filter media/paper. This kind of dual scintillator-based detector assembly is mounted by facing the filter paper and performs the counting of alpha and beta particulate activity simultaneously in CPM/CPS/Bq on continuous basis. For any continuous air monitoring system, the volume air sucked through the glass fiber filter paper is used for estimation of air activity which is expressed in terms of Bq/m^3.

Air-sampler detector assembly consists of a filter holder 60 mm in diameter, a suction chamber with air inlet and outlet & detector housing fabricated with stainless steel. A dual scintillator-based detector assembly measures simultaneously both alpha and beta activity present in a given sample separately. It uses ZnS(Ag) scintillator deposited on a plastic scintillator of 50 mm diameter and coupled to a PMT along with a pre-amplifier. To minimize the gamma background, adequate lead collar/brace shielding has been provided. Detection efficiency achieved 25% for alpha with Am-241 and 25% for beta, it is about with Sr-90/Y-90. Collection efficiency for glass fiber filter paper should be better than 97% in any type of equipment design. A dedicated suction pump and provision for collection chambers are included in the design to enable the user to easily replace the filter paper periodically as defined in the air monitoring guide.

Electronic module: Continuous Air Monitors measure air concentration/activity of air samples collected on the filter paper and displayed in terms of CPM or CPS or Bq, or Bq/m^3. Besides, this system has several built-in electronic sub-systems including HVPS module, Simulated Mode of Power Supply, controller card, Electromagnetic compatibility filters, current loop circuit, relay, and relay driver circuit. The front panel has a Thin-film-transistor display with touch screen, alarm indication on the TFT display, and an audio buzzer. Enclosure has connectors for connecting to the scintillation probe, a pin Input/Output connector, test sockets, an Ethernet connector, an AC mains switch, and a fuse holder. This electronic module is capable of triggering both visual and aural alarms when the activity exceeds the pre-set level. CAM output parameters are communicated to the data logging system via the Ethernet port for better visualization of parameters. The electronic unit comprises LVPS, HVPS, Pre-amplifier & Amplifier, Count rate meter, and Alarm generation module. The electronic unit and the detector assemblies are mounted inside a single floor-mounted trolley. Sampling lines are made up of SS-304L to avoid deposition on the lines. The manufacturer has designed and fabricated the air sampler to achieve the particle collection efficiency better than 97% for air particles down to 0.3 μm size using glass filter paper.

Calibration and maintenance

Air-monitoring equipment should have a valid calibration for measuring the detector and flow meter prior to use. Frequency of calibration should be performed in conformance with national regulations. Further guidance is also provided in National Physical Laboratory UK-NPL Good Practice Guide 82 [10]. Calibration should be performed with a radioactive source having the same physical & radiological characteristics as the source used during type testing (size, radionuclide, and method of construction) in accordance with the manufacturer's instructions.

Installed radiation monitoring systems and equipment in nuclear facilities should conform to the requirements by demonstrating through type tests, inspections, and tests carried out at the manufacturer's site. This should also comply with the facility's on-site acceptance tests, the installation inspection, pre-operational testing, and the commissioning inspection. The licensee shall oversee the manufacture of the equipment and the quality management during manufacture.

Electronic criteria for operation and calibration: The following requirements have to be tested while commissioning the continuous air monitoring systems according to ANSI A42.17B [11]. They are high radiation alarm, alarm reset, alarm activation delay, false alarms, audible alarm intensity, visual alarm visibility, response time, and coefficient of variation.

1.8.5 Tritium and Its Monitoring

Background: Tritium is the only radioactive isotope of hydrogen atom and is produced in Pressurized Heavy Water Reactor (PHWR) type of reactors due to neutron activation of heavy water. Tritium is presents in the workplace in the form of tritium water vapor (HTO) and a low-energy beta emitter could enter the human body either by inhalation or by skin absorption. Tritium is classified as low toxic according to the IAEA classification scheme (Group 4). The extent of internal dose due to tritium gas exposure is about 10,000 times less than the total effective dose from an equal exposure to airborne HTO. The committed dose of any HTO exposure depends on its biological half-life.

Tritium is a low-energy beta emitter ($E_{max} - 18$ keV) and it needs special techniques for measurement [12]. Tritium exists in workplaces as DTO (PHWR), HTO (Fusion reactors), and as organically bound tritium (OBT) . Tritium is also formed as by byproduct of nuclear ternary fission (0.01%), produced due to neutron capture by Li-6 and B-10. The objectives of tritium monitoring are:

- To measure the tritium activity in air samples in the workplace and in systems.
- To recommend suitable PPE to control the occupational exposures.
- To monitor the airborne concentrations of HTO and HT in the workplace to improve the dose estimate.
- To provide data for planning the work from a radiological perspective.

Purpose: Workplace monitoring of tritium accomplished either by real-time measurement or by air sampling using an appropriate measurement assembly like a bubbler. In case of the presence of HT in a workplace, there is a possibility that a fraction of HT may be converted into HTO. Owing to the significant radiological risk due to HTO is 10000 times greater than HT, the accurate determination of HTO and HT fractions is necessary, which will provide a more accurate assessment of the potential radiological hazard [13].

Techniques: Typically, there are two types of real-time measurement of airborne tritium levels. In one case, a flow through ionization chamber is used, whereas in another case, the proportional counters are employed. Both detectors use a sealed container in which charge carriers are created when the beta particle expends its energy in the gas inside the chamber. Real-time monitors can be installed or portable and may be fitted with an alarm function.

Sampling and equipment used for sampling: The sampling of tritium may be dynamic or passive. For sampling, the tritium in air is collected in an appropriate medium: tritium-free water or water absorbent, then the activity is measured by liquid scintillation. With dynamic samplers, the air sampled circulates through one or more bubblers filled with clean water. In some cases, solid absorbers such as zeolite or silica gel are used as a tritiated water trap. In this case, a desorber is also required to remove the water from the absorber and collect it in a low-temperature collector. In some circumstances, passive samplers can be used. This consists of a vial containing tritium-free water or water absorbent as an HTO trap, the top of the vial being fitted with a diffusion orifice. Both a sampling and a measuring assembly are required. For a dynamic sampling the assembly is equipped with a sampling pump, a pre-filter placed upstream of the bubblers to trap any particulate and iodine (if applicable), a flow meter, one or more bubblers containing tritium-free water (or columns containing a solid absorber such as zeolite or silica gel) and an oven catalyzed by copper or palladium to separate HT and HTO.

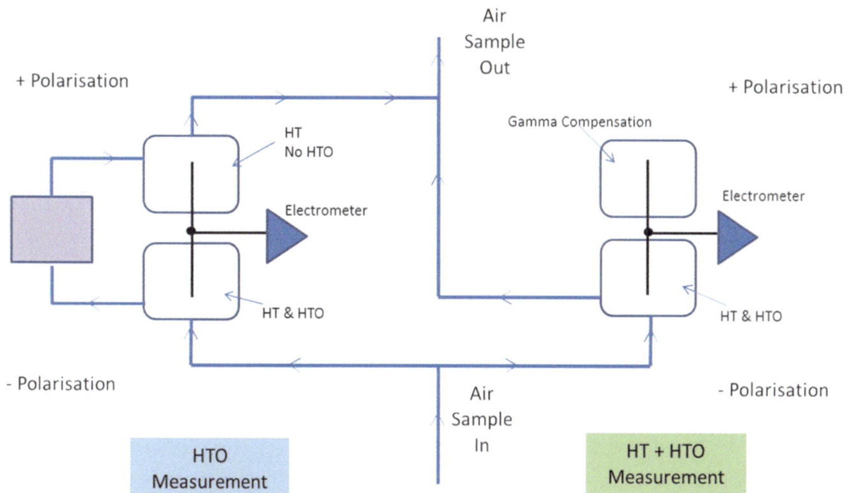

Fig. 1.2 Real-time tritium measurement (Simultaneous ion chamber measurement of HTO and total tritium)

Real-time monitors for tritium: A typical tritium monitor consists of an air sampling circuit, measurement cells (one or two) separated by a tritiated water trap, associated electronics, data processing software, and alarm devices.

Particulate radiation which may be present in samples may be retained using a pre-filter, and to remove the iodine, a charcoal filter paper is also provided. However, the pre-filter is chosen in such a way that it does not affect the performance of the monitor, nor shall it trap or temporarily retain tritium or decrease the flow rate or the pressure inside the measurement cell over the limits specified by the manufacturer.

The choice of instrument depends on the detection capability and the response time. The majority of tritium monitors use ionization chambers. The gamma compensation chamber is set up to compensate at a defined temperature and pressure and is sealed. An example of a real-time monitor is shown in Fig. 1.2.
Ion chambers for low-level tritium measurement have some limitations including:

- Offset drift occurring due to insulators; It is necessary to provide high insulation resistance to avoid drifting of the electrometer.
- Magnitude of ionization produced by two alphas emitted by radon gas is equal to the magnitude of ionization produced by tritium concentration of 37,000 Bq/m^3 and therefore radon discrimination is essential, which is present in the workplace environment.
- Background gamma; Ambient gamma compensation is required for cosmic radiation and other gamma radiation sources. Moisture due to condensation on the surface of the collector can create drift in instrument response.

The first series of bubblers (or columns containing a solid absorber) trap the tritiated vapor, then the tritium gas is oxidized in the oven at a typical temperature of 400 °C in

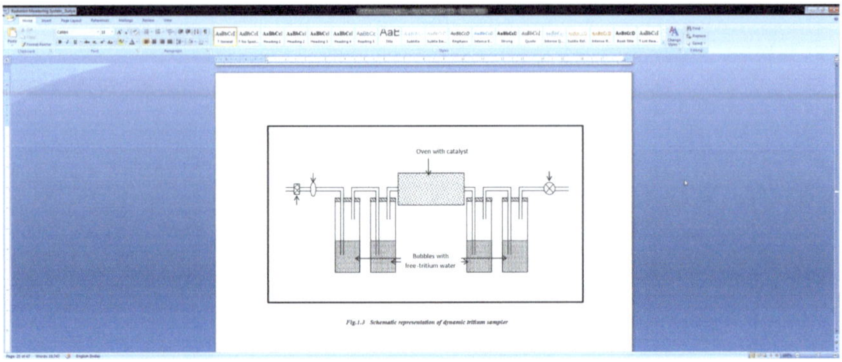

Fig. 1.3 Schematic representation of dynamic tritium sampler

the presence of a catalyst, and the tritiated vapor produced in this manner is trapped in the second series of bubblers (or columns containing the solid absorber) as shown in Fig. 1.3.

Calibration and maintenance of Real Time Monitors

Calibration of tritium in air monitors using tritium is a very demanding process. It requires a facility which can circulate known concentrations of tritium through the instrument's measuring chamber. Monitors cover a wide tritium concentration range, and it is difficult to produce a wide range of concentrations for circulation on a routine basis. There is also the problem of memory. The interior surfaces and components in a well-designed tritium in air monitor are chosen to have a low tendency to adsorb tritium onto the surfaces in the charge collecting volume but it is inevitable that testing with high concentrations will lead to some minimal adsorption onto the surface of the ion chamber components which will lead in turn to an increase in the background. Further details of the process can be found in the Good Practice Guide (GPG) issued by National Physical Laboratory (NPL-GPG) tritium in air monitor testing [14].

Type testing should be conducted in accordance with IEC 60761-1 [15] and IEC 62303 [16] standards. A well-defined calibration programme and calibration regime are mandatory for air monitoring systems in conformance with regulatory guidelines. Functional testing of each instrument should be defined by the manufacturer and usually includes testing with a radioactive source.

It is recommended to follow the maintenance timing and process indicated by the manufacturer. The maintenance operation shall include changing the pre-filter and any water absorbent before saturation.

Liquid scintillation counting (LSC) is a method generally used to determine the radioactivity in low-energy beta-emitting radionuclides. This is a sensitive radiometric technique which requires scintillation cocktails that are capable of converting the deposited energy into light pulses to enable detection by PMT. To realize the efficient scintillation in liquid medium, the LSC cocktails should be made up of a mixture of two components, namely an organic compound and scintillating flour.

LSC is based on the principle that, upon excitation by low-energy beta or alpha the solvent molecules get excited and transfer energy to scintillator molecules. The energy absorbed through the organic scintillators causes electronic excitation, which, upon returning to the ground state, emits light in the visible region. A photomultiplier tube (PMT) detects the light emitted from the process of scintillation. A Multi-Channel Analyser (MCA) is a memory card that stores the electrical pulses from the PM tubes and segregates according to energy.

Like in the GM counter, LSC has the problem of quenching. Quenching process derails the efficiency of scintillation and energy loss in the liquid scintillation solution. This quenching process has a deleterious impact on the energy spectrum of the radionuclide by shifting the peak to lower energy. Photon quenching, chemical quenching, and optical quenching are major types considered. Photon quenching occurs with the inadequate transfer of beta particle energy to solvent molecules. Chemical quenching occurs due to the presence of impurities, causing energy losses while transferring energy from solvent to solute. Optical quenching reduces the intensity of photons produced by solute.

Conventional liquid scintillation counters are integrated with one or two PMT detectors, whereas to enhance the counting geometry, certain LSCs have three PMTs (120° angles to each other), and two different coincidence outputs are made available. This arrangement enables higher counting efficiencies, providing quenching correction by triple-to-double coincidence ratio (TDCR) and luminescence free counting [17]. LSC equipped with three photomultipliers is suitable for a triple-to-double coincidence ratio counting. TDCR is an absolute counting method for obtaining the counting efficiency of the active samples without external or internal standardizing sources. The TDCR method was originally developed for the direct determination of the absolute activities of beta and EC-decaying radionuclides in liquid scintillator medium. TDCR combines experimental data with theoretical calculations of the detector efficiency. The knowledge of the radionuclide decay scheme data is a precondition [17].

1.8.6 Contamination Monitoring

The presence of radioactive materials in undesired places is called contamination. This poses a risk to occupational workers who are involved in various assignments as part of their routine job in the NPP. In a nuclear power plant, radiation monitors and surface contamination detectors are installed at various elevations of controlled areas. Both the personnel contamination monitoring and equipment contamination monitoring are performed to detect and remove the surface contamination. The main types of surface contamination monitors include Hand contamination and Foot contamination monitors, Frisker contamination monitors that measure the surface contamination on a worker's body. Dedicated contamination monitors are made available for the detection of contamination on the surface of equipment ranging from small-sized portable tools used by workers to large scaffolding boards. Active laundry contamination monitors measure the presence of contamination on protective clothing worn

Table 1.4 Specifications of hand and foot contamination monitor

Equipment	Specification
Detector	Pancake GM detector
Number of detectors	15–18 detectors
Detector sensitivity	For beta 200 ± 10 cps/(10^{-4} μCi/cm^2) For gamma 300 ± 10 cps/(mR/h)
Back ground	0.1 μSv/h
Measurement time	10 s
Calibration source	Sr-90 planar source 10 × 10 (cm)
Recommended distance	Hands & foot: close contact (1–3 mm) Head: 50 mm Chest: 100 mm
Distance	Hands & feet: close contact Head: 50 mm Other: 100 mm

by workers who worked inside the controlled area. Contamination monitor's sensitivity for hand and foot is 4 Bq/cm^2 for beta and 0.4 Bq/cm^2 for alpha contamination. A large array of pancake-based GM detectors are employed for this purpose. Contamination measurement is generally performed within a fixed period to facilitate the minimization of processing time.

Foot detector assembly

This consists of two detachable detector assemblies mounted in a tray (one each for the Left and Right foot). Each of the above is provided with 8 numbers of end window Pancake GM detectors of 45 mm diameter covering a sensitive area of about 12 cm × 30 cm. A Mylar sheet and grill plate shall be provided to prevent physical damage and to protect from the dust during operation. Optical switches shall be provided on this assembly to sense the presence of the user. The operation of these switches controls the signal processing. Each grill plate shall be designed to take a load of 120 kg. The distance between the detector window and grill plate shall be optimized to provide the required beta sensitivity as given in Table 1.4. EHT for the detectors shall be provided by the EHT module mounted in the electronic processing and display unit.

Hand detector assembly

This consists of two shielded detector boxes, one for each hand, with detachable detector assemblies. Each box has a vertical slot for the insertion of the hand. 10 numbers of end window pancake GM detectors of diameter 45 mm covering sensitive areas of 12 cm × 36 cm on each side of the hand in such a way to get the sensitivity requirement given in Table 1.4. Optical switches shall be provided to sense that hands are inserted and properly positioned in the slots. The operation of these switches controls the signal processing. A Mylar sheet and grill plate shall be provided to protect the detectors from any physical damage and dust. Also, detectors shall be shielded with 6 mm thick lead to reduce the background.

Signal counting

Background counting shall be inhibited when a person steps on the Foot detector assembly or inserts their hand in the Hand contamination detector assembly. If both the feet and hands are steady, signal counting will start and record the reading. The ready window shall flash to indicate that signal counting is in progress. At the end of the signal counting time, the monitor shall acquire and compute the contamination level in CPS for each of the detector assemblies individually for the left and right foot, left and hand. The instrument is designed in such a way that background contamination levels are automatically subtracted, and the contamination level is expressed in terms of CPS. Based on the magnitude of this count rate, the monitor displays either 'clean' as indicated by the Green LED light. Whereas the 'contaminated' signal is indicated by a Red LED light, which prompts the personnel to take appropriate actions.

1.9 Environmental Radiation Monitoring

Background

Environmental radiation monitoring is carried out to ascertain the radiological conditions prevailing in the vicinity of nuclear facilities and assess the impact of the operation of nuclear facilities, including NPP, both in normal operating conditions as well as during emergency conditions. Environmental radiation monitors installed measure the ambient gamma dose rate, the concentration of discharged gaseous radioactive material, and the concentration of airborne radioactive material by the installed monitors around the NPP, which continuously measure and communicate the radiation levels. During the routine radiological surveillance around the nuclear power plants/facilities, the radioactivity levels due to target nuclides such as Cs-137, Sr-90, Pu-239, Cs-134, I-131, and H-3 are identified and estimated as part of environmental monitoring. As part of confirmatory monitoring, the samples are collected from soil, sediments, river soil, air, underground water, surface water, sea water, cow milk, crops, vegetables, fishes, and fruits are also analyzed in the laboratory. The objectives of environmental monitoring depend on prevailing conditions. For example, in the case of routine monitoring conditions, the early detection of abnormal environmental radiation, provides data on ambient levels of radiation in the environment whereas during nuclear or radiological emergencies, it provides radiological data for appropriate emergency response arrangements including communicating to the general public, public officials and adjusting the protective actions. This monitoring also helps to disseminate the information on the status of the environment to the general public to have a better understanding and reassurance on the safe operation of nuclear power plants. Regulatory body demands operationalization of environmental monitors applicable for both routine conditions and emergency scenarios.

Environmental radiation monitoring is defined for the activities that involve measurement of external dose rate, activity concentration in soil, air, and water samples. The fundamental objective is to collect and provide information to protect people and the environment. In many countries, the real-time radiation monitoring instruments are installed at critical locations in the environment for rapid detection and preliminary estimates of routine and unanticipated releases. Ground-based environmental radiation monitoring serves the purpose of early notification due to an unexpected increase in the ambient level of local gamma radiation. The Comprehensive Test Ban Treaty Organization (CTBTO) located in IAEA headquarters in Vienna, has established a worldwide monitoring network. CTBTO has a network of 80 radionuclide measurement stations for the detection of airborne radionuclides. In addition, 40 of these stations are equipped with systems for the measurement of radioactive xenon. These systems have demonstrated their capability of detecting the accidental radiological release that occurred in the aftermath of the INES level-7 nuclear accident in Fukushima, Japan, in 2011.

Objectives

Real-time evaluation of ambient radiation levels in and around the nuclear sites is a regulatory requirement. Hence, appropriate monitoring strategies should be developed and deployed to continuously measure the releases of radioactive material into the environment. The dispersion of radioactive materials is sometimes strongly affected by wind direction and velocity. The radiological characterizations of ambient air, water bodies, soil, and vegetable/fruit samples are necessary.

Radiation exposures to the general public from controlled discharge due to the operation of nuclear power plants may arise from the direct discharge of radioactive effluents and dispersion of radionuclides in the environment. In order to continuously measure the radiation levels in and around the nuclear facilities, large numbers of radiation monitors are installed and commissioned. They intend:

- To provide online information about current radiation levels at various locations/stations to the emergency control room,
- To facilitate the environmental impact assessment of nuclear emergencies.
- To establish background environmental radiation levels and monitor long-term shifts in them.

Broadly, two classifications of environmental monitors are (a) fixed monitors and (b) field monitors.
The minimum capabilities required for such monitors are, they must be in a position to measure the gamma dose rate level and airborne activity levels. In order to ensure uninterrupted transmission of data, each station is equipped with:

- Radiation monitor equipped with GM Counter
- Radiation monitor equipped with an Ion Chamber
- High Volume Air Samplers for air activity estimation and identification.
- Data Acquisition and Communication System.

Environmental conditions

The operating system should be capable of working in the temperature range of 8–45 °C with a relative humidity ranging from 55 to 85%. Occasionally, due to changes in weather conditions, the temperature range may vary from 5 to 45 °C and the relative humidity may vary from 50 to 95%. Environmental monitoring is also performed by measuring low-level gamma activity using state-of-the-art gamma ray spectrometry by analyzing typical samples like foodstuffs, soil samples, well water, and runoff water from different locations around the country. Airborne particulates are collected periodically on filters in each station and analyzed by gamma ray spectrometry to identify the contributing nuclides. In this way, even lower levels of environmental contamination can be detected and identified compared to other portable radiation monitor detectors. Environmental radiation monitors are installed in the monitoring post that houses a dose rate monitoring device for continuous measurement of environmental gamma dose rate. This assembly is also equipped with both a dose rate meter and an airborne activity monitor in order to measure the concentration of airborne contamination is nothing but a monitoring station. Generally, a nuclear facility has a group of several monitoring posts, and monitoring stations are generally installed at 5–10 locations around the security perimeter of a nuclear power plant (on-site domain), and at 10–20 locations in the off-site domain.

Specific requirements

Gamma measuring systems have a scintillation detector optically coupled with a Photo Multiplier tube and shall be suitable for detecting gamma radiation having the energy from 60 keV to 3 MeV. Detector shall be of integral assembly consisting of 2 2-inch diameter and 2-inch thick NaI(Tl) scintillator with bi-alkali photomultiplier tube. An appropriate pre-amplifier shall be provided, and this shall be powered from the DC power supply unit of the electronic processing and display unit. The detector, along with the PM tube and pre-amplifier, shall be suitably housed in an enclosure as per IP 56. This shall be mounted on the top of the enclosure which houses the electronics processing and display unit.

The following are the specifications generally adopted for environmental monitoring in selected locations around the exclusion zone. It consists of a NaI(Tl) based detector which directly measures gamma dose rates from the radionuclides that emit gamma within the energy range of 30 keV–3 MeV. A detector with a dimension of 50.8 mm in diameter and 50.8 mm in thickness is optically coupled with a Photomultiplier Tube. The detector and corresponding electronic accessories are well protected from environmental conditions by an IP56 enclosure. The low intensity gamma dose rate monitoring system is fabricated with a hermetically sealed NaI(Tl) based scintillation detector. This system has a PMT, an amplifier circuit, a high voltage circuit, and a temperature compensating circuit, and provides output in the form of standardized pulse signals. These monitors have a four-decade range covering from 0.1 to 1000 μSv/h. Appropriate electronic display units are also provided within the monitoring post, which communicate the readings to the main control room. Environmental radiation data is critically important in planning response action in case of an emergency.

significant difference between open window and closed-window readings indicates plume shine. In both these cases, air samples are not required.

During the intermediate phase of a nuclear emergency, the gamma dose rate measuring instruments, along with gamma identification of deposited radionuclides, are used to determine the boundaries of the restricted zone, wherein the people who were not already evacuated must be relocated. Isotopic analysis can be efficiently carried out by a High-Purity Germanium (HP-Ge) Detector. Air samplers are deployed to assess the re-suspension of ground-deposited radioactive material that contributes significantly to the inhalation dose.

Geiger Muller nuclear instruments designed for measuring low dose rate (few tens of μSv/hr) may underestimate the high-exposure rates (few tens of mSv/h), because at high dose rates they become saturated. In view of this challenge, it is necessary to have a calibrated monitoring system that is capable of measuring high dose rates as well.

The pulse height produced by the radiation detector must be sufficiently high so that it can be reliably sensed by the detector. GM detectors inherently deficient in having long dead time, i.e, the time during which the detector is rendered unresponsive to incoming radiation. This dead time is especially true in the case of high gamma intensity. Radiation is no longer detected during the recovery time period, and it prevents the detector from fully recovering. This process is called saturation. During saturation, the instrument displays essentially that background levels of radiation are present, even though the operator is in a high radiation area. This is potentially a high-exposure situation. In order to overcome this, low-range GM instruments must be accompanied by high-range survey meters during accident conditions.

For the applications in intermediate phase emergency operations, the micro-R survey meters, in addition to the milli-R and R meters, are also deployed. Most micro-R meters are inorganic scintillation detectors having the dimension of $1'' \times 1''$ [NaI(Tl)]. Care should be taken not to damage them while they are in use, since they are equipped with PMT.

In order to detect the various air contaminants due to fission products, noble gases, and radioiodine, air samplers with calibrated air pumps are used. These cartridges are loaded with the appropriate particulate filter and cartridge-type absorber filter for selective collection of radioiodine in the presence of fission products and noble gases. Both battery-operated air samplers and AC-supplied air samplers may be used.

Air samples are collected in such a way that air is allowed to pass through the pre-filter, filtering particulates such as caesium, strontium, and iodine present. The remaining air stream may have a fraction of iodine, and large fission product noble gas is allowed to pass through the pre-filter and enter the iodine cartridge absorber, where the iodine gas is collected. The noble gases are then exhausted through the other end of the air sampler.

Air Sampler

During the early phase of emergency exposure situations, air monitoring is required for taking protective action. Suitable air collection media is required [19].

- **Absorption**—A high efficiency particulate air (HEPA) has higher collection efficiency for 0.3 micron particle size. One can also use glass fiber filter paper to collect the particulate activity. Few inorganic polymers are suitable inorganic absorber materials that can minimize adsorption of fission product noble gases vis a vis radioiodine.
- **Charcoal**—Activated charcoal, which has an enhanced surface area, is an organic adsorbing medium to collect the radioiodine efficiently. However, it has a tendency to adsorb a significant portion of the radioactive noble gases. Charcoal impregnated with Triethylene diamine (TEDA) is always recommended. It acts as a chelating agent to chemically bind the iodine in the cartridge and reduce desorption loss.
- **Inorganic absorber media**—Silver activated zeolite, silver silica gel, and silver alumina are commercially available inorganic absorber media that are suitable for adsorption. Silver zeolite and silver alumina have higher rejection efficiency for FPNG/Ar-41 gases and higher radioiodine retention efficiency.

1.9.2 Fixed Monitors

Fixed radiation monitoring systems are pegged in identified locations that continuously measure the radiation exposure rate. These types of environmental monitors are networked to keep track of the external dose rate due to release. In contrast to this, field monitoring systems are deployed to accumulate samples of air and water and subject them to radiological analyzes. During the normal operation of nuclear power plants, the release of radioactivity is controlled and monitored. The monitors are capable of providing real-time dose rates directly to the control room, both routine data and emergency data.

Requirements

Radiation monitoring systems should be proficient in giving out a real-time display of the external gamma dose rate, which ranges from natural background dose rate [$<$10 micro-Sievert per hour (μSv/h) to emergency levels ($\geq$100 Sievert per hour (Sv/h)]

- Detectors are pegged at 1 meter above the ground
- They can be used for continuous operation and unattended operation, and are capable of communicating results
- Provision to record the measurement and recording frequency every minute to provide near real-time hourly dose rate data
- Should withstand outdoor weather conditions
- Capable of automatic uploading and integration with the data management system and visualization system.

Radiation detectors

Environmental monitoring requires special instruments. Generally, gas-filled detectors and solid-state scintillation detectors are deployed. Geiger Muller detectors, Plastic Scintillators, and High-Pressure Ionization Chambers like HPXe detectors are highly popular. They are filled with inert gases like xenon or air inside the cylindrical or spherical chambers. Detection of ionizing radiation is realized by ionizing the gas, which produces a measurable electrical pulse. In order to increase the sensitivity of ionization chambers, the pressure of gas inside the chamber volume is increased to several atmospheres. Solid state sodium iodide (NaI), caesium iodide (CsI), Cerium (III) bromide ($CeBr_3$), or Lanthanum (III) bromide ($LaBr_3$), produces scintillation of light photons that are converted into a measurable electric pulse by PMT. In emergency exposure situations, the dose rates and contamination levels are expected to be on the higher side, and therefore a wide range of measurements covering from background radiation to emergency levels is required. For this purpose, the systems with several detectors, like two GM tubes or a scintillator and a GM tube, are preferable.

Plastic scintillators are capable of detecting alpha, beta, gamma, and neutron radiation. They are primarily organic scintillators in solid form. Polystyrene and polyvinylbenzene are two types of scintillators widely deployed for detection purposes. By incorporating a suitable proportion of a neutron-sensitive element like B-10, this scintillator is made neutron sensitive. Many such detectors are commercially available, and they are well known for the detection of surface contamination and air contamination as part of dual phosphor detectors. Though its light output is less than inorganic scintillator, the plastic scintillator has a very short lifetime, thereby enabling fast detection of radiation. In a nuclear power plant, the iodine monitoring is being done by using the plastic scintillator, whose sensitivity is high for low-energy gamma.

High-pressure ionization chambers are widely used [20] to make accurate gamma dose rates due to natural background radiation and, in particular, to measure the stack effluent discharges within the exclusion zone of a nuclear facility. Detectors with enlarged dimensions and with energy-compensated GM counters are also being used for this purpose.

The salient feature of the high-pressurized ionization chambers is that they exhibit better isotopic response as compared to the Geiger Muller counter. Besides, from the experience it is known that the HPICs (High-Pressure Ionization Chamber) measure exposure more accurately than the energy-compensated Geiger Muller instruments [21]. It is technically desirable to have required knowledge of the response of radiation detectors that are deployed for measuring the environmental radiation field to obtain accurate measurement results.

Energy-compensated, halogen-quenched GM-based monitoring systems show acceptable energy response between 50 keV and 2 MeV and are suitable for a simple pulse-counting circuit. Significant directional dependence is possible from long GMs, which requires further correction, particularly for environmental radiation measurement. These detectors do have the physical dimensions of 400 mm in length and a diameter of 40 mm. They have a desirable plateau length of 100 V and an operating voltage of 500 V. The background radiation level is reduced drastically to 400 cpm by using a lead shield with a thickness of 40 mm (Pb). The response of the GM detector at low energy, around 100 keV photons, is higher compared to 1MeV photons because at low photon energy, there exists enhanced photoelectric interaction. This shortcoming is overcome by providing a metal sheet around the GM detector which filters away the low-energy photons. Copper with perforated lead wrapper is used for this purpose [22]. The use of filters significantly reduces the over-response at low energies. The metallic shielding filters are sufficient to eliminate any contribution from beta radiation. An uncertainty level of more than ±20% over a range of 50 keV–1.3 MeV is the one usually quoted by manufacturers for such energy-compensated devices. Functionally, GM tubes produce a large output pulse height of 1 volt, and therefore, low electronic noise is sufficient.

The electronics provide the required high voltage to the GM tube and the pulse amplification and shaping.

Environmental enclosures

The components of environmental monitors are sensitive to weather. In view of this, the components shall be placed inside weather-proof housing while installed outdoors. These enclosures shall comply with relevant IP (ingress protection) rating as recommended by the International Electrotechnical Commission (IEC-60529). This ingress rating category is based on the degree of protection offered by the enclosure in order to safeguard the internal components, including detectors. Such requirements are outlined in the manufacturer's instrument specifications.

Data analysis

Software programmes for handling and analysis of emergency data are essential for fixed radiation monitoring systems. Emergency managers require real-time measurement data or compare the data from different locations and different monitors, and review the previous measurements. Application software for the user interface may be adopted for the monitoring system either by compact disc or through internet/intranet download. This has various features like desktop-only or web-based access, an interactive data display, mapping, temporal or spatial analysis, alarm, or fault notifications. For routine monitoring, we use data management software for trend analysis and graphical representation.

Calibration

Calibration is required to verify the performance of monitoring instruments. Gamma background radiation levels and radiation sources require different and variable environmental conditions. While performing the detector calibration, one needs to consider the complex nature of gamma ray spectra due to the background energy spectrum up to 3 MeV. Some salient features of the detectors, like large size and high sensitivity, along with the need for extreme stability of the associated electronics, call for special care in instrumentation design and in calibration techniques and procedures.

1.10 Post-accident Monitoring Systems

Background: Radiation monitoring is pertinent not only during normal operation of NPP but also during and post-accident conditions in which the gamma dose rate up to 10^4 Gy/h and 10^{12} Bq/m^3 of air activity levels are expected.

The nuclear accident at Fukushima Daiichi nuclear power plant in Japan caused due to failure of both on-site and off-site power failure has triggered the emergency situation. Consequently, several monitoring instruments were deprived of electricity, leading to core melting and a hydrogen explosion. Several important functions of the instrumentation systems were lost. A challenging situation prevailed, which made it an uphill task for the operator to perform the essential monitoring during prevailing emergency conditions in the plant [23]. In such a situation, it is necessary to have a set of reliable and robust monitoring instrumentation systems. These instruments are designed and commissioned in such a way that even in the case of escalation of an event into a severe accident, it is capable of mitigating the consequences of a severe accident. These arrangements will ensure a safe restoration of the nuclear plant. The radiation monitoring systems designed for post-accident monitoring must maintain operability even in the case of a single failure. They must be accurate enough in measuring the parameters over a wide range that is possible during nuclear accidents. To the extent possible, the measurements are so planned that the operators will easily recognize in case of failure of measurement or in case the measurement range is exceeded. There are multiple barriers that try to impede radiological releases, namely, are (i) fuel cladding, (ii) nuclear reactor coolant system, and (iii) reactor containment. Most of the Post-Accident Monitoring parameters are selected to identify the conditions that indicate a threat to the integrity of the barriers.

Fuel melting causes the release of fission products due to overheating of fuel elements during the accident, but measurement of the large releases into the containment is rather difficult. Radiation damage to equipment that is located inside the containment during Design Extension Conditions (DEC) with core melting would be much higher. The radiation exposures are unlikely to be uniform inside the locations of the controlled area during the accident. High temperature, high pressure, the presence of combustible gases, and sodium aerosols are some of the considerations taken into account in designing the instrumentation. Nuclear instrumentation systems used for the implementation of Severe Accident Management Guidelines (SAMG) are required to be designed, installed, and commissioned to withstand local anticipated transient conditions.

The radiological assessment and monitoring in emergency exposure situations are quite different from the monitoring during normal operational situations. For an effective emergency mitigative accident management, a set of dedicated nuclear monitoring instruments called post-accident monitoring systems (PAMS) that include radiation monitors are stipulated depending upon the hazard potentials of the nuclear facility. From the experience of the Fukushima Daiichi accident, the need for reliable accident monitoring instrumentation is emphasized, which indicates the safety functions during the progression of the accident as well as the aftermath of the accident in NPP. In a Nuclear Power Plant, the Post-Accident Monitoring Systems (PAMS) are integrated with other monitoring systems in the design stage itself to manage the detection, measurements, and indicating the critical safety functions for effective implementation of Emergency Operation Procedures (EOP) and SAMG.

The PAMS provides information on the critical parameters of the reactor containment building. PAMS aims to monitor the plant conditions and perform its role in bringing the plant to a safe configuration during an accident. Accident monitoring systems support the functions envisaged in level 3, level 4, and level 5 of the defence in depth concept.

- It provides the information for verification of reactor shutdown, integrity of the containment, intensity of activity release, for carrying out actions to mitigate the consequences of postulated Core Disruptive Accident (CDA) in fast reactors [24]. It provides vital information to the operators to indicate the potential breach or the actual breach of one or more fission product barriers.
- It ensures the supply of vital information to indicate the performance of the essential safety systems for the mitigation of DBE.
- It continuously monitors the release of radioactive materials to the environment, which may be in the form of particulates or gases.
- It minimizes the on-site and off-site releases and their adverse radiological consequences.
- It assesses gamma radiation levels and radioactivity inside the reactor containment building and fuel building are monitored in order to support emergency preparedness and response plan.
- It measures radiation levels and activity concentration in the Main Control Room and other controlled areas of the plant where access may be needed for plant recovery.

For Sodium cooled fast reactor, the instrumentation for PAMS measures and provides the operator with the necessary information during severe accident conditions, by monitoring and displaying critical safety parameters such as core neutron flux, pressure, temperature, radioactivity, presence of sodium aerosols in the Reactor Containment Building (RCB), and Main vessel temperature.

All instrumentation for PAMS is qualified to survive the seismic conditions associated with severe accidents. It provides information to operators to implement the manually controlled actions if no automatic control is provided. It communicates the information to the emergency control center. PAM supports the informational needs about verification of reactor shutdown, radioactivity discharges, assessment of radiological conditions, and assistance in carrying out recovery actions. Normal process deviations and any events such as transient over power, pump trip and seizure, primary pipe rupture, uncontrolled withdrawal of absorber rods, and loss of off-site power supply are not considered as design basis events for the design of PAM. Some of the above-mentioned initiation conditions are generally anticipated during normal operational plant states and do not require post-accident monitoring. Certain common mode failures, such as fire and flooding, are not considered as post-accident monitoring events since the plant is normally designed for such common mode events by complementary safety design principles and guidelines, such as grouping and separation requirements and fire protection requirements.

Instruments for PAM have ranges that extend to the maximum values that selected parameters can attain under worst-case conditions, and the instrumentation components are qualified to withstand the higher level of environmental conditions in which they are intended to function. Signal processing units for all the above systems are located outside RCB. The equipment should be easily testable and maintainable. The components and modules are provided with unique and easy identification to distinguish them from other process systems. All the sensors placed inside RCB are qualified to ensure satisfactory operation after the accident. The equipment is designed to survive the specified accident conditions and operate satisfactorily for at least 100 days after the CDA (conditions such as temperature, pressure, radiation level). A high-level alarm set point is provided for all the parameters. Local indications are provided in the control building (CB) local control centre (LCC) (where the signal processing units are located) for maintenance and calibration during normal plant operation. The signals are connected to Alarm annunciating windows and indicating meters provided in the control room for all the parameters in the control room. Alarm outputs of all the parameters are fed to the Safety Class-2 data highway for centralized display and data acquisition. Good Operation Trip (GOT) facility is provided to all the instruments, when there is any problem (power supply voltages are low, modules not properly inserted in the electronics, high voltage not applied to the sensor) in the system. A built-in test facility is provided for calibration, which can be authorized from the control room by the operator with a key-operated switch. Redundant channels are independent, physically and functionally, so that the operation of one system in no way is affected by a failure of the other system and ensures that the elements do not encounter similar conditions due to common extraneous causes like power failure.

1.10.1 Gamma Dose Rate and Air Activity Monitoring System

During CDA, the gamma intensity is estimated to reach a high level of 10^3 Sv/h. In order to measure gamma dose rate, two ionization chambers (ranging from 1 μSv/h to 1000 Sv/h) with the sensitivity of 10^{11} Amp/(R/h) are pegged on RCB walls. This constantly measures and displays the readings in the main control room. Furthermore, the other dominant radiological hazard during CDA is the release of radioactive sodium aerosol into the RCB atmosphere due to the combustion reaction of sodium with air/moisture. NaI(Tl) based inorganic Scintillation detector and associated electronics are provided to detect the radioactive sodium aerosol by measuring the Na-22 and Na-24 activity. For this purpose, ambient air is drawn from the containment building at two different locations in the control building through 25 mm stainless steel tubing by means of a dry carbon vacuum pump, and the air is returned back to RCB due to its high specific activity. Sampling arrangement and detection schemes are depicted in Fig. 1.4. Design considerations should be such

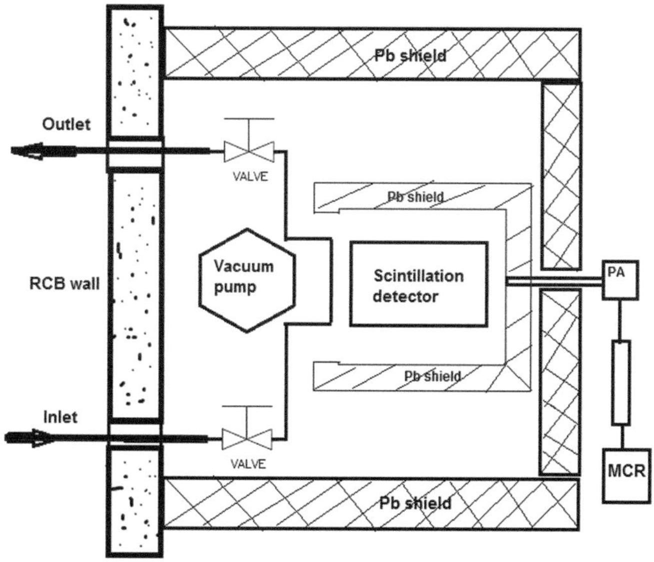

Fig. 1.4 Sampling of aerosol and monitoring of radioactive sodium in RCB [25]

that, even in case of severe accidents, there shall be provision to assess radionuclide concentrations in the containment gas plenum by sampling or some other method. In these cases, a range of iodine isotope and radioactive aerosol concentrations shall extend to a minimum of 10^{15} Bq/m^3. The sensitivity and range of the post-accident monitoring system are given in Table 1.6.

High Range Area Gamma Monitors (HRARM) equipped with ionization chambers are installed inside the RCB, and corresponding monitors are placed outside the RCB by design in order to reduce the probability of exposure and contamination. The performance criteria for PAMS are defined with respect to its range, accuracy, and response time. The calibration of these monitors is carried out using one curie Cs-137 source every year. The survivability and specifications of PAMS have been defined in such a way that, the detectors, cables and other associated electronics withstand extreme conditions including high temperature (>200 °C) due to sodium fire and high radiation field [10–10^3 Sv/h] due to intense gammas from Na-24 having gamma energies at 1.37 and 2.4 MeV and Na-22 gamma energies peaking at 0.511, 1.275 MeV.

High Temperature Fission chamber (HTFC) having the sensitivity of 0.1 cps/nv is deployed for core neutron flux monitoring with the range of 1E3 nv to 2E7 nv. This is a type of ionization chamber coated with enriched U-235 and suitable for measuring neutron flux in the presence of an intense gamma background and high temperature.

The expected gamma dose rate inside the RCB of the fast reactor may go up to a maximum of 10^3 Sv/h during CDA. Gamma monitors based on ionization chambers

Table 1.6 Sensitivity and range of PAMS

Sl.No.	Detector	Sensitivity	Range
1.	Ionization chamber (HRARM)	10^{-11} A/(R/h)	1 μSv/h–1000 Sv/h
2.	NaI(Tl)	1 cps/(Bq/cc)	10–10,000 Bq/cc
3.	High temperature fission chamber (HTFC)	0.1 cps/n/cm^2/s	10^3 to 2×10^7 n/cm^2/s

with the sensitivity of 0.1 pA/R/h and measurement range of 1 μSv/h–1000 Sv/h are suitable for measuring the intense gamma radiation levels. The signals are fed to the main control room for emergency response actions.

1.11 Personnel Dosimetry

Background: Thermoluminescence dosimetry (TLD) is a versatile tool for the assessment of external exposures. Since the invention of the Thermoluminescence phenomenon by Daniels and his colleagues from the University of Wisconsin-Madison, there are plenty of phosphors that have been prepared and investigated for their desirable dosimetric properties. Thermoluminescence is a phenomenon of emission of light in the visible region due to heating of crystalline materials. These materials are capable of essentially measuring the radiation dose due to beta, gamma, X-rays, and sometimes due to neutron radiation. The desirable characteristics of a dosimeter are: it should be preferably tissue equivalent, should exhibit low fading, should have high sensitivity for beta, gamma, X-rays, and neutrons, should exhibit high temperature glow peak, and it should exhibit linear dose response. TLD-based badge is recommended for personnel dosimetry in many countries. Among them, $CaSO_4$: Dy is highly sensitive and therefore for assessing the occupational exposures [26]. Table 1.7 explains the various commercially available thermoluminescence dosimetry. They have desirable sensitivity, a glow peak, and less fading properties. Figure 1.5 depicts the typical example of the physical configuration of a TLD dosimeter having three discs of $CaSO_4$: Dy with and without a cassette. Metallic covering is provided to moderate the enhanced efficiency for low-energy gammas.

Regulatory requirements of personnel dosimetry

National regulatory requirements stipulated by the Atomic Energy Regulatory Board (India) demand that the radiation professional working in the controlled area of a nuclear power plant or any other nuclear facilities shall be periodically monitored for their exposure to radiation in order to control the stochastic effects and to prevent the deterministic effects. Every facility shall have well equipped and accredited thermoluminescent laboratory and a whole body dosimetry laboratory to assess both external dose and internal dose. Besides, the other purposes of personnel monitoring are: it helps to ascertain the effectiveness of radiation protection practices in the workplace, to detect changes in radiological intensity and levels in the workplace,

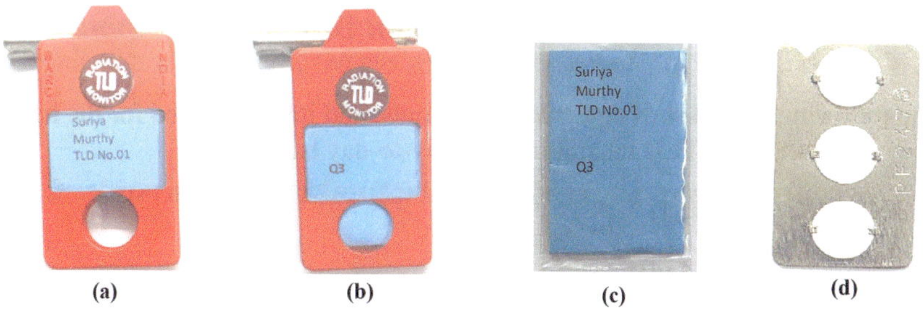

Fig. 1.5 TLD badge: (**a**) Anterior view, (**b**) rearmost view, (**c**) TLD dosimeter loaded in wrapper and polythene pouch, and (**d**) bare TLD card without pouch

and to confirm or supplement static workplace monitoring. Further, it is also used to identify working practices that minimize doses and provide information in the event of accidental exposure.

Regulatory requirements further underline that the occupational radiation exposure of individuals arising from various practices should be subject to dose limits to ensure that no individual is exposed to a risk that is judged to be unacceptable.

The International Commission on Radiological Protection (ICRP) downwardly reduced the occupational dose limit from 50 to 20 mSv per year, averaged over 5 years for and from 5 to 1 mSv for the public [27].

Table 1.7 Details of some commercially available thermo luminescent dosimeters [26]

TLD type	Effective atomic number (Z_{eff})	Main peak (°C)	Emission maximum (nm)	Relative sensitivity	Fading (at 25 °C, samples kept in dark)	Useful dose range
LiF: Mg, Ti	8.14	200	400	1	5%/year	20 μGy–10 Gy
LiF: Mg, Cu, P	8.14	210	368	40	5%/year	0.2 μGy–10 Gy
LiF: Mg, Cu, Si	8.14	240	384	55	Negligible	1 μGy–20 Gy
$Li_2B_4O_7$: Mn	7.3	220	605	0.40	4%/month	0.1 m Gy–3 Gy
$Li_2B_4O_7$: Cu	7.3	205	368	8	10%/2 months	10 μGy–10^3 Gy
MgB_4O_7: Dy/Tm	8.4	190	490	6–7	4%/month	5 μGy–500 Gy
BeO	7.1	190	330	1.00	8%/2 months	0.1 m Gy–0.5 Gy
Mg_2SiO_4: Tb	11	200	380–400	40–53	Very slight	10 μGy–1 Gy
$CaSO_4$: Dy	15.3	220	480–570	30	1%/2 months	2 μGy–10 Gy
$CaSO_4$: Tm	15.3	220	30	30	1%/2 months	2 μGy–10 Gy
CaF_2: Mn	16.3	260	500	5	16%/2 weeks	10 μGy–10 Gy
CaF_2 (natural)	16.3	260	380	23	Very slight	10 μGy–50 Gy
CaF_2: Dy	16.3	215	480–570	15	8%/2 months	10 μGy–10 Gy
Al_2O_3: C	10.3	190	420	60	5%/year	0.1 μGy–10 Gy

The response time of the detector used for scanning is not dependent on the type and amount of radionuclide present inside the body. The technique has advantages like: good sensitivity, response is independent of dispersal of radionuclides; shielded room is not required since the detector is well shielded, and hence reduces the cost. This type of arrangement is useful for transportable whole body counters since it does not have massive shielding.

Indirect Measurements

Biological samples like urine, faeces, and blood are used for the estimation of internal body burden through indirect measurement. This involves the determination of concentrations of radionuclides in materials separated from the body. These bioassay samples are urine, faeces, breath, and blood. For suspected intake, particularly of concern due to intake of radiotoxic Pu-239, the nose blow and nasal swab are considered as early estimates of contamination levels.

The selection of urine or faeces samples depends on the biokinetic model of radionuclide and the physico-chemical form of intake. For soluble radionuclides, urine samples are preferred, whereas for insoluble radionuclides, faeces samples are preferred.

Ideally, a 24 h urine sample is used for routine monitoring. Where 24 h samples are not easily collected, the first morning voiding is preferable for analysis. The alpha-emitting radionuclides are radiochemically separated, and individual radioisotopes are electrodeposited on a stainless steel disc and then subjected to alpha spectrometry for qualitative and quantitative analysis. Similarly, the beta-emitting radioisotopes like Sr-90, Y-90, etc. can be radiochemically separated from the urine sample and then subjected to liquid scintillation counting or any other suitable radiometric method for qualitative and quantitative analysis.

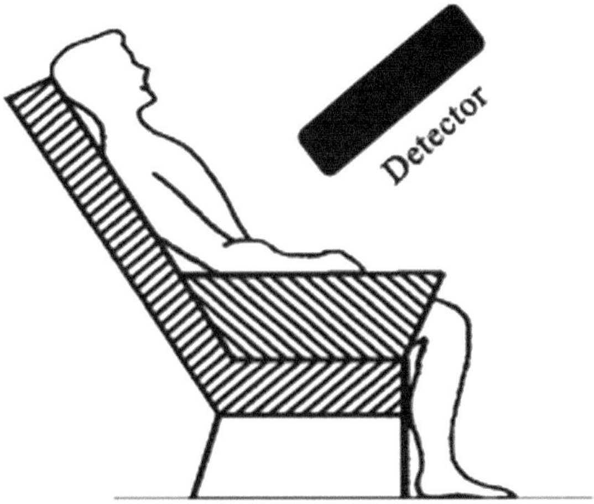

Fig. 1.6 Chair geometry for internal monitoring

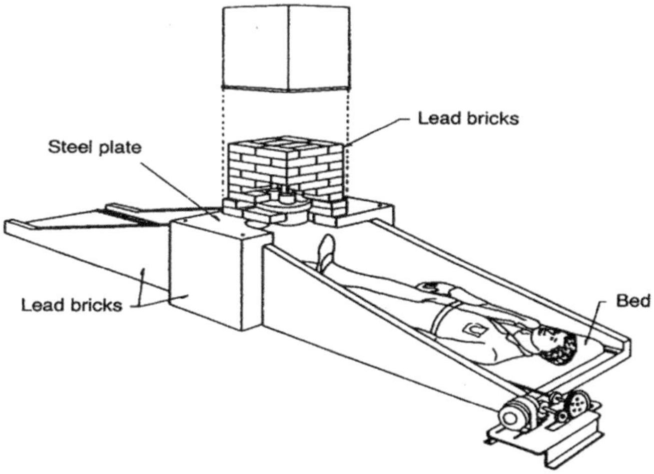

Fig. 1.7 Shadow shield bed geometry for internal monitoring

1.12 Quality Assurance in Radiation Monitoring

A sound Quality Assurance (QA) programme should be established to ensure that precise radiological measurements and assessment are carried out by the laboratory. The principal objective of incorporating QA in radiation monitoring is to achieve the highest standards in radiological safety practices. The QA programme includes all quality control tests and proficiency tests required to ensure a consistent high level of reliability and accuracy in all measurements carried out by the HP Unit.

For effective implementation of a quality assurance program applicable for radiation protection selection of appropriate standards for calibration and testing is required. All the QA procedures shall be documented, reviewed by competent personnel. The calibration record should show the unbroken traceability to primary standards which is necessary for reference purposes. An approved QA manual with all details of Quality Control (QC) tests, procedures, and records that are to be maintained. There are various standards from American National Standards Institute (ANSI), International Standard Organisation (ISO) and International Electro technical Commission (IEC) that are applicable for radiation monitoring systems.

Quality Assurance should start from the procurement stage. Here, the objective is to assure that the purchase is done from reputed, qualified, and appropriate sources. All the quality requirements relating to design, manufacturing, inspection, and testing are suitably included in the procurement documents and also followed in all the stages so that the product satisfies the intended requirements consistent with safety, reliability, and availability. The details of QA program elements that should be developed and implemented for obtaining satisfactory quality of data and results for radiological monitoring programs [30].

- Organizational Structure, roles and responsibilities of Managerial & Operational Personnel
- Specification of education, skills, and competency of Personnel
- Standard Operating Procedures and Instructions
- Records and record maintenance
- Quality Control in air and liquid sampling
- Quality Control in the Nuclear Counting Laboratory
- Calibration and Quality Control of Instruments, Measuring Devices, and Test Equipment
- Internal Quality Control Samples and Analysis
- Performance Evaluation Program
- Verification and Validation
- Assessments and Audits
- Preventive and Corrective Actions.

1.12.1 General Performance Criteria for Instruments

Radiation monitoring systems are subjected to periodic testing and calibration to ensure the suitability of the instrument for continued operation. Therefore, acceptance criteria are defined, and the standards are satisfied by undertaking the proficiency testing. The performance criteria for the calibration and testing of instruments needed to conduct these programs are described in ANSI N317-1980 [31]. Performance specifications are also given in ANSI N-323 (ANSI 1997b), ANSI N42.17A (ANSI 1988b), and ANSI N42.17C-1989 (ANSI 1987c) for portable health physics instrumentation and IEC Publication 325 (IEC, 1981) for alpha and beta contamination meters and monitors. Criteria for air monitoring instrumentation are contained in ANSI N13.1 [32]. Criticality alarm systems are discussed in ANSI/ANS-8. 3-1986 (ANSI 1986a). The performance criteria that are discussed in the following subsections are specified in these standards as referenced.

1.12.1.1 Portable Radiation Survey Instruments

Radiation survey instruments should satisfy the calibration and performance verification as recommended in the nuclear facility in tune with ANSI N-323 [33] (ANSI 1997b). The instruments should satisfy the following requirements:

- The overall accuracy shall be within $\pm 20\%$, and the precision shall be within $\pm 10\%$ at the 95% confidence level.
- The detector response time (i.e., the time for the instrument reading to go from zero to 90% of full scale) shall be <10 s on the most sensitive scale and <2 s at readings of 100 mSv/h, 1000 μGy/h, and 500 dpm or greater.

- The instrument shall be capable of maintaining accuracy and precision for a minimum of 24 h of continuous operation.
- The instrument shall be equipped with a battery having a lifetime of 200 h of continuous operation.
- The instrument system shall function within specifications over all anticipated combinations of temperature and humidity (e.g., 15–65 °C, 40–95% relative humidity).

Gamma survey meters should meet the accuracy requirements stated in ANSI N317 [30] over the energy range of 0.01 to 1.25 MeV. The angular response of this gamma survey meter should be within ±15% over a 2π steradian frontal direction using at least two photon sources with energies ranging from 0.06 to 1.25 MeV. Based on the experience, it can be stated that this response specification is not satisfied by most instruments at lower energies due to attenuation of the gamma photon. The energy dependence should be within ±15% over the range of very low energy to 1.25 MeV, and the operating range should be from 5 μSv/h to at least 50000 μSv/h. Experience has indicated that ±20% over very low energy to 1.25 MeV is more realistic and achievable.

1.12.2 Performance Criteria for Installed (Fixed) Monitoring Instruments

Installed radiation monitoring systems, such as particulate airborne contamination monitors, surface contamination monitors, photon area monitor, and neutron area monitors, and post-accident monitoring systems should show adherence to the following standard performance criteria.

1.12.2.1 Particulate Airborne Contamination Monitors

Detection of airborne contamination due to dispersal of particulate radiation in air is generally detected by continuous air monitoring (CAM) systems. These monitors should meet the following criteria according to ANSI N317 [30]. The primary objective of any continuous air monitor is to detect the presence of particulate airborne radioactivity and activate an alarm to alert the personnel in the controlled area so that actions can be taken to minimize personnel radiation exposures. The selection of CAM should be based on the fact that it is capable of performing rapid functioning as quickly as possible and should have the lowest minimum detectable level of radioactive airborne concentration. The concentration of airborne radioactivity that can cause alarm within a given time interval is defined in units of DAC-h for a particular radionuclide. DAC-h is a function of the nuclide's airborne concentration in DACs, the sampling rate, the minimum detectable level of the instrument, and the

(vi) Generation of Logarithm and period signals
(vii) Introduction to nuclear systems for safety, safety related and monitoring: Start-up, intermediate, power range instrumentation, Reactor Regulating system, Flux Mapping system, Failed Fuel Detection & Localization systems, Stack Monitoring system.

2.2 Reactor Instrumentation is Different

Reactor instrumentation is different from conventional instrumentation. The reason for this is as under. Nuclear power plants are loaded with enough fuel to last for years, compared to continuous fuel feed in conventional power plants. So, while in the case of conventional power plants, to stop the production of heat, all that is needed is to stop the fuel feed, in the case of a nuclear power plant, increasing nuclear reaction rate due to equipment failure or malfunction cannot be stopped by exhaustion of fuel. In other words, the system is not self-limiting. While the concentration of heat energy in the fuel is 40×10^9 BTU/lb of pure nuclear fuel, the value is 14×10^3 BTU/lb of coal, a factor of $\sim$3 million. This means that nuclear fuels can generate intense heat, exceeding the capabilities of the coolant to remove heat. Further, the presence of high nuclear radiation fields and high temperature in core regions of a nuclear reactor causes deterioration in signal and system performance.

In view of the above, an approach of "Defence-in-depth" is the guiding principle of nuclear reactor instrumentation. This is illustrated in Fig. 2.1. The philosophy of the approach is "create, maintain, and update multiple independent and redundant layers of protection to compensate for potential human and mechanical failures. No single layer is exclusively relied upon".

This philosophy is incorporated in the following manner:

- Design & Construction of the reactor vessel and building.
- Multiple barriers against radioactive leaks.
  - Primary: Fuel, Cladding.
    The fuel is clad in suitable material like Zirconia, and this provides the first barrier to the release of radioactivity in the environment.
  - Secondary barrier: Reactor Vessel.
    In case of a breach in the fuel clad, the reactor vessel provides the next barrier.
  - Tertiary barrier: containment building.
    The reactor vessel is in turn housed in a concrete building built to withstand the rigors of natural calamities like earthquakes and provides the third boundary.
- **Redundancy and diversity in the instrumentation**
  The Instrumentation and Control (I&C) systems deployed in a nuclear reactor are based on diverse principles and employ sufficient redundancy (two out of three logic). This ensures that monitoring and control of the reactor is possible for several

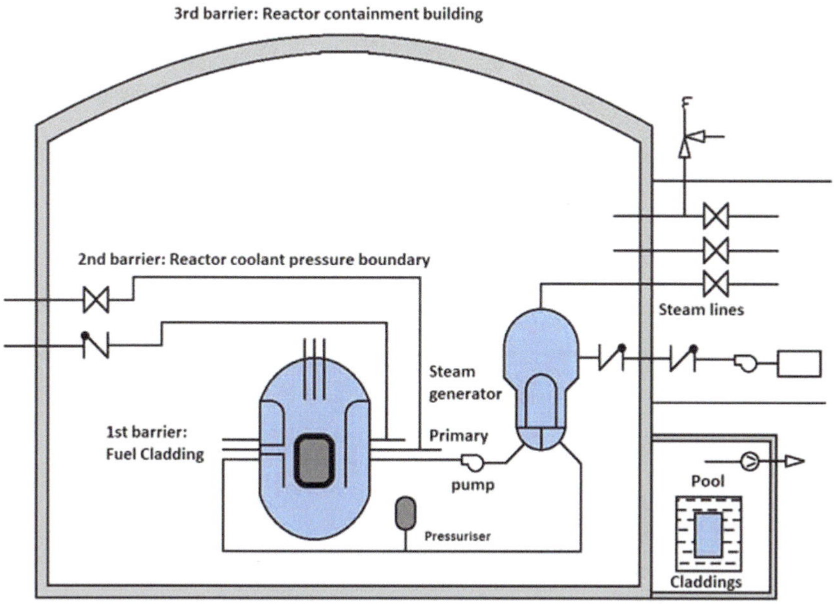

Fig. 2.1 Principle of "Defence in Depth"

types of postulated Design Basis Accidents (DBA) as well as Beyond Design Basis Accidents (BDBA).

- **Maintenance & operation**
  The I&C systems undergo regular periodic maintenance while the reactor is operating, to weed out discordance and failures. The nuclear power plants maintain the necessary inventory of spares to minimize the Mean Time To Repair (MTTR).
- **Physical security**
  The nuclear power plants deploy multiple levels of Physical Security and access controls to ensure that only authorized personnel have access to safety-critical systems and thus prevent any intentional sabotage.
- **Emergency preparedness**
  The emergency preparedness drill is carried out periodically at all nuclear installations to ensure that all the personnel are aware of Standard Operating Procedures (SOPs) in the event of any emergency.

2.3 What to Measure

When heavy nuclei such as U, Pu are bombarded with neutrons, they absorb the neutrons and split into two smaller nuclei in a process called nuclear fission. In addition to the two "fission products", the reaction also produces a few neutrons, gamma rays, and neutrinos in addition to releasing a copious amount of energy resulting from the difference in the binding energies of the parent nuclei and the fission products, as shown in Figs. 2.2 and 2.3. Some nuclei like U-235 fission with neutrons of any energy, with high probability of fission when impacted by "slow" neutrons (neutrons of less than 0.05 eV energy), while some nuclei like U-238 fission only with high-energy (>1 MeV) neutrons. Be that as it may, the availability of more than one neutron following one fission essentially means that a system can be built in which the fissions, once started, can be sustained as a "chain reaction" by judicious use of the neutrons.

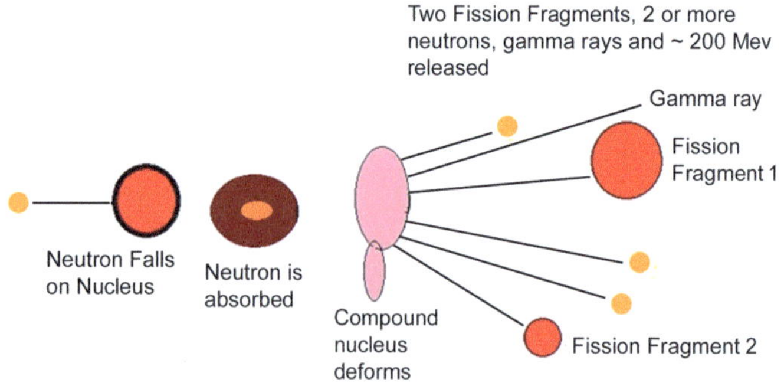

Fig. 2.2 The fission process

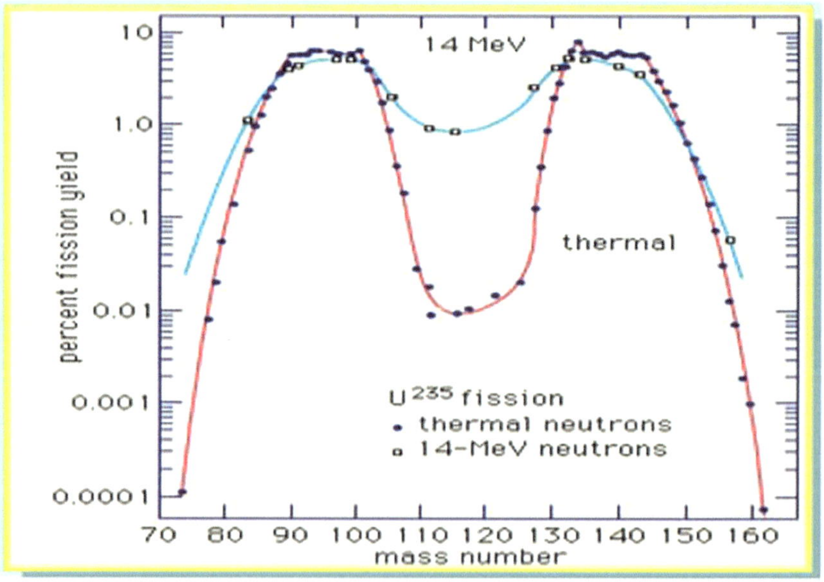

Fig. 2.3 Mass distribution of fission fragments [https://www.britannica.com/science/nuclear-fission/The-stages-of-fission]

Depending on the intention, one can have an uncontrolled chain reaction, such as in a Nuclear Bomb, in which the emphasis is on maximizing energy release, i.e., high yield, and minimizing duration of energy release: viz, an explosion. One can have a controlled chain reaction, the principle on which Nuclear Reactors operate. Here, things are engineered to ensure that exactly 1 neutron remains available after every fission event by using neutron absorbers & moderators (to slow down the neutrons). A steady-state operation with constant power is achieved by using control rods for the regulation of power.

The control system of a reactor has two main functions: (i) maintain/regulate the neutron flux and (ii) regulate its rate of change within the permissible limits. This is as demanded by the power requirements. On the other hand, the safety system generates ALARM and TRIP signals to restore the neutron flux and its rate of change to safe values whenever the regulation deviates beyond permissible limits. This is also necessary if and when there is a breach of integrity in some part of the reactor system and when the required response time is beyond the capabilities of the regulating system.

2.3.1 Parameters for Reactor Control and Safety

There are a number of parameters that are important for the control and safety of a reactor. These include Neutron flux, Primary coolant inlet & outlet temperature, Coolant channel temperature, Moderator temperature, Fuel surface temperature, Primary system pressure, Steam pressure, Primary coolant flow rate, Steam flow, Feed water flow, Control and absorber rod position etc. This chapter deals exclusively with neutron flux.

2.4 Principles of Neutron Detection

Gas-filled type neutron sensors are the preferred choice for reactor applications. This stems from the inherent advantages that gas-filled detectors have for the application at hand. These include: Wide dynamic range, long-term stability, resistance to radiation damage, and inherent gamma-ray discrimination.

Neutrons are detected by making them interact with a suitable target material, which would emit charged particles that can be detected by the ionization they produce. This two-step process is necessitated by the fact that neutrons, being neutral, do not ionize directly. Commonly used materials in gas-filled detectors for reactor instrumentation are $^{10}B(n, \alpha)7Li$ and the fission reaction in ^{235}U. The movement of the charges released for the interaction under the influence of an electrical field constitutes the signal. In detectors using the ^{10}B reaction, gas multiplication is used to increase the number of ions from each event, especially so when used in the pulse counting mode.

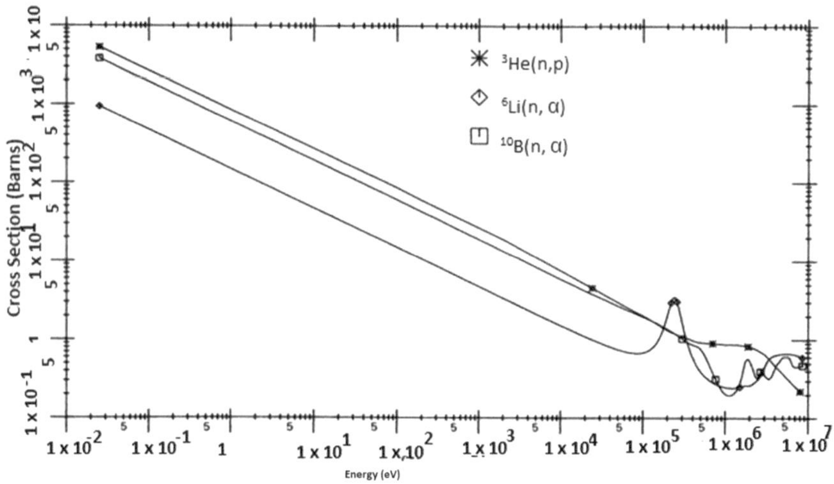

Fig. 2.4 Interaction cross section vs energy for the nuclear reactions of ^{3}He, ^{6}Li and ^{10}B

The nuclear reactions of neutrons with various isotopes are given below:

$$^{10}_{5}\mathrm{B} + ^{1}_{0}n \rightarrow \begin{cases} ^{7}_{3}\mathrm{Li} + ^{4}_{2}\alpha + 2.792\,\mathrm{MeV\ (ground\ state)} \\ ^{7}_{3}\mathrm{Li}^{*} + ^{4}_{2}\alpha + 2.310\,\mathrm{MeV\ (excited\ state)} \end{cases}$$

$$^{6}_{3}\mathrm{Li} + ^{1}_{0}n \rightarrow ^{3}_{1}\mathrm{H} + ^{4}_{2}\alpha + 4.78\,\mathrm{MeV}$$
$$^{3}_{2}\mathrm{He} + ^{1}_{0}n \rightarrow ^{3}_{1}\mathrm{H} + ^{1}_{1}\alpha p + 0.764\,\mathrm{MeV}$$

The cross sections for the above reactions also vary with energy as shown in Fig. 2.4.

The various materials that are employed as neutron sensing materials and their relative merits and demerits are included in the Table 2.1.

2.5 Gas-Ionization Detectors-Operational Characteristics

Ionization chambers, the simplest and the oldest of gas-filled detectors, are even today the workhorses of any nuclear reactor instrumentation. In its simplest form, the ionization chamber consists of an outer sealed housing inside which are dispensed two or more electrodes held insulated from each other and the outer housing. One set of electrodes serves as the High Tension (HT) electrode, while the remaining are used as signal electrodes. The volume between these electrodes constitutes the so-called sensitive volume of the chamber. This is filled with a suitable gas at an adequate pressure depending on the application at hand. Radiation interacting with

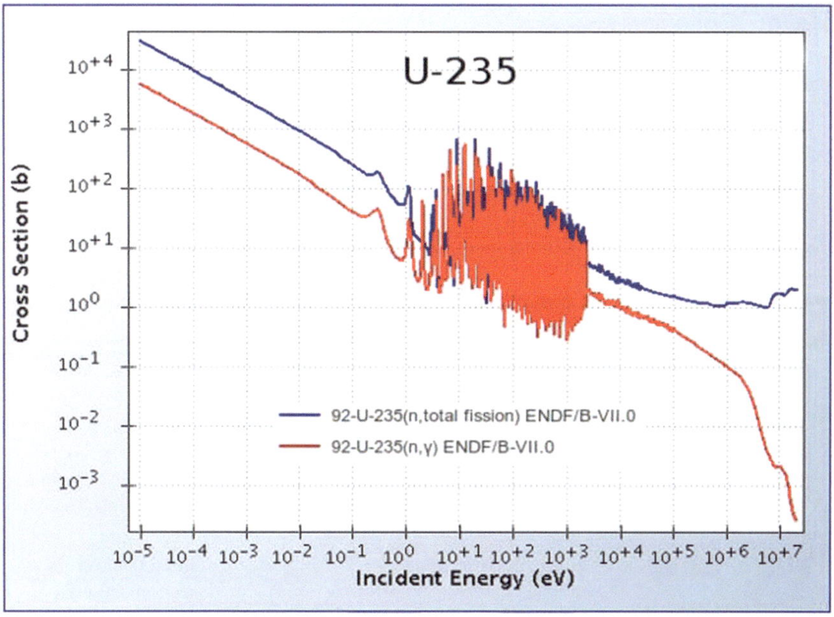

Fig. 2.5 Interaction cross section vs energy for the nuclear reactions of ^{235}U (https://whatisnuclear.com/fast-reactor.html)

this volume generates ion pairs, and these drifting in opposite direction under the influence of an applied electric field constitute the signal output from the chamber.

As the applied voltage to the chamber is increased, the positive and negative ions (or electrons) begin to drift toward the opposite electrodes with increasing drift velocity. This separates the ions of either polarity and reduces the equilibrium concentration of ions within the gas, thereby reducing the recombination of the ions in their path to the collecting electrodes. As a result of the reduction in the charge that is lost, the measured current increases with increasing voltage. When the voltage applied is sufficient, the electric field is sufficiently high to effectively reduce the recombination to negligible levels, all the primary charges produced get collected, and any further increase in voltage does not lead to an increase in the output current. This is the saturation region, the region in which the DC ionization chambers are conventionally operated. This ensures that the current measured in the external circuit faithfully indicates the rate of formation of ions in the active or sensitive volume of the chamber and hence, indirectly, a measure of the incident radiation flux. The voltage required to obtain saturation current is called saturation voltage, and its value depends mainly on the electrode gap, the type of gas used, and its pressure.

At higher voltages, charge multiplication takes place. The electrons released in the primary ionizing events are accelerated sufficiently to produce secondary ionization,

Table 2.1 Relative merits and demerits of neutron sensing materials

Isotope and natural abundance (%)	Cross section (barns)	Energy per reaction (MeV)	Remarks
^{10}B, 19	3840	2.792/2.310	High cross section and low energy per reaction. Useful in cases where low neutron flux is to be measured while the gamma field is not too high (<1 kR/h, typ)
^{6}Li, 7.6	940	4.78	Low cross section and low energy. Rarely used in reactors
^{3}He, 0.00014	5330	0.764	High cross section and low energy. Useful at <1 R/h background and for initial start-up only
^{235}U, 0.7	540	200	Reasonable cross section and high energy. High gamma tolerance (up to 1 MR/h or more) general work-horse

and the pulse height is proportional to the energy of the radiation. Proportional counters operate in this region. At still higher voltages, the charge is amplified to a value that is limited by the characteristics of the chamber and the external circuit ultimately leading to electric breakdown of the gas when the space charge inside the avalanche is strong enough to shield the external field. This results in the recombination of ions producing photon emission and in secondary ionization with new avalanches beyond the initial one. When the process propagates backwards from the avalanche tail, eventually an ion column links the anode and cathode, and a spark discharge will happen. The chamber or counter is then said to operate in the Geiger-M*ü*ller mode.

When this discharge is stymied by adding quenching agents to the gas, the region is called limited Geiger mode. Here, output pulses at the anode are much higher than in the proportional mode. The process of spark discharge can also be stopped by altering the electric field: application of short (a few ns) pulses of high voltage results in short discharges (streamers) from the ion trail of a crossing particle, and a track image can be obtained by photography as in a streamer chamber. An effect similar to limited Geiger mode can be obtained using thick, against the usual 20–30 μm) anode wires without using quenchers. This, limited streamer mode, is advantageous in view of a higher mechanical reliability due to the thick wires. The various regions of operation of a gas-filled detector are shown in Fig. 2.6, where I is the recombination region, II is the saturation region, III is the proportional region, IV is limited proportionality region, and V is the Geiger region.

Ionization chambers have been used for the detection of virtually every form of nuclear radiation, such as gamma rays, X-rays, neutrons, alpha particles, beta rays, etc., except neutrinos. The advantages of ionization chambers include their ruggedness, long-term stability, wide-range capability, good resistance to radiation damage, ease of fabrication, etc. They are ideally suited for applications involving gross measurement of radiation intensity. It must be mentioned, though, that specially designed ion chambers have been used for spectroscopic applications, especially those involving alpha particles. The schematic diagram of an ion chamber is shown in Fig. 2.7.

Several factors can cause a loss of saturation in an ion chamber, the chief among them being recombination losses. This can be minimized by ensuring that a sufficiently high electric field exists everywhere within the sensitive volume of the chamber. Columnar recombination is most significant in cases where the initial ion density is very high, as in the case of tracks of heavy ionizing particles like alpha particles or fission fragments, and low ionizing particles like electrons or gamma radiation at high gas pressures. The recombination rate is given by

$$\frac{dn_+}{dt} = \frac{dn_-}{dt} = -\alpha n_+ n_- \tag{2.1}$$

The magnitude of the ion current or the intensity of radiation also affects the degree of recombination. At lower levels of radiation intensity, the number of ion pairs is less, and as seen from Eq. 2.1, the recombination rate will be less significant. As the radiation intensity goes up, the rate of ion formation and recombination goes up. The effect on the voltage-current (V/I) characteristic is illustrated in Fig. 2.8.

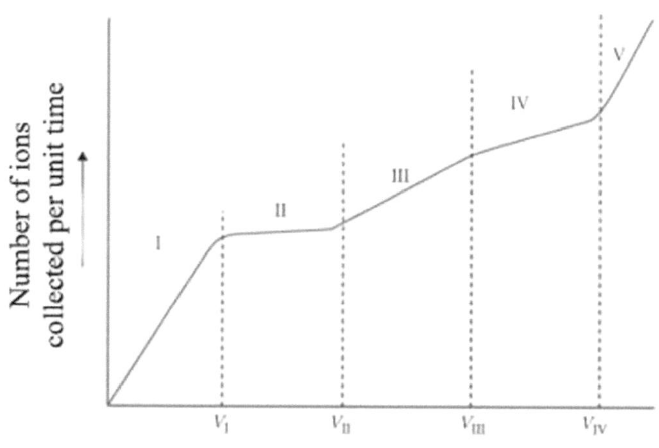

Fig. 2.6 Regions of operation of a gas-filled detector

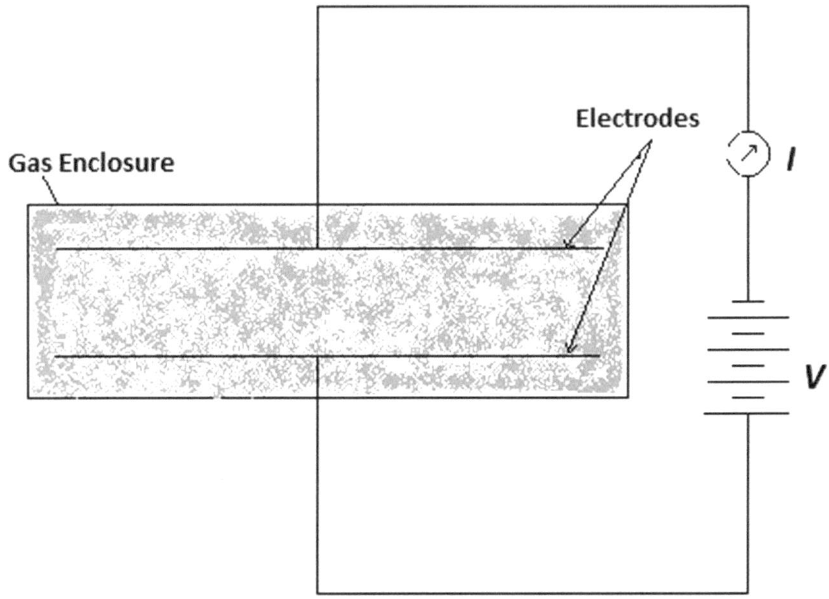

Fig. 2.7 Schematic diagram of the basic components of a DC ion chamber

If the production of ions is uniform throughout the volume of the ion chamber, the drift of the ions toward their respective electrodes will result in an imbalance in their equilibrium concentration, thus creating a gradient in their concentrations. This leads to diffusion in the direction of decreasing concentration. This direction is opposed to the direction of charge flow and can cause a reduction in the measured current. The resultant perturbation in the measured current for parallel planar geometry is given by

$$-\frac{\Delta I}{I} = \frac{\varepsilon k T}{eV} \tag{2.2}$$

electrodes have been assumed. Here ε—ratio of the energy of the charge carrier in the presence of electric field to that in the absence of electric field, k—Boltzmann constant, T—absolute temperature, e—electronic charge and V—voltage applied between the electrodes. The value is nearly unity for ions, and hence the loss due to this effect is negligible for ions. For electrons, however, the value of ε is of the order of a few hundreds, and hence the losses due to free electron diffusion can be quite significant at low voltages. Operating the chamber at appropriate high voltages minimizes the loss.

The combined effect of recombination and diffusion determines the shape of the V/I characteristic of an ion chamber. An empirical relationship between the saturated current I_s, the electrode spacing d, and the measured current I at an applied voltage of V is given by

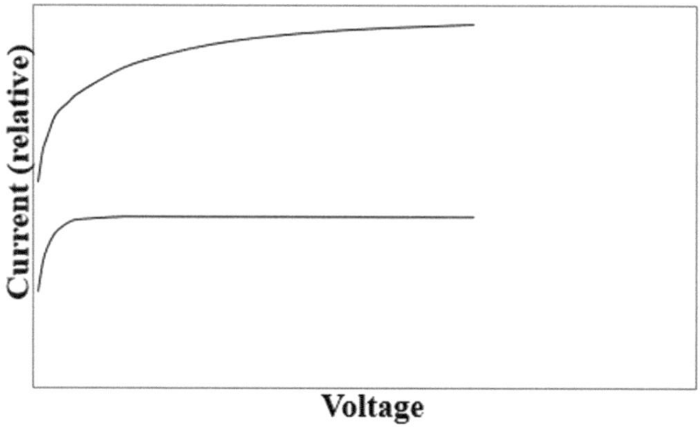

Fig. 2.8 Effect of radiation intensity on the V/I characteristics of an ion chamber

$$\frac{1}{I} = \frac{1}{I_{\text{sat}}} + \frac{k}{V^2} \tag{2.3}$$

From Eq. 2.3, it is seen that the plot of $1/I$ versus $1/V^2$ should be a straight line. The value of saturation current I_s can be obtained by extrapolating the line to the zero of $1/V^2$ (infinite voltage).

2.5.1 *Modes of Ion Chamber Operation*

Ion chamber output can be processed either as individual interactions or the time average taken over many individual events. The former, called the pulse mode, is used when the number of events is small in number. The latter called the current or DC mode is used when the number of events is too high for the pulse mode operation. In the pulse mode, the ratio of the time constant **RC** (τ) of the external circuit and the charge collection time t_c leads to two extreme modes of operation. For $\tau << t_c$, the current flowing through the load resistance, **R**, is practically the prompt value of the detector current. The voltage then approximately follows the time dependence of the current. Detectors are used in this mode when event rates are very high. This is termed the current-sensitive mode of pulse operation. However, detectors are operated more commonly in the other extreme, $\tau >> t_c$, such that a small current flows in the external circuit during the charge collection and the detector current is integrated briefly on the capacitance. If the timing between the events is large, the capacitance will discharge through **R**, returning the voltage to zero (Fig. 2.9).

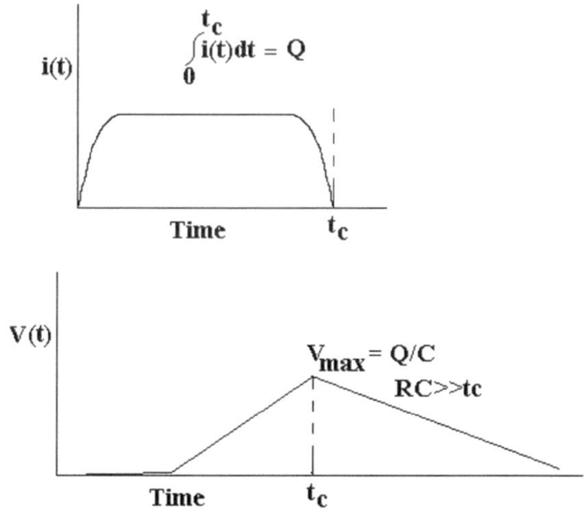

Fig. 2.9 The output from an ion chamber as a function of time

2.6 Types of Neutron Detectors: In-Core and Ex-Core

Detectors for reactor instrumentation are located either as in-core or out-of-core detectors. In-core detectors, as the name indicates, are located inside the core of the reactor. They are generally located in the narrow coolant channels in the reactor core and provide detailed information about the neutron flux distribution within the core. By contrast, out-of-core detectors are located outside and at some distance from the core. They, hence, respond to the neutron flux integrated over the whole core. These are located either inside or outside the reactor pressure vessel and see a less severe environment compared to the in-core detectors. The high-energy neutrons arising from fission reactions are slowed to thermal energies by the moderator atoms in the core. So neutron sensors also detect and measure neutrons of low energy, where the cross section for neutron interactions of interest is high.

2.7 Measurement Ranges of Reactor Neutron Flux

As explained earlier, neutron flux is proportional to the instantaneous reactor power and is a true indicator of the operating status of the reactor. Since the reactor power ranges from watts to hundreds of megawatts, it is impossible to use one set of detectors with one circuitry to cover the entire range. So, as a convenience, the reactor power is divided into three ranges, viz., source, intermediate, and power ranges.

Each set of detectors is used to measure a limited part of the range and is complemented by additional sets of detectors. To ensure safety and reliability, one set of detectors have a minimum of 2 decades range overlap with the other set of detectors. This overlapping (typically 2 decades) makes for the smooth transfer of control and safety from one set of detectors to the next. It must be mentioned that with fission detectors, it is possible to use one detector to cover the entire range of reactor neutron flux by operating the device in three modes of operation, viz., pulse counting, Mean Square Voltage (MSV), or Campbell and DC mode.

2.7.1 Source Range

The permissible counting rates in counts per-second dictate the limits of operation of the detectors in pulse mode (Source range). The minimum counting rate for a safe condition, as specified by reactor safety norms, is typically 1–10 cps. This sets the lower end of the counting range. On the other hand, the upper limit of neutron flux that can be measured with good accuracy depends on the maximum counting rate. Maximum counting rate is dependent on the pulse width or the system resolving time (the longer of the two is the limiting factor) and the maximum counting loss that is acceptable. As a rough rule of thumb, for a counting loss of less than 10%, the maximum counting rate would be $1/10\tau$, where τ is the resolving time. So, a system with 200 ns resolving time would have a maximum counting rate of 5×10^5 cps.

Gamma background is invariably present at most detector locations, and the selection of detectors for source range monitoring is governed by the ability of the device to discriminate against the gamma flux, i.e., measure "few" neutrons in the presence of "many" gammas. Using pulse height discrimination, the unwanted and relatively short gamma-induced pulses can be eliminated. Here, fission detectors are ideal candidates because of their superior gamma tolerance capability. ^{10}B lined proportional counters are also used in this range.

2.7.2 Intermediate Range

The intermediate range overlaps with the source range, and here the influence of the gamma background is less severe. The higher neutron flux means that individual pulses pile-up, and the signal is somewhat akin to a DC current. Due to this, the signal becomes indistinguishable from the gamma-induced signal. To overcome this situation, ^{10}B-lined gamma compensated ion chambers are the most commonly used detectors in this range. Another means of measuring the neutron flux is the Mean Square Voltage or Campbell mode.

In the DC mode, the mean value of the generated current is considered. It can be shown that the standard deviation σ of the fluctuating component of this mean current is proportional to the square root of the neutron flux. If the resulting signal is employed for detection, we get the so-called Campbell's method or Mean Square Voltage (MSV), usually employed in the intermediate range with fission detectors. For a chamber connected to a capacitor C and leak resistance R is given as

$$\sigma = q(nR/2C)^{1/2} \tag{2.4}$$

Here n is the mean number of events per unit time. The main advantage of this method is its gamma discrimination capability.

$$I = \Sigma nq = n_n q_n \left(1 + \frac{n_\gamma q_\gamma}{n_n q_n}\right) \tag{2.5}$$

and

$$\sigma^2 = k\Sigma nq^2 = kn_n q_n^2 \left(1 + \frac{n_\gamma q_\gamma^2}{n_n q_n^2}\right) \tag{2.6}$$

We thus gain a factor of 1000 in the neutron-gamma discrimination ratio. The sensitivity in the MSV mode is given by

$$S_{\text{MSV}} = \frac{1}{2}\overline{N} q_e^2 Z_{12}^2 \frac{\omega_H^2}{\omega_H + \omega_L} \tag{2.7}$$

Here, S is in volts^2, is the average pulse rate (number per sec), qe is the charge times the fraction of potential field through which it falls (C), Z_{12} is the mid band pass transfer impedance and ωH and ωL are the upper and lower cut-off frequencies of the system through the a-c amplifier.

The intermediate range extends into and completely overlaps the power range. The power range covers from 1 to 150% of full power to provide allowance for small power excursions. ^{10}B-lined chambers with or without gamma compensation or fission chambers are employed in this range.

2.7.3 Power Range

In the power range, the mean value of the generated current is considered by operating the detectors in the DC mode.

2.8 Considerations for Deploying Detectors at Different Stages of Reactor

The selection of a particular type of detector based on one of the above reactions alluded to earlier is necessary to ensure that the purpose of neutron detection, viz, control and safety of the reactor, is not compromised. The basic considerations are Signal amplitude, Response time, Tolerance to interfering radiations, and Temperature.

2.8.1 Signal Amplitude

The signal amplitude refers to the output from the detector in terms of either charge or current. Signal amplitude should be far higher than the amplitudes of background and electronic noise put together. There are several contributions to the background: the Alpha background in Fission counter, DC leakage current in Ion chambers, contributions from other interfering radiations like gamma, etc. Contributions to electronic noise include those inherent to the processing circuit and EMI from the surroundings. Ideally, a Signal/Noise (SNR) > 5 is recommended for reliable detection.

2.8.2 Response Time

In the event of a reactor accident or major component/system failure, the detection system should respond rapidly to initiate shutdown of the reactor before damage can occur. The rapidity with which this is achieved depends on the response time of the detector-electronics system. Typically, the detector response time $\sim$100 ns are achievable, especially with Fission Counters, and the processing instrumentation limits the system's response time

2.8.3 Tolerance to Interfering Radiation

The location chosen for deploying the detector need not always be ideal, and there may be interfering or competing radiations like gamma rays from structural material and fission product gamma rays. So, it is necessary to choose a detector with care to ensure that the detector is tolerant to gamma background. Also, special processing techniques like fast current pulse amplifiers or Campbell mode operation are used to ensure that neutrons are detected with adequate accuracy.

2.8.4 Operating Temperature

Since the temperature obtained in reactors can vary from room temperature to 600 °C or more, the ambient temperature at the detector location becomes crucial in the design and choice of detectors. As an example, Fast Breeder Reactors and gas-cooled reactors often require the use of high-temperature detectors. The high temperature has the following effects:

- Reduced insulation resistance of ceramic insulators
- At high voltages, possible discharges on the surface of the ceramic insulators at high voltage resulting in pulses that mimic neutron pulses

- Contamination of the filling gas due to out-gassing from the chamber body
- Loss of fill some gases due to the combination with the materials of the chamber
- Reduction in the strength of the metal body and electrodc support.

The problems are overcome by the following solutions:

- Use of high-density high-purity alumina
- Specially contoured Ceramic
- Special chemical cleaning and degassing methods
- Selection of Inconel-600Z as construction material
- Use of special Mineral Insulated cables.

2.9 Important Specifications of the Neutron Detectors

2.9.1 Sensitivity

This parameter, which is measured either in cps/nv (Pulse mode) or in Ampere/nv (DC mode), informs us about the response of the detector to a unit flux of neutrons falling on the detector. It combines the geometric and intrinsic efficiencies of detection as well as the response of the detecting medium. Here, it may be mentioned that the intensity of neutron radiation is determined by the rate of flow of neutrons. The neutron flux value is indicated as the neutron density (n) multiplied by neutron velocity (v), where n is the number of neutrons per cubic centimeter (expressed as neutrons/cm^3) and v is the distance the neutrons travel in 1 s (expressed in cm/s). Thus, neutron flux (nv) is measured in neutrons/cm^2/s.

The sensitivity of a detector for a given application is chosen keeping in mind the upper and lower ranges of the flux to be measured. In case of pulse counters, which are used invariably during the start-up range of the reactor power, the sensitivity is so chosen as to provide (at the lower limit) a minimum of 3 cps/nv, at a slightly sub-critical condition of the reactor. The upper limit is usually 5 decades above this lower limit. In case of fission counters employed with fast current amplifiers, the range can be 6 decades. In all other ranges, the sensitivity is chosen so as to provide a sufficient signal at the lower limit and not saturate at the upper limit.

2.9.2 Average Charge Produced Per Event

This parameter is a physics design parameter and depends on the energy deposited by the reaction products of the neutron interaction and the energy required to produce an ion pair. This depends on gas mixture, electrode spacing, and gas pressure and is described in the section on Gamma Tolerance.

2.9.3 Charge Collection Time

This is again dependent on the parameters described in the previous paragraph and will be detailed later.

2.9.4 Capacitance and Insulation Resistance

The capacitance of the detector should be kept to a minimum, as this plays a significant role in determining the pulse width. However, the capacitance is a function of the area of the sensitive volume and the electrode gap, and as such, there are competing and often opposing requirements which have to be reconciled to optimize the detector design (see section on Gamma tolerance). The insulation resistance is determined by the quality of the insulators used and the care taken during every stage in the development of the detector. Insulation resistance determines the leakage current (that is, the current produced by the detector in the quiescent state) and shall be as high as possible (tens of gigohms or more).

2.9.5 Operating Voltage Range

The operating voltage is a design parameter that depends again on gas mixture, electrode spacing, and gas pressure.

2.9.6 Maximum Operating Temperature

This factor depends on the materials used in the construction of the detectors including the insulator and the cables used thereof.

2.9.7 Permitted Range of Neutron Flux Measurement

This is also a design parameter that depends again on gas mixture, electrode spacing, and gas pressure. The practical limits on the lower end of the measurement range are imposed due to the statistical fluctuations at low count rates for an acceptable counting time. The higher end of the measurement range is dictated by the counting losses occurring due to pile-up at higher incident neutron flux levels.

2.9.8 Gamma Tolerance

The gamma discrimination capability of a pulse detector is improved by reducing the charge collection time. The charge collection time depends on the following four variables: gas mixture, electrode spacing, gas pressure, and applied voltage. Of these, four, two, the electron drift velocity and the electrode spacing, dictate the collection time. The four parameters mentioned above have an impact on the elector drift velocity. The electrode spacing decides the distance traveled by the electrons. The charge generated is proportional to the energy deposited in the gap. This, in turn, is proportional to the product of pressure and gap [20]. So,

$$q = kpd$$

Here q is the charge liberated (in C) in an electrode gap of d (mm) at a gas pressure of p atmospheres and k is a constant. It is thus obvious that maintaining a constant pressure-gap product provides a constant charge. The charge collection time is given as

$$\tau = k_2 d^2 p / V$$

Even though the higher pressure is needed in order to maintain a sufficiently high charge per event for smaller electrode gaps, the collection time is reduced by a reduction in the electrode spacing as shown in the above equation.

In the equation above, τ is the collection time (in ns) and V is the applied voltage. k_2 is a factor that depends on the gas-fill, pressure, and the value of E/p. For a cylindrical geometry, the electric field at a point r, $E(r)$ is given by

$$E(r) = V / r \log(b/a)$$

Thus, for a system operated in the current mode, a reduction in electrode spacing increases the voltage pulse at the input of the preamplifier even though the charge per event is less. For an ideal current pulse, that is, for zero detector capacitance

$$I_{\text{in}} = q/\tau$$

Here I_{in} is the peak current at the input.
So

$$I_{in} = qV / k_2 d^2 p$$

Reducing the gap by a factor of two reduces the charge by the same amount. But the collection time would be reduced by a factor of four. Hence, resulting in an increase in the current into the preamplifier by a factor of two. Another factor that needs to be considered in selecting the electrode gap, gas-fill, and pressure is the maximum voltage that can be applied to the detector. This is limited by the breakdown noise pulses (BNP) in the Mineral Insulated (MI) cable at elevated temperatures. These noise pulses simulate the neutron pulses, resulting in spurious counts.

The fill-gas should be chemically inert, have good thermal conductivity, low neutron cross section, and suitable ionization properties. The gas should be chemically inert so that it does not combine with the chamber materials and thus reduce the

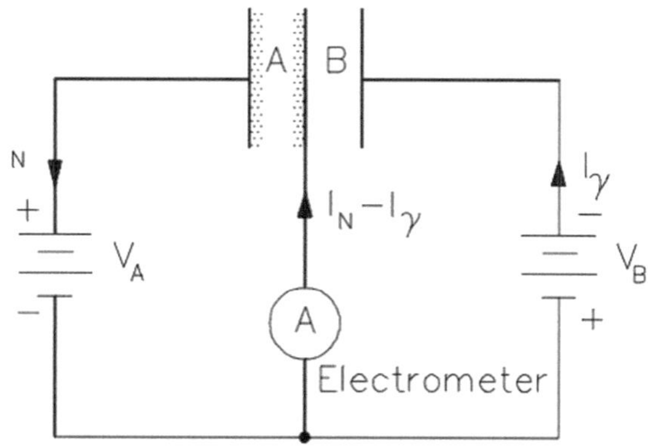

Fig. 2.10 Schematic of a gamma compensated ion chamber

available gas for ionization. If the neutron cross section is high, the fill gas would be depleted by transmutation. Almost all the noble gases and nitrogen meet the above requirements. None of these pure gases, however, gives sufficiently fast collection times. Gas mixtures such as argon-carbon dioxide or argon-methane provide fast collection times. However, polymerization of methane at high irradiation and the possibility of breakdown of CO_2 at high temperatures make these choices unattractive.

In the case of DC chambers, the gamma tolerance is achieved by providing an additional volume, nearly identical to the neutron-sensitive volume, but without the neutron-sensitive material. This ensures that the gamma DC signal is nearly identical from both volumes. By applying opposing polarities, the net current would be the neutron current as illustrated in Fig. 2.10.

2.9.9 *Burn Up/nv-t Lifetime*

This is also a design parameter that depends on neutron-sensitive coating material and the operating region of the detector. In general, the proportional counters have lower nv-t lifetime while the ion chambers & fission counters have higher nv-t lifetime.

2.10 Detector Design

The design of the detector proceeds on the following lines:

1. The size of the detector is dictated by the space available at the location of the detector.
2. The second design input is the environmental parameters, like the temperature and ambient gamma background.
3. The next design input is the range of neutron flux to be measured.

2.11 Detector Signal Processing

2.11.1 Signal Processing for Pulse Mode Operation

In pulse mode operation, each individual current pulse is recorded as a separate event. The charge content of each current pulse (Q) is proportional to the energy of the incident particle. Pulse mode operation can be deployed for either energy spectroscopy or for pulse counting requirements.

In the energy spectroscopy requirement, both the energy of individual pulses and their rate of occurrence are recorded, and a histogram of energy versus rate of occurrence is generated. This histogram/energy spectrum is analyzed to identify the particular radionuclide as well as to quantify the percentage of the particular radionuclide present in the given sample. Typically, energy spectroscopy requirements emphasize more on achievable FWHM (Full Width at Half Maximum) and hence are limited to counting rates below 10000 cps.

In the pulse counting requirement, the rate of occurrence of pulses is the only quantity of interest, and emphasis is more on count linearity over a given dynamic range of occurrence. Typical input dynamic ranges are $\sim$1 Mcps, and this is achieved at the expense of achievable FWHM.

It may be noted that for the Neutron Flux Monitoring System (NFMS) in a nuclear reactor, the parameter of interest is their flux rather than their energy. Thus, the NFMS deploys pulse counting instrumentation wherein the rate of occurrence is recorded with a high degree of linearity over a wide dynamic range covering 6 decades of incident neutron flux and count rates up to 2×10^6 cps.

In this chapter, the discussion of pulse mode instrumentation is limited to that deployed for pulse counting applications as required for NFMS for reactor applications.

The various functional blocks shown in Fig. 2.11 are briefly described in Sects. 2.11.1.1 through 2.11.1.7.

2.11.1.1 Neutron Detectors in Pulse Mode Instrumentation

Pulse mode instrumentation is deployed for fission counters and various types of proportional counters. The fission counters have neutron sensitivity of 0.1–1 cps/nv, extremely fast charge collection times ($\sim$80 ns), and require operating HV in the range of 400–600 V. The proportional counters have neutron sensitivity of 1–200 cps/nv, collection times $\sim$300 ns, and require operating HV in the range of 1000–2500 V. The type of instrumentation deployed depends largely on the response time of the neutron detector as well as the distance between the detector and the preamplifier.

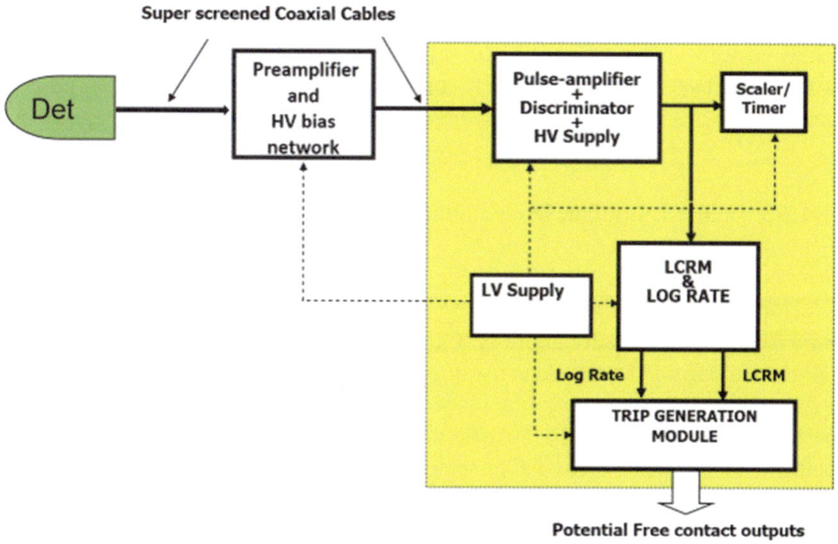

Fig. 2.11 Simplified block diagram of pulse mode instrumentation

2.11.1.2 Preamplifier Module

The preamplifier is the first and the most crucial functional block of the entire signal processing chain. A charge-sensitive type preamplifier is commonly used along with neutron detectors operating in pulse mode, and it works satisfactorily provided the required count rates are below 15000 cps and the cable length between the neutron detector and the preamplifier is ~30–50 m. The charge-sensitive preamplifier has relatively lower bandwidth and is more tolerant to electromagnetic interference (EMI) .

If, due to operational reasons at the installation site, the preamplifier has to be installed more than 50 m away from the neutron detector, or if the required count rates are in excess of 30000 cps, then a current-sensitive type preamplifier has to be deployed. The current-sensitive preamplifiers are high-gain, wide-bandwidth amplifiers that can process count rates up to 2 Mcps with acceptable linearity. However, due to their wide bandwidth, they are extremely susceptible to EMI, and several additional components, viz., common-mode filters, differential mode filters, soft ferrite beads etc., have to be incorporated in the design to achieve the desired electromagnetic compliance (EMC) .

The signal cable between the neutron detector and the preamplifier is carefully chosen to provide sufficient shielding to the low-level pulse signals that are carried by it. In most instances, a "Super-Screen" type coaxial/triaxial cable with two or more u-metal foils and an anti-microphonic semi-conducting layer is used between the neutron detector and the preamplifier. Such super-screen cables have a very low "surface transfer impedance" and provide extremely good grounding and shielding, and minimize susceptibility to radiated and conducted EMI. In some instances, these cables have to withstand very high temperatures/background gamma radiation or Loss of Coolant Accident (LOCA) conditions, and accordingly, special-purpose "mineral insulated" cables have to be used between the detector and the preamplifier.

2.11.1.3 Pulse Amplifier, Discriminator and HV Supply Module

This module generates the High Voltage supply required for biasing the neutron detector and further amplifies/"shapes" the pulses received from the preamplifier. For a Charge Sensitive Preamplifier, shaping is essential before converting to logic pulses, while for a Current Sensitive Preamplifier, no shaping is required. Baseline restoration is essential in both cases as pulse heights are measured w.r.t. baseline and the input pulses occur with a Poisson distribution. Commonly used pulse shaping methods include quasi-Gaussian or semi-Gaussian pulse shaping.

The shaped analog pulses are "discriminated" against a set threshold voltage called as "discriminator bias" to generate a logic pulse (TTL) for each valid event from the detector. The discriminator bias has to be optimally set so that the pulses due to electronic noise or gamma background are "discriminated out" and "loss of valid events" generated by the incident neutrons are minimized.

Before deploying the pulse mode instrumentation for a reactor, the H.V. plateau characteristics and discriminator bias characteristics of the neutron detector, along with the instrumentation channel, is studied by carrying out a series of experiments in a research reactor. The baseline data is analyzed to determine the optimal operating point (operating H.V. & Discriminator Bias (D.B.)) & sensitivity (cps/nv), and to validate the count rate linearity, count rate stability, etc., over the desired operating range of the instrumentation channel.

2.11.1.4 LCRM and Log Rate Generation Module

The logic pulses (TTL) generated by the pulse amplifier are further processed by the log count rate meter (LCRM) and Log rate generation module. The generated Log(Count_Rate)/Log.P signal is proportional to instantaneous reactor power, while the Log.Rate signal gives the information about the e-folding Period of the reactor. The Log.P signal and Log.Rate signals are monitored for the generation of ALARM and TRIP signals for the control and safety of the reactor.

It must be noted that the Log P signal should maintain linearity over 6 decades of input count rate and should have excellent temperature stability and long-term stability. These stringent design requirements are conformed to by using special circuit techniques as well as careful selection of active and passive components for the LCRM module.

The LCRM and Log.Rate module has several built-in calibration and diagnostic features, including the safety control rod axe man (SCRAM), to enable periodic calibration and preventive maintenance of the instrumentation channel. Typically, preventive maintenance is carried out once every quarter to weed out possible failures of the instrumentation channel.

2.11.1.5 Scaler/Timer Module

The logic pulses (TTL) generated by the pulse amplifier are also provided as input to the Scaler/Timer module, which counts them for a preset time duration and displays the counts on the front panel. The counting duration can be user-settable (Manual mode) or internally set (Auto mode) with typical values of counting time between 0.1 and 999 s. The scaler/timer modules have 6–7 digit display on the front panel and have a typical pulse pair resolution of 50 ns. In some instances, the Scaler/Timer modules are located on the control panel of the nuclear reactor, in which case the logic pulse generated by the pulse amplifier has to be transmitted from the channel room all the way up to the main control room. To transmit these fast, logical pulses over long distances while maintaining the fidelity of count rates, optical pulse transmission is used in such instrumentation channels.

2.11.1.6 TRIP/ALARM Generation Module

The TRIP module generates an ALARM signal when Log.P signal falls below the threshold value set by user, generates TRIP signals when Log.P and/or Log.Rate signals exceeds the threshold values set by user. The TRIP module also generates ALARM signal if any module in the pulse channel malfunctions and the channel becomes "un-healthy". All the TRIP/ALARM signals are potential free, fail-safe relay contact outputs (Normally Open contacts) to ensure galvanic isolation of the nuclear-instrumentation channels. Thus, the TRIP module is the "Watch Dog" of the neutronic instrumentation channel as it performs all these safety actions without manual intervention.

2.11.1.7 Low Voltage Power Supply Module (LVPS)

The LVPS module receives safety class-II or class-I station power supply which could be 230 V AC mains or 24/48 V DC supply respectively and generates the required Low Voltage power supplies (typically 24, +15, –15 and 5 V) for all the modules in the instrumentation channel. All the generated low voltages should have 1.5X to 2X current capability, fold-back protection, Short circuit protection, overload protection, built-in diagnostics, and fail-safe NO type failure contact outputs. The LVPS module should have low output noise, excellent temperature stability, and long-term stability. They are based on AC to DC or DC to DC converters to provide galvanic isolation between the input and output. The LVPS module has to be qualified as per stringent conducted and radiated EMI/EMC requirements, which are met by incorporating special-purpose components such as Transient Voltage Suppressor/Metal Oxide Varistors, common-mode filters, snap ferrites, high Self Resonant Frequency (SRF) capacitors, etc.

2.11.2 Signal Processing for Campbell/MSV Mode Operation

The Campbell mode or Mean Squared Voltage mode (MSV) operation is deployed for extending the useful range of pulse mode instrumentation to the intermediate and power range of nuclear reactors. In this mode, the derived signal is proportional to the square of the charge generated (Q^2) by the incident particle, and thus it provides higher signal output for neutron vis-à-vis gamma radiation. The simplified block diagram of the MSV mode instrumentation is shown in Fig. 2.12 and the basis is given by Campbell's theorems.

r = event rate
$Q = Eq/W$ = charge produced for each event
E = average energy deposited per event
W = average energy required to produce a unit charge pair
$q = 1.6 \times 10^{-9}$C.

Let $f(t)$ be the response of the channel to each pulse produced in the detector. Here, it is assumed that all the input pulses are of similar shape and their rate of occurrence follows a Poisson distribution. Let $(\overline{N})$ be the average rate of pulses and $\overline{E(t)}$ be the average response of the circuit to N input pulses, then Campbell's first theorem states that

$$E\overline{(t)} = \overline{N} \int_0^\infty f(t)dt \tag{2.8}$$

Similarly, Campbell's second theorem states that

$$\left[E(t) - \overline{E(t)}\right]^2 = \overline{N} \int_0^\infty \left[f(t)\right]^2 dt \tag{2.9}$$

But,

$$E(t) - \overline{E(t)} = \sigma \ldots \text{Variance}$$

Hence, we can re-write Eq. 2.9 as

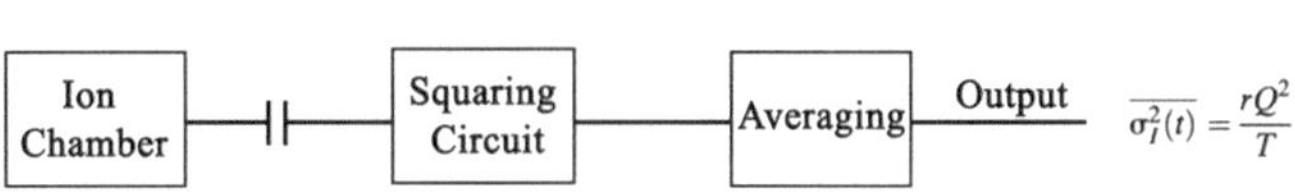

Fig. 2.12 Campbell/mean squared voltage mode of operation

$$\sigma^2 = \overline{N} \int_0^\infty [f(t)]^2 dt \tag{2.10}$$

These two Campbell's theorems enable us to measure higher order moments of the neutron-induced signal pulses.

2.11.2.1 Mean Squared Voltage Mode: A Case-Study

Let us study the implications of Campbell's theorem for a Single Time Constant Circuit comprising a neutron detector as shown in Fig. 2.13. The output voltage $f(t)$ can be given as:

$$f(t) = \frac{q}{C} \exp^{\frac{-t}{RC}} \tag{2.11}$$

By substituting Eq. 2.11 in 2.8, we obtain the average output voltage as:

$$\begin{aligned} E\overline{(t)} &= \overline{N} \int_0^\infty \frac{q}{C} \exp^{\frac{-t}{RC}} dt \\ \overline{E(t)} &= \overline{N} \times R \times q \end{aligned} \tag{2.12}$$

Similarly, by substituting Eq. 2.11 in 2.9, we obtain variance in the output voltage as:

$$\left[E(t) - \overline{E(t)}\right]^2 = \frac{\overline{N} \times R \times q^2}{2 \times C} \tag{2.13}$$

By comparing Eqs. 2.12 and 2.13, we can see the advantage obtained by measuring variance in output voltage which is proportional to square of input charge, instead of measuring the average output voltage which is proportional to input charge. The charge q generated by incident neutron is much higher as compared to that due to gamma, hence the Campbell mode provides a very good n–γ discrimination.

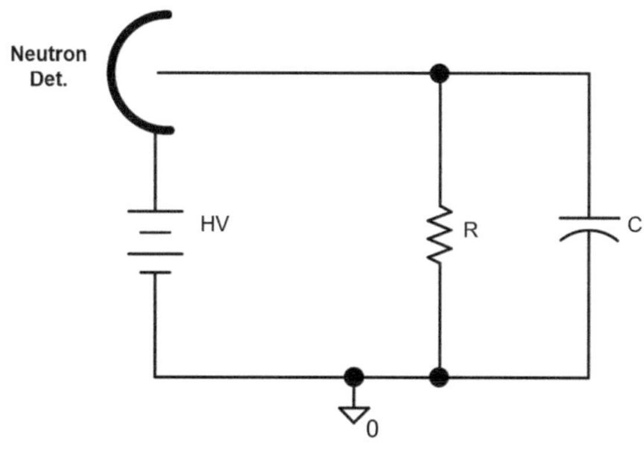

Fig. 2.13 Mean squared voltage mode: a case-study

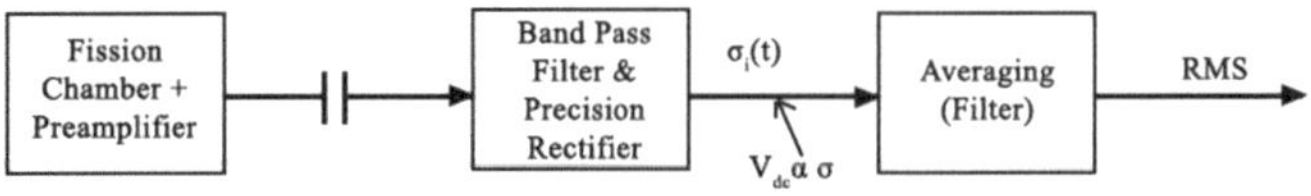

Fig. 2.14 MSV channel: a simplified block diagram

From Eqs. 2.12 and 2.13, we have:

$$\overline{E(t)} = k\left[N_\gamma q_\gamma + N_n q_n]\right] \quad (2.14)$$

$$\left[E(t) - \overline{E(t)}\right]^2 = \sigma^2 = k_1\left[N_\gamma q_\gamma^2 + N_n q_n^2\right] \quad (2.15)$$

For neutronic instrumentation, we are interested in neutron's signal only and gamma's signal is considered to be error, thus the error factors are:

$$\text{Error factor in } (14) \ldots \frac{N_\gamma}{N_n}\left(\frac{q_\gamma}{q_n}\right)$$

$$\text{Error factor in } (15) \ldots \frac{N_\gamma}{N_n}\left(\frac{q_\gamma}{q_n}\right)^2$$

Comparing the error factors in average signal and variance signal, we can see that the error factor in variance is reduced by a factor proportional to $\frac{q_\gamma}{q_n}$.
It may be noted that:

$$\frac{q_\gamma}{q_n} \text{ for } B^{10} \text{ lined proportional counter is } \frac{1}{10}$$

While

$$\frac{q_\gamma}{q_n} \text{ for } U^{235} \text{ fission counter is } \frac{1}{100}$$

Hence, by using Fission Counter in MSV mode, one can obtain the variance signal for incident neutrons, which will have about 1% error due to gamma background.

A simplified block diagram of the MSV channel is shown in Fig. 2.14. Analog pulses from the preamplifier are processed by the Band Pass Filter and precision rectifier. The band pass filter has a 3 dB cut-off frequency in the range of 10–400 kHz. The rectified output is further averaged out to obtain the MSV output. The MSV channel covers the intermediate and power range of the reactor.

2.11.2.2 Advantages of Campbell or MSV Technique

a. MSV mode provides higher Neutron to Gamma discrimination, which can be improved further by using Fission counters.
b. MSV mode is AC-coupled signal processing technique, and Insulation Resistance, leakage current-related issues become insignificant in this mode of instrumentation.
c. MSV mode also allows for operating the detector at elevated temperatures, as leakage current does not affect the Campbell output.

d. Using MSV mode in conjunction with pulse mode operation, a single detector can be used (usually an FC detector) for processing up to 10 decades of neutron flux measurement.

2.11.3 Signal Processing for DC Mode Operation

In DC mode operation, the incident event rate is so high that the individual current pulses overlap with each other, and as a result, an average DC current flows through the neutron detector. This average current is a product of event rate and average charge generated per event, and hence it is proportional to the incident field/neutron flux. Generally, DC channels come "on-scale" 3–5 decades after criticality and are used for safety and monitoring of the core in the intermediate and power range of a nuclear reactor. A simplified block diagram of DC mode instrumentation, along with an overview of basic parameters governing the signal generation, is given in Fig. 2.15.

2.11.3.1 Neutron Detector for DC Mode Operation

DC mode instrumentation is deployed for Uncompensated Ion Chamber (UIC) or Compensated Ion Chamber (CIC) . These ion chambers have a ^{10}B coating/lining on the inside of the detector, which interacts with the incident neutrons and produces charge particles which are detected in the sensitive volume of the ion chamber. Typical neutron sensitivity is 10^{-14} Amp/nv while their gamma sensitivity is $\sim 10^{-12}$ Amp/R/h. The CICs provide better linearity in the presence of gamma background, and the useful range can be extended by 1–2 decades on the lower end of the range.

2.11.3.2 High Voltage Supply Module

This module generates the required operating HV supply for biasing the detector. The CIC detector requires +HV and –HV operating supplies, while the UIC detector requires only +HV operating supplies.

Before deploying the DC mode instrumentation for a reactor, the H.V. plateau characteristics of the neutron detector along with the instrumentation channel are studied by carrying out a series of experiments at various power levels in a research reactor. The baseline data is analyzed to determine the optimal operating point (+H.V. & –H.V.) & sensitivity (Amp/nv), and to validate the linearity, stability, etc. over the desired operating range of the DC instrumentation channel.

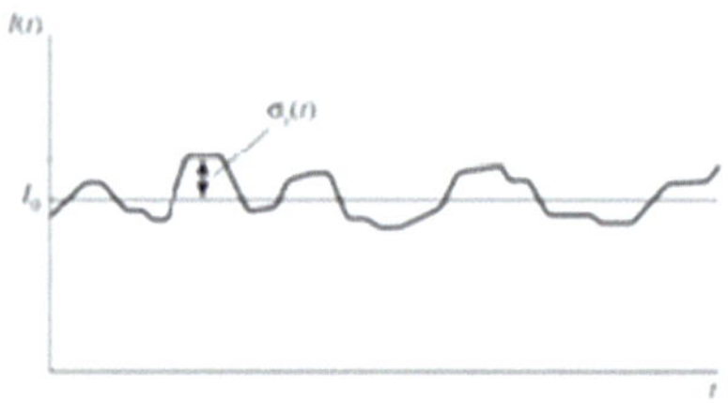

$$I_0 = rQ = r\frac{E}{W}q$$

r = event rate
$Q = Eq/W$ = charge produced for each event
E = average energy deposited per event
W = average energy required to produce a unit charge pair
$q = 1.6 \times 10^{-19}$C

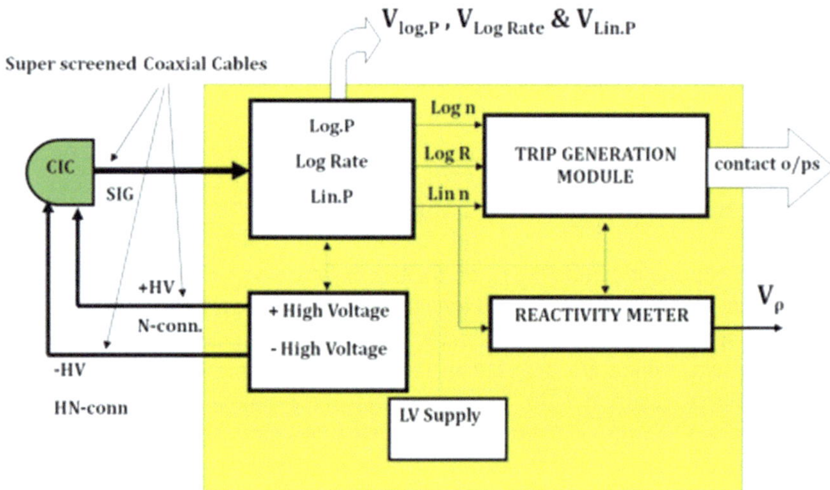

Fig. 2.15 Simplified block diagram of DC mode instrumentation

2.11.3.3 Log.P, Lin.P and Log Rate Generation Module

The average DC current for the ion chamber is directly processed by the logarithmic frontend amplifier in this module, and the Log.P, Lin.P as well as Log.Rate signals are generated by this module. The generated Log.P signal is proportional to the instantaneous reactor power, while the Log.Rate signal gives the information about the e-folding Period of the reactor. Additionally, a linear power signal (Lin.P) is also generated for precise control of the reactor power over the highest two decades.

Typically, the Log.P signal covers 5–7 decades of neutron flux (intermediate and power range) over 0–10 V full scale, while the Lin.P signal covers only the last two decades (1–150% of nominal power) linearly over 0–10 V full scale, with a better accuracy. The Log.P, Lin.P, and Log.Rate signals are monitored for the generation of ALARM and TRIP signals for the control and safety of the reactor.

It must be noted that Log.P signal should maintain linearity over 5–7 decades of incident neutron flux and should have excellent temperature stability and long-term stability. These stringent design requirements are conformed to by using special circuit techniques as well as careful selection of active and passive components for the Log.P, Lin.P, & Log.rate module.

The Log-Linear/Log.Rate module has several built-in calibration and diagnostic features, including SCRAM, to enable periodic calibration and preventive maintenance of the instrumentation channel. Typically, preventive maintenance is carried out once every quarter to weed out possible failures of the instrumentation channel.

2.11.3.4 TRIP/ALARM Generation Module

This module generates an ALARM signal when Log.P signal falls below the threshold value set by the user, generates TRIP signals when Log.P, Lin.P, and/or Log.Rate signals exceed the threshold values set by the user. The module also generates an ALARM signal if any other module in the DC channel malfunctions and the channel becomes "un-healthy". All the TRIP/ALARM signals are potential free, fail-safe relay contact outputs (Normally Open contacts) to ensure galvanic isolation of the nuclear-instrumentation channels. Thus, thc TRIP/ALARM module is the "Watch Dog" of the neutronic instrumentation channel as it performs all these safety actions without manual intervention.

2.11.3.5 Reactivity Meter Module

This is a special-purpose module in the DC instrumentation channel, which further processes the Lin.P signal and computes the reactivity of the core/reactor. The reactivity computation is generally implemented on a programmable device such as a microprocessor/FPGA (Field Programmable Gate Array). It solves the point kinetic equations for six groups of delayed neutrons

$$dn/dt = n(\rho - \beta)/l + \Sigma\lambda_i C_i \text{ and}$$
$$dC_i/dt = k\beta_i n/l - \lambda_i C_i$$

where n is neutron density, ρ is reactivity, C_i is the i-th delayed neutron concentration, β is effective delayed neutron fraction, l is prompt neutron generation time, λ_i is decay constant of the i-th delayed neutron and β_i is the fraction of the i-th delayed neutron. The reactivity (ρ) is calculated online, close to criticality of the reactor. The output signal is represented in fraction of β or per cent mille (PCM) or mK. The reactivity meter is used during the cold-physics/phase-B experiments to compute and validate the safety margins of the reactor core.

2.11.3.6 Low Voltage Power Supply Module (LVPS)

The LVPS module receives a safety class-II or class-I station power supply, which could be 230 V AC mains or 24/48 V DC supply, and generates the required low-voltage power supplies (typically 24, +15, –15 & 5 V) for all the modules in the DC instrumentation channel. Generally, the LVPS module in DC mode instrumentation is identical to that used in pulse mode instrumentation, and all features of the LVPS supply are identical to those outlined in Sect. 2.11.1.7.

2.12 Generation of Logarithm and Period Signals

In general, the neutronic instrumentation for a reactor is required to monitor the core over more than 10 decades of power. To process such a wide dynamic range of incident neutron flux, logarithmic amplifiers are used in both pulse mode instrumentation as well as DC mode instrumentation. The main advantages of logarithmic signal processing are

a. Logarithmic signal processing allows for monitoring 5–7 decades of neutron flux automatically without requiring manual intervention.
b. It provides an automatically varying time constant which is high at lower incident neutron flux and reduces as the neutron flux increases.
c. Differentiation of log amplifier output (Log.P) gives Log.rate signal, which gives information about the e-folding period reactor.

2.12.1 Generation of Log.P Signal

As outlined in earlier sections, the Log.P signal should maintain linearity over 5–7 decades of incident neutron flux and should have excellent temperature stability and long-term stability. These stringent design requirements are met by using special circuit techniques. The simplified block diagram of a practical logarithmic amplifier is shown in Fig. 2.16.

The signal current (I_{Det}) and the reference current (I_{Ref}) are processed by two different logarithmic amplifiers and their outputs V_A and V_B are subtracted by an instrumentation amplifier Instrumentation Amplifier (INA). In this way, the temperature-dependent term due to reverse saturation current of the transistors Q1 and Q2 (I_{ES}) is eliminated at the output of the instrumentation amplifier (V_C) and we get a transfer function of $\sim$59 mV/decade change in the signal current at node C. This signal is scaled up as per the desired input dynamic range by amplifying it through an inverted summing amplifier as outlined in the Fig. 2.17. The gain elements include a thermistor R_T having a positive temperature coefficient, carefully chosen to cancel out

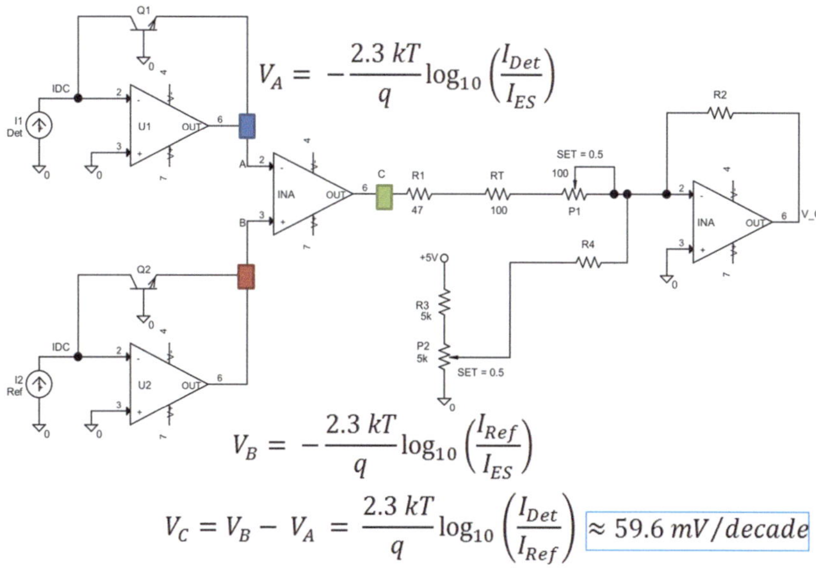

Fig. 2.16 Simplified block diagram of a logarithmic amplifier

the temperature-dependent term kT/q present in V_C at node C. Thus, the final output voltage V_{OUT} has excellent temperature and long-term stability.

In most instances, the instrumentation channel is required to process detector current signal from 10 pA and upwards, which puts a constraint on the op-amps that can be used as the logarithmic amplifiers U1 and U2. It is desirable that U1 and U2 should have ultra-low input bias current. Similarly, the logarithmic feedback elements Q1 and Q2 should be matched transistor pairs packaged in a single TO-8 metal header so that their IES are precisely matched.

The calibration of the log amplifier is a crucial step and is generally done in two steps as outlined in Fig. 2.18. The node voltage V_C should be zero when $I_{Det} = I_{Ref}$ is given as input to the instrumentation channel using a precision current source. For this condition, the OFFSET-ADJ trimpot is adjusted so as to get a voltage signal V_{OUT} corresponding to the input current. In the next step, the test input from the current source is made equal to the full-scale value, and the GAIN-ADJ trimpot is adjusted to obtain the output voltage V_{OUT} corresponding to the full-scale value.

2.12.2 Generation of Reactor Period Signal

The instantaneous neutron flux $N(t)$ in a nuclear reactor is given as:

$$N(t) = N_O \exp^{\left(\frac{t}{\tau}\right)}$$

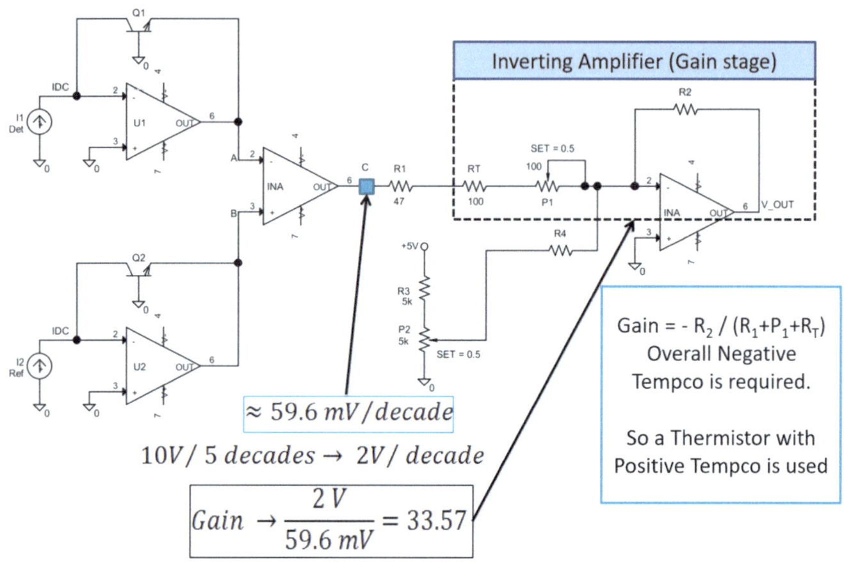

Fig. 2.17 Case-study: setting the gain of log amplifier

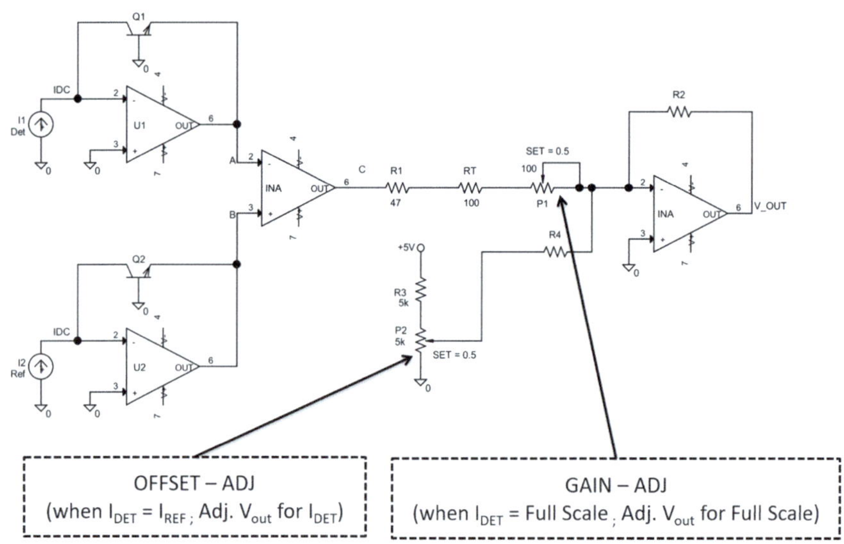

Fig. 2.18 Case-study: calibrating the log amplifier

where N_O is the initial neutron flux at time $t = 0$ and τ = time period after which the neutron flux increases e-fold multiples. The detector signal current $I(t)$ is proportional to the incident neutron flux and hence it follows an identical relationship as given below:

Table 2.2 Relation between different period meter calibrations

e-folding period	τ (s)
Doubling time	0.693τ
Nepers/s	$1/\tau$
%/s	$100/\tau$
Decades/min	$26/\tau$
Decades/s	$0.43/\tau$

$$I(t) = I_O \exp^{\left(\frac{t}{\tau}\right)}$$

Taking $\log_e$ on both sides gives

$$\log_e I(t) = \log_e I_O + \frac{t}{\tau}$$

Differentiating w.r.t. time on both sides, we obtain

$$\frac{1}{I(t)} \frac{dI(t)}{dt} = \frac{1}{\tau}$$

Thus, it is evident that the time derivative of the logarithmic amplifier output is inversely proportional to the e-folding period of the reactor.

There are instances where the period meters in instrumentation channels are calibrated to display "doubling time" or log.rate % per-second, etc. These are correlated to the e-folding period as given in the Table 2.2.

2.13 Nuclear Systems for Safety, Safety Related and Monitoring

In a nuclear reactor, it is mandatory to monitor the state of the core in all states, i.e., initial fuel loading and first approach to criticality, source range, intermediate range, power range, and shutdown state. The dynamic range over which neutron flux has to be monitored is more than 10 decades, and it is impossible to use only a single detector or instrumentation channel to monitor the state of the reactor core. Generally, this is achieved by deploying three or more types of instrumentation channels, viz., the source range/start-up instrumentation channel, intermediate range instrumentation channel, and power range instrumentation channel, as shown in Fig. 2.19.

2.13.1 Sources of Neutrons in a Reactor

In a nuclear reactor, neutrons can be produced either by neutron-induced fission or by other nuclear reactions. The neutrons produced by reactions other than neutron-induced fission are called *source* neutrons. Source neutrons are important because

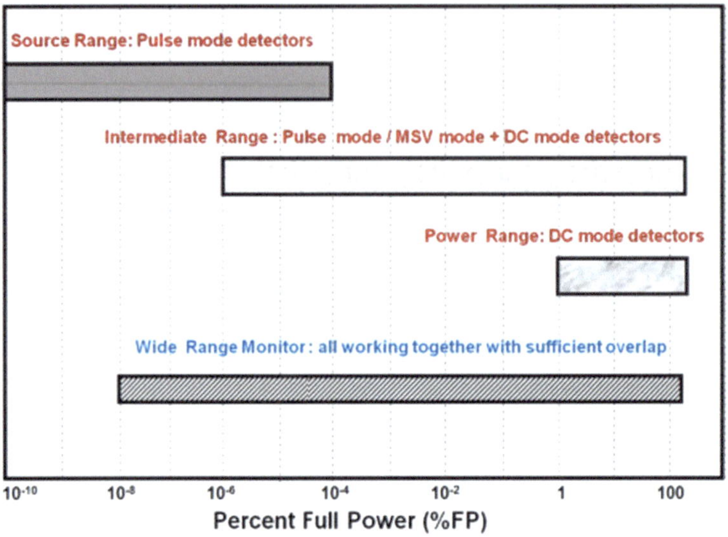

Fig. 2.19 Range and overlap of various neutronic instrumentation channels

Table 2.3 Neutron production by spontaneous fission

Nuclide	$T_{1/2}$ (Fission) (years)	$T_{1/2}$ (α-decay)	Neutrons/sec/gram (years)
^{235}U	1.8×10^{17}	$6.8x10^8$	8.0×10^{-4}
^{238}U	8.0×10^{15}	4.5×10^9	1.6×10^{-2}
^{239}Pu	5.5×10^5	2.4×10^4	3.0×10^{-2}
^{240}Pu	1.2×10^{11}	6.6×10^3	1.0×10^3
^{252}Cf	66.0	2.65	2.3×10^{12}

they ensure that the neutron flux remains high enough and the start-up instrumentation is "on-scale" and provides adequate monitoring of the core during initial fuel loading, first approach to criticality, subsequent start-up and shutdown states. Source neutrons can be classified as either intrinsic or installed neutron sources as listed in Table 2.3.

Intrinsic neutron sources are those neutron-producing reactions that always occur in reactor materials, e.g., spontaneous fission of ^{238}U nuclei present in the natural Uranium based fuel element of PHWR, nuclear reaction with ^{11}B, photo-fission of deuterium nuclei etc.

$$^{11}_{5}\text{B} + ^{4}_{2}\alpha \xrightarrow{\text{gives}} ^{14}_{7}\text{N} + ^{1}_{0}n \quad (\alpha - n \text{ reacton})$$

$$^{2}_{1}\text{H} + ^{0}_{0}\gamma \xrightarrow{\text{gives}} ^{1}_{1}\text{H} + ^{1}_{0}n \quad (\text{Photo-fission or Photo-neutrons})$$

Many types of reactors, viz., PWRs, LWRs , Sodium cooled reactors, have artificial sources of neutrons installed inside the core, which ensures that the shutdown neutron flux remains high enough. Installed neutron sources include Californium-252, Beryllium ($\alpha - n$) reaction-based sources, or Beryllium ($\gamma - n$) based photoneutron sources.

$$^{9}_{4}\mathrm{Be} + ^{4}_{2}\alpha \xrightarrow{\text{gives}} ^{12}_{6}\mathrm{N} + ^{1}_{0}\mathrm{n}\ (\alpha - n \text{ reaction used in Am-Be, Pu-Be sources})$$

$$^{9}_{4}\mathrm{Be} + ^{0}_{0}\gamma \xrightarrow{\text{gives}} ^{8}_{4}\mathrm{Be} + ^{1}_{0}n\ (\gamma - n \text{ reaction used in Sb-Be source})$$

2.13.2 Start-Up/Source Range Instrumentation

This includes a set of duplicated/triplicated pulse mode instrumentation channels processing the neutronic signal generated by the respective in-core/ex-core neutron detector. The neutron flux in the source range is low, and hence pulse mode neutron detectors are deployed to cover this range. The neutron detectors can be located in-core during fuel loading and FAC stage, while they are typically located ex-core for source range instrumentation. For beginning of equilibrium core (BOEC) the gamma background is low and ^{3}He or ^{10}B-lined gas proportional counters are used, which provide higher sensitivity and low to medium gamma tolerance. The pulse mode instrumentation channel is a charge-sensitive-amplifier-based pulse channel. For start-up after long shutdown (special start-up) wherein the core/structural gamma background is high, ^{235}U lined fission counters are used because of their higher gamma tolerance. The pulse mode instrumentation channel used for special start-up is a wide-band current-sensitive-amplifier-based pulse channel for ensuring satisfactory n-gamma discrimination.

The source range instrumentation generates the Log.power and Log.Rate signals which are monitored for the generation of ALARM and TRIP signals for the control and protection of the nuclear reactor.

2.13.3 Intermediate and Power Range Instrumentation

The intermediate range includes a set of duplicated or triplicated DC mode instrumentation channels processing the neutronic signal generated by the respective ex-core neutron detectors. The neutron flux in the intermediate and power range is sufficiently high and hence uncompensated or compensated ^{10}B lined ion chamber operating in DC mode are deployed for covering these ranges. In the intermediate range, the Log.Power and Log.Rate signals are generated over 5–7 decades, and at least a decade of overlap with source range instrumentation is ensured by design. This is implemented by keeping a safety interlock for checking that before the source range instrumentation enters into count-loss regime, the intermediate range instrumentation is "on-scale" and the reactor control can be safely transferred to intermediate range instrumentation. Additionally, in some reactors, the MSV mode instrumentation is also deployed to implement a wide-range safety channel.

The power range instrumentation includes a set of triplicated DC mode instrumentation channels processing the neutronic signal generated by the respective ex-core uncompensated or compensated ^{10}B lined ion chamber operating in DC mode. The power range instrumentation covers the last two decades of nominal reactor power (1–150% N_{nom}) and generates Linear_Power and Reactivity signals for control and safety of the reactor core.

2.13.4 Reactor Regulating System

The reactor regulating system (RRS) processes the signals generated by the power range instrumentation for automatic maneuvering of reactor power by driving the control rods in/out of the reactor core. The RRS ensures uniform burn-up of the fuel present in the reactor core.

2.13.5 Flux Mapping System

The flux mapping system (FMS) processes the signals generated by specialized in-core neutron detectors viz. the Self Powered Neutron Detectors (SPND). The FMS generates a 3-D map of the neutron flux inside the reactor core for efficient fuel utilization and prevention of localized hot-spots inside the core.

2.13.6 Failed Fuel Detection and Localization Systems

The failed fuel detection (FFD) system monitors the "healthiness" of the fuel sub-assemblies (FSA) present inside the reactor core. It generates a TRIP signal for safe shutdown of the reactor when the release of fission product noble gases (FPNGs) exceeds the safe design limit, signifying rupture of one or more fuel sub-assembly. The failed fuel localization (FFLM) system identifies the particular FSA which has failed by detecting the delayed neutrons and/or the FPNGs being released by the ruptured FSA. Once the failed FSA has been identified, it is replaced by a fresh FSA, and the reactor can be re-started.

2.13.7 Stack Monitoring System

The stack monitoring system monitors the Fission Product Noble Gases (FPNGs) being released by the stack column of a nuclear reactor and ensures that the release of radioactivity by the nuclear power plant is within acceptable limits.

2.14 Major Design Features of I&C Systems for Reactor

The I&C systems deployed in a nuclear reactor have several unique features to ensure a very high level of reliability. There is one set of triplicated instrumentation channels that provide signals for protection of the nuclear reactor (safety-channels) and another set of triplicated instrumentation channels which provide signals for regulation of reactor power (control channels). All these triplicated safety and control channels are completely independent and galvanically isolated to ensure that common cause failures are avoided. Optimum redundancy in instrumentation channels is ensured at all times, and any safety or control action is initiated only when at least two out of three channels call for such action.

Engineered safety features are incorporated, such as diversity in I&C systems to avoid common cause failures. For instance, there are Primary Shutdown system, Secondary Shutdown system, and Tertiary Shutdown system in any reactor, all of which are based on diverse safety principles. Failsafe design philosophy is followed, e.g., TRIP contacts are normally open (NO) relay contacts, and failure of the power supply will cause the relay to OPEN and trigger a TRIP condition.

Online discordance monitoring is implemented, and an ALARM is annunciated if a given parameter from two different channels shows widely different values. Online testing (FIT/SCRAM) is carried out on a periodic basis to weed out discordance. Periodic maintenance work is carried out on one channel at a time while the reactor is under operation, so that results are immediately known.

The I&C systems should have high reliability, high mean time between failure (MTBF), low mean time to detect failure (MTTD) , and low mean time to repair (MTTR). This is especially true of Safety Class-IA systems having higher safety implications, e.g., neutron flux monitoring systems, shutdown systems, etc. Typical failure rate for the Shutdown system is 10^{-6} failures/demand.

To meet such stringent design requirements and to achieve high reliability, the I&C systems are qualified for environmental, seismic, and EMI-EMC conditions anticipated during installed conditions at the site. The prototype as well as production batches of I&C systems undergo thorough functional testing, including environmental stress tests (dry heat, dry cold, damp heat, burn-in, etc.) to weed out infant mortality. Additionally, one set of I&C systems is subjected to type tests, which include seismic/vibration/shock tests as well as EMI-EMC tests as per MIL-461E/IEC-61000 standards.

2.15 Electro Magnetic Compatibility of I&C Systems

2.15.1 EMI-EMC Qualification Tests

The I&C systems for a nuclear reactor are qualified to work satisfactorily under the anticipated electromagnetic interference (EMI) during their operation on site. The severity of EMI is stated in MIL-461E (for mobile platforms) as well as IEC-61000 standards (for land-based reactors), and the applicable Electro Magnetic Compatibility (EMC) tests are conducted for the I&C systems. EMI-EMC tests can be classified broadly under four different categories

i. Conducted Emission tests
ii. Radiated Emission tests
iii. Conducted Susceptibility tests
vi. Radiated Susceptibility tests.

The emission tests ensure that noise emitted by the I&C systems is within acceptable limits and they don't become the source of EMI in the installed environment. The susceptibility tests ensure that the I&C systems are able to work satisfactorily despite the anticipated EMI in the installed condition.

2.15.2 EMI Mitigation Techniques

Ensuring electromagnetic compatibility (EMC) of I&C systems is a major challenge, and it calls for the incorporation of special techniques for mitigating anticipated EMI right from the design stage up to the commissioning stage.

2.15.2.1 Design Techniques for EMI Mitigation

The design techniques for ensuring EMC include:

a. Incorporation of special-purpose components such as common-mode and differential mode noise filters & soft-ferrites, transient overvoltage suppressors, Y-capacitors, etc.
b. Power supply isolation, separation of input and output grounds by using AC-DC or DC-DC converters appropriately.
c. Ensuring galvanic isolation between the I&C channel and output signals by using 4–20 mA isolated current loops for driving the output devices or by using electromagnetic relays for driving output loads.

2.15.2.2 Commissioning Techniques for EMI Mitigation

The commissioning of I&C system at the site includes:

a. Implementation of a proper grounding scheme for the entire site, including the channel room where the I&C systems are installed.
b. Provision for separate instrumentation earth-pit for heavy electrical loads (viz. trubines, pumps, etc.) and for I&C systems.
c. Implementation of proper shielding scheme for all the interconnecting cables, laying of proper and isolated cable-conduits, μ-metal enclosures, etc.
d. Laying of suitable super-screen cables and separate cable-conduits for the interconnecting cables carrying sensitive signals, such as those between neutron detectors and their instrumentation channels.

2.16 Conclusions

The subject of Instrumentation and Control of Nuclear Reactors is very vast and encompasses several branches of science, engineering, and technology. In this chapter, the authors have tried to rouse the interest of graduate-level students by giving a brief overview of neutron detectors and their instrumentation channels. For further reading, the interested readers may refer to the textbooks mentioned in the references.

Acknowledgements Authors would like to thank and acknowledge the encouragement and support provided by Shri P.V. Bhatnagar, Head, Reactor Instrumentation Section, Smt. Anita Behere, Director E&I Group, BARC, and Shri P.P. Marathe, Ex-Director E&I Group, BARC. The authors further wish to place on record their sincere thanks to Dr. Kallol Roy, Ex CMD, BHAVINI, for going through the manuscript and providing many critical comments which went a long way in improving the quality of the same.

Bibliography

1. J.M. Harrer, J.G. Beckerley, *Nuclear Power Reactor Instrumentation Systems Handbook*, vol. 1 (Argonne National Laboratory, 1973)
2. Defence in depth in nuclear safety. INSAG-10 report by the International Nuclear Safety Advisory Group, IAEA
3. G.F. Knoll, *Radiation Detection and Measurement* (Wiley, 2010)
4. S. Glasstone, A. Sesonske, *Nuclear Reactor Engineering* (Springer, 1998)
5. Evaluated Nuclear Data Files (ENDF) by IAEA, https://www-nds.iaea.org/exfor/endf.htm
6. W.D. Allen, *Neutron Detection* (George Newnes Ltd., London, 1960)
7. G. Grosshoeg, *Neutron Ionisation Chambers* (North-Holland, Amsterdam, 1979)
8. B.B. Rossi, H.H. Staub, *Ionisation Chambers and Counters* (McGraw-Hill, 1949)
9. F.H. Attix, E. Tochilin, in *Radiation Dosimetry*, Chap. 9 ed. by J.W. Boag (Academic Press, 1969)

10. R.A. DuBrigde, IEEE Trans. Nucl. Sci. **NS-14**(1), 241 (1967)
11. H.A. Thomas, A.C. McBride, IEEE Trans. Nucl. Sci. **NS-15**(1), 15 (1968)
12. L.D. Philipp, N.C. Hoitnik, W.G. Spear, Nucl. Tech. **20**(1973)
13. L.C. Wimpee, GEAP10611
14. General principles of nuclear reactor instrumentation, IEC 60231:1967
15. R.A. DuBridge, Campbell theorem-system concepts and results. IEEE Trans. Nucl. Sci. **NS-14**(1), 241–146 (1967)
16. Nuclear reactor instrumentation-Wide range neutron fluence rate meter—Mean square voltage method. IEC 61501 (1998)
17. Nuclear power plants—Instrumentation and control systems important to safety—Safety logic assemblies used in systems performing category a functions. IEC 60744 (2018)
18. MIL-STD-461 series, https://www.nts.com/industries/defense/mil-standards/mil-std-461-revisions-a-d/
19. Electromagnetic compatibility (EMC). IEC 61000 Series

Further Reading

20. V. Balagi, K.R. Prasad, S.K. Kataria, Effect of high gamma background on neutron sensitivity of fission detectors. Indian J. Pure Appl. Phys. **42**, 390–397 (2004)
21. G.F. Knoll, *Radiation Detection and Measurement* (Wiley, 2010)
22. J.M. Harrer, J.G. Beckerley, *Nuclear Power Reactor Instrumentation Systems Handbook*, vol 1 (Argonne National Laboratory, 1973)
23. S. Glasstone, A. Sesonske, *Nuclear Reactor Engineering* (Springer, 1998)

Chapter 3
Instrumentation for Sodium Cooled Fast Reactors

Sethumadhavan Sridhar

3.1 Characteristics of Fast Reactors

Nuclear reactors are of different types, based on the nature of technology deployed for extracting nuclear energy. Though the first classification is based on whether the energy released is due to nuclear fission or fusion, at present, only fission reactors are found to be practically viable. Of the different types of fission reactors, one major classification is based on the type of coolant used: light water (H_2O), heavy water (D_2O), gas, or liquid metals. Fission reactors can also be classified based on the energy of the neutrons used for producing fission. Thermal reactors are those which use moderator to slow down the neutron speed to increase the probability of fission. The average neutron energy that causes the self-sustaining fission reactions is low in thermal reactors (around 0.025 eV) whereas fast reactors use highly energetic neutrons (around 10 keV). It may be noted that probability of fission reaction, known in the nuclear parlance as fission cross section and measured in barns (1 barn = 10^{-23} cm^2), is higher in the case of less energetic neutrons. Those reactors which use sodium as the coolant operate in the fast region and are generally known as SFR (Sodium Cooled Fast Reactors). When the fission is caused by fast neutrons, the number of neutrons released per fission is high. Hence, more neutrons are available in the core over and above those required for maintaining the chain reaction. This amount of excess neutrons can be used for breeding more fuel (converting fertile nuclides to fissile nuclides) than is consumed. Hence the term breeder reactors. However fast reactor needs high enrichment of fissile material to overcome the low fission cross section due to fast neutrons. A comparison of fast and thermal reactors is presented in Table 3.1.

S. Sridhar (✉)
Bharatiya Nabikia Electricity Limited (BHAVINI), Kalpakkam, India
e-mail: sethu@igcar.gov.in

U. Mangalanathan et al. (eds.), *Nuclear Instrumentation and Control*,
https://doi.org/10.1007/978-981-97-1283-0_3

Table 3.1 Comparison of thermal and fast reactors

	Thermal reactors	Fast reactors
Neutron energy	Low	High
Fission probability	High	Low
Fuel enrichment	Low	High
Coolant	Water, heavy water	Sodium
Moderator	Water, heavy water	Not required
Breeding of fuel	Possible	Possible
Type of refueling	Online and offline	Offline only

3.2 Process Flow Diagrams

The 235 MWe PHWR uses natural uranium as the fuel. Heavy water acts both as a moderator and a coolant. The primary pumps circulate the heavy water into the pressure tubes placed in the caldarium, which is filled with heavy water to act as a moderator. The fuel bundles are inside the pressure tubes, in which cold heavy water enters around 249 °C. The 306 pressure tubes housing the fuel are designed to produce 756 MWt. The hot D_2O exiting the pressure tubes around 290 °C transfers the heat to the light water (H_2O) in the Steam Generator, which generates saturated steam at 40 kg/cm^2 and 251 °C. The process flow sheet for a typical PHWR type thermal reactor is shown in Fig. 3.1.

Fast Breeder Test Reactor (FBTR) located at Kalpakkam, India, uses sodium as the coolant and a mixture of uranium carbide and plutonium carbide as fuel. The fuel is housed in the reactor vessel to produce 40 MWt. There are a total of four sodium loops: two primary sodium loops and two secondary sodium loops. Primary sodium enters the bottom of the reactor vessel at around 380 °C and leaves it at around 515 °C. The heat from the primary sodium system is transferred to the secondary sodium system in the IHX, which is a sodium-to-sodium heat exchanger introduced to prevent radioactive sodium from leaving the reactor containment building. The steam generator produces superheated steam at 125 kg/cm^2 and 480 °C. The steam runs a turbine to produce 13 MWe. The process flow sheet for a typical fast reactor (FBTR) is shown in Fig. 3.2.

3.3 Instrumentation and Control for Nuclear Power Plants

The nuclear power plants, irrespective of the technology used, use extensive Instrumentation and Control (I&C) Systems to help in the smooth start-up, normal operation to produce thermal power and convert the same to electrical power in the turbine, and for safe shutdown of the reactor in case of any abnormal process condition. Apart

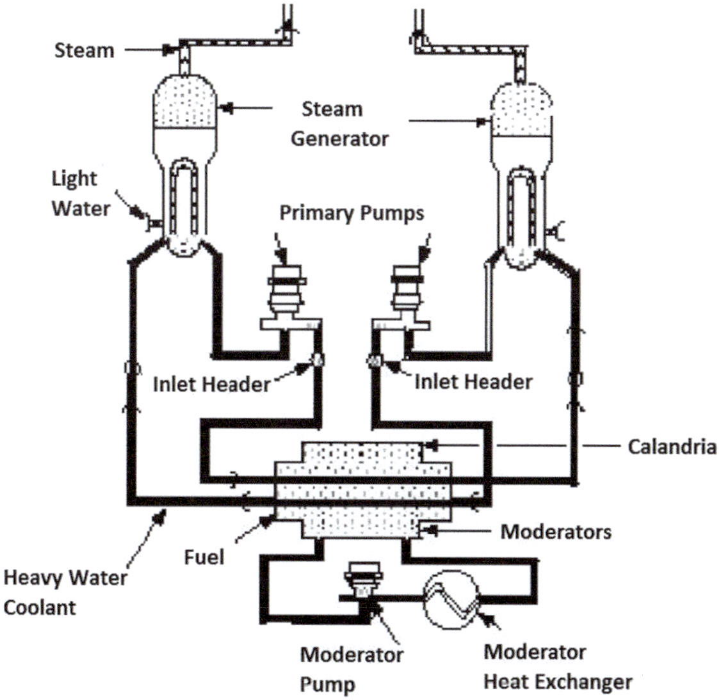

Fig. 3.1 Process flow sheet of PHWR

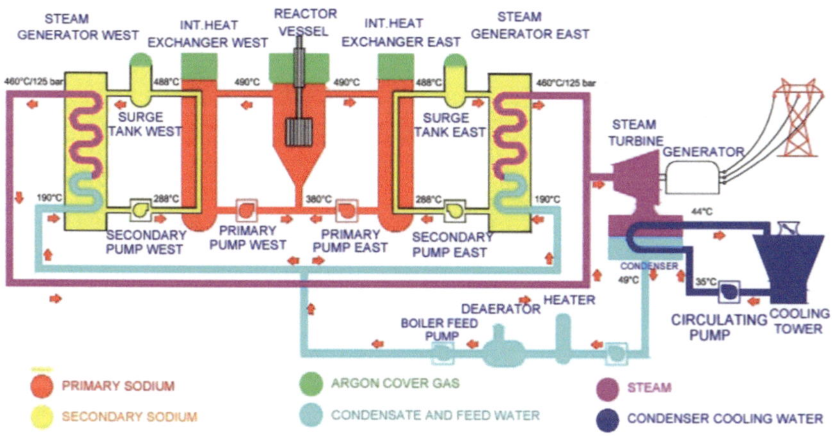

Fig. 3.2 Process flow sheet of FBTR

from this, it is also required to monitor the status of the reactor in case of any accidents. Extensive instrumentation is required to monitor any unsafe condition in the plant, including seismic events, and also to monitor radioactivity release to the public.

In order to ensure that I&C systems meet the stringent requirements in their design and implementation, several mechanisms exist. First, the role of the I&C system should be clearly defined. Then the design should follow the globally established standards, commensurate with the safety classification of the I&C system. Third, the design and its implementation should be validated with extensive testing. Fourth, well-established maintenance schedule should be in place to ensure that I&C systems perform as intended throughout their lifetime in the NPP where it is deployed.

Requirements of I&C systems can be met by either using a purely Hardware system or a combination of both hardware and software. The choice depends on the complexity of the system specifications, availability of components based on market survey, reliability, involved verification of validation process, time to deliver, maintainability, human machine interface, etc. Also, it is required to give a wise thought on whether the system can be developed from scratch or using a pre-developed system, i.e., an indigenously developed generic system or a commercial off-the-shelf system. It is mandatory for I&C system not only to meet functional requirements but also to meet regulatory requirements, as laid by the regulatory body based on the classification of the system, before deployment into NPP.

3.3.1 Role of the Regulatory Body

In India, the Atomic Energy Regulatory Board (AERB) is the central agency responsible for issuing safety guides and standards that help the designer of I&C systems to design the systems so as to meet the requirements of the NPPs. The safety guides listed below in Table 3.2 are of importance.

Table 3.2 List of safety guides applicable to I&C systems of Indian NPPs

Sl. no.	Safety guide	Description
1	AERB/NPP-PHWR/SG-D-1	Safety classification and seismic categorization for structures, systems and components of pressurized heavy water reactors
2	AERB/NPP-PHWR/SG-D-10	Safety systems for pressurized heavy water reactors
3	AERB/NPP-PHWR/SG-D-20	Safety related instrumentation and control for pressurized heavy water reactor based nuclear power plants
4	AERB/NPP-PHWR/SG-D-25	Computer-based systems of pressurized heavy water reactors

3.3.2 Safety Functions as per SG-D-1

This safety guide defines the important safety function of structures, systems, and components (SSCs) in an NPP. Any SSC (civil, mechanical, or I&C component) is considered a safety system if it is associated with the following operations: normal reactor operation, preventing plant events from leading to accident conditions, and mitigating the consequences of accident conditions (e.g., emergency core cooling system).

3.3.3 Safety Classification of I&C Systems as per SG-D-1

Safety class IA is assigned to those I&C systems and SSCs which perform a principal role in the achievement of reactor safety or in the maintenance of the same. For example, the reactor protection system, core temperature monitoring system, seismic instrumentation system, as these systems initiate a reactor trip to protect the reactor from unsafe conditions. I&C systems which play a role in the removal of decay heat and in preventing release of radioactive material to the public, are also classified in this category. For example, the pressure switches meant for detecting high pressure in the reactor containment structure are classified as Safety Class IA. The instrumentation system that detects low sodium flow in the primary sodium loop is another example.

Safety class IB is assigned to those I&C systems which perform a complementary or auxiliary role in the achievement of reactor safety or in the maintenance of the same. Failure of these systems to perform their functions may lead to a demand on class IA systems. For example, the window alarm annunciation system, which alerts the operator regarding any unsafe condition, does not directly play a safety role, but informs the operator of any unsafe condition and hence plays a complementary role in safety, and is classified as safety class IB. Other examples are the computer-based systems that implement the interlocks for the control rod drive mechanisms, the instrumentation system that monitors radioactivity inside the plant, etc.

Safety class IC is assigned to those I&C systems that perform some safety functions. However, such functions will not directly fall under either safety class IA or safety class IB. For example, a reactor vessel displacement measurement system can be classified as safety class IC.

Safety class NINS (Not Important to Nuclear Safety) is assigned to those I&C systems that do not play any direct role in reactor safety. However, these systems may play an important role in industrial safety. For example, a fire-alarm system or a CCTV surveillance system can be classified as NINS.

3.3.4 Seismic Considerations

Another important aspect that needs to be considered while designing I&C systems for NPPs is the seismic category of the system. NPPs are designed for a life span of 40 years, during which the site in which the NPP is located can face seismic event/events of different magnitudes. Depending on the magnitude of the seismic events, two maximum ground accelerations are defined: Operation Basis Earthquake (OBE) and Safe Shutdown Earthquake (SSE), as per AERB Code of Practice on Safety in Nuclear Power Plant Siting (AERB/SC/S).

The OBE (also known as S1 level earthquake, as per IEC 60980) is the maximum ground motion the NPP site may experience once during the operating life of the nuclear power plant. The SSE (also known as S2 level earthquake) is a very low probability level of ground motion for which the NPP has to be designed to remain safely shut down.

The SSC in a NPP necessary for the continued operation of the plant shall be designed to remain functional during and after OBE. Necessary repair works may be carried out to resume normal operation after an OBE. The NPP is expected to be safe after an SSE, but the damage is likely to render the NPP non-operational. Typical values of the maximum ground acceleration expected at the site of a NPP are given in Table 3.3.

3.3.4.1 Seismic Categorization of SSC

All SSCs, including I&C systems should be categorized into one of the followings:

Seismic Category-1

Seismic category-1 shall include all I&C systems which are required for

i. shutting down the reactor
ii. maintaining the reactor in a safe shutdown condition by removing decay heat
iii. preventing radioactive release to the public

Systems whose failure could directly or indirectly cause accident conditions also come under this category. All seismic category-1 SSCs should be designed or qualified for both OBE and SSE. Examples of Seismic category-1 I&C systems: Nuclear

Table 3.3 Typical values of maximum ground acceleration at the site of a NPP

Direction	SSE	OBE
Horizontal acceleration (g)	0.156	0.078
Vertical acceleration (g)	0.104	0.052

Flux Monitoring System for generating reactor trip on positive reactivity, reactor protection system, etc.

Seismic Category-2

Seismic category-2 shall include all I&C systems which play a similar role in reactor safety as described above, but to a lesser degree. All seismic category-2 SSCs should be qualified to withstand the effects of OBE. Examples of Seismic category-2 I&C systems: Reactor Power Monitoring System, sodium leak detectors of primary sodium system, etc.

Seismic Category-3

Seismic category-3 includes I&C systems not important to reactor safety and those not covered in category-1 or 2. Examples of Seismic category-3 I&C systems: Surface thermocouples of primary sodium purification system piping, sodium flow meters of secondary sodium system, etc.

3.3.5 Guidelines for the Design of I&C Systems Performing Safety

The fundamental requirement for the design of any safety I&C system is the presence of clear, comprehensive, and unambiguous functional requirements and design specifications.

Guidelines specific for the design of safety class IA I&C systems:

a. The design shall be based on applicable codes/guides/standards.
b. The design shall be simple so that the same can be easily verified.
c. Redundancy shall be provided to meet the single failure criterion.
d. Appropriate segregation of components shall be ensured to reduce internal hazards that can affect the redundancy.
e. Online testing facility shall be provided to detect unsafe faults.
f. If common cause failures are known to be present, there shall be diversity.
g. The I&C system shall be subject to formal qualification by testing, or analysis or both.
h. Multi-layer Security against unauthorized access and modification shall be provided, for example: Access control.

Guidelines specific for the design of safety class IB I&C systems:

a. The design shall be based on applicable codes/guides/standards.
b. The design shall be simple so that the same can be easily verified.

c. Redundancy is preferable. However, if redundancy is not provided, analysis should be done to ensure that the consequences of failure of the I&C system in any mode are safe. If not, redundancy shall be provided.
d. The design shall meet the single failure criterion.
e. Appropriate physical and electrical segregation of components shall be ensured to reduce the effect of internal hazards that can affect the redundancy.
f. The I&C system shall be qualified by testing, analysis, or both.

Guidelines specific for the design of safety class IC I&C systems:

a. Redundancy shall be considered to meet the overall reliability requirements.
b. The I&C system may be accepted based on commercial design standards.

3.3.6 Development and Qualification of Computer-Based Safety Class I&C Systems

Many I&C systems are now being implemented with the help of programmable devices, which include any I&C system using personal computers, industrial computers, PLCs, microprocessors, micro-controllers, FPGAs, CPLDs, PALs, etc. Such systems fall under the broader classification of Computer-Based Systems (CBS). The guidelines for the design, development, and qualification of CBS are covered in the AERB safety guide SG-D-25.

Based on the requirements, derived from the Preliminary Safety Analysis Report (PSAR) or system control note, and market survey, a decision has to be taken whether CBS shall be a newly developed system or commercial pre-developed system or certified Pre-developed system based on the component availability and their suitability to meet requirements. Generic design certification from regulators allows to design of hardware or software for a specific purpose, which can be used as a ready-to-use building block for the design and implementation of CBS. Pre-developed system can be further divided into a certified pre-developed system or a commercial pre-developed system.

One of the main aspects of this process is to ensure that the software does what is required to be done, and does not do what is not required. This is essential because any program consisting of code written in higher level languages/machine language can contain inadvertent errors (like representing years in two-digits instead of four digits, as happened in the Y2K issues), or even malicious code. The code is normally not accessible to the user. Even if it is so, the user may have the expertise to understand what the code is doing. Hence, in order to address the issue of design, development, and qualification of software-based systems, extensive guidelines are provided that need to be followed, especially for safety class IA and IB systems, where programming is involved, one of the main requirements is to ensure that the program/software is independently verified and validated.

a. Verified: all the system requirements are implemented in the software
b. Validated: the implemented software is evaluated by testing and ensured to meet the system requirements

This process, known as the Independent Verification and Validation (IV&V) process, is to be done by those who are independent from the design and development of the concerned CBS.

The IV&V is required to be done for both hardware and software. The system development methodology and the IV&V process are explained in the form of a flow chart in Fig. 3.3 and Fig. 3.4 respectively.

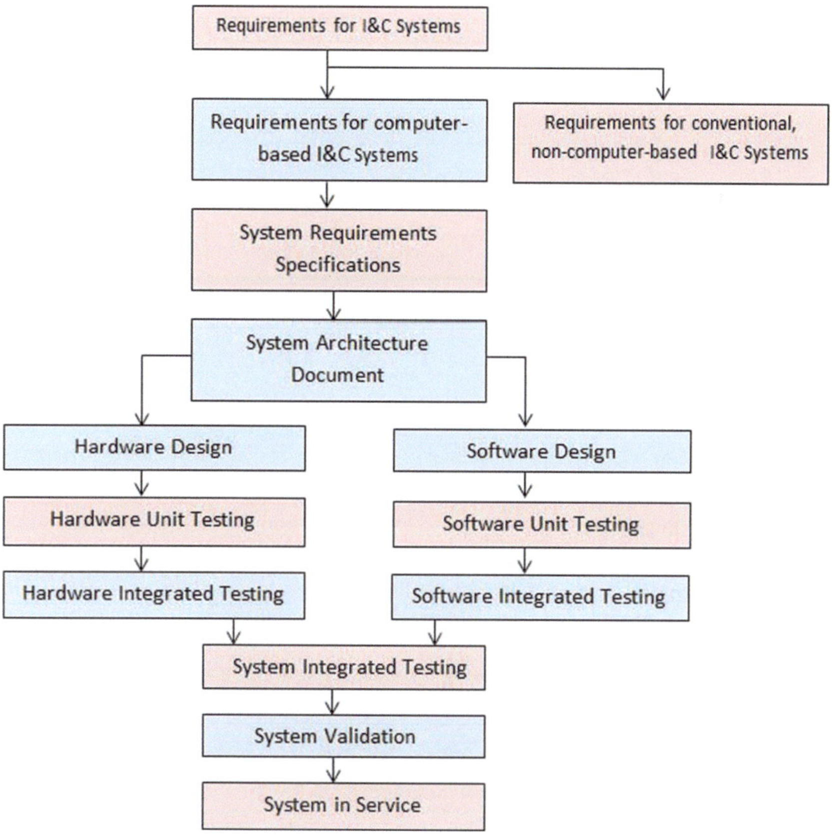

Fig. 3.3 The system development methodology

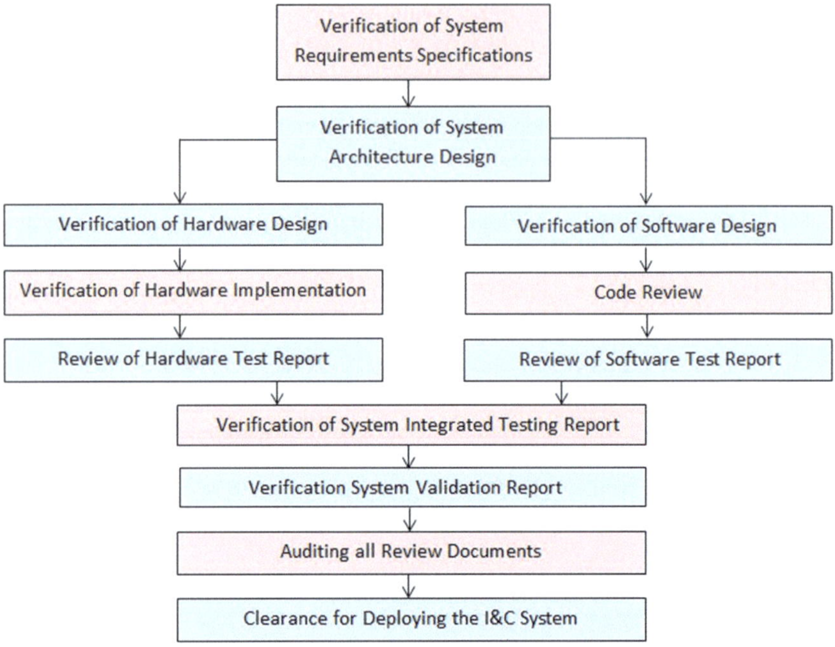

Fig. 3.4 The IV&V methodology

3.3.6.1 Documents Required for IV&V

The documents required for IV&V can be classified as follows:

(a) System requirements
(b) Plan Documents
(c) Design Outputs
(d) Analysis documents

The list of documents under each category is also listed in Table 3.4.

3.3.6.2 The IV&V Process

The IV&V process is a means to qualify the computer-based I&C system for a safety application. It is basically a document-centric process to ensure that the computer-based IV&V system is built to meet the requirements, as well as to ensure that good design practices are followed so that nothing that jeopardizes safety is present in the system, especially in the form of code. A number of documents, depending on the safety classification, is required to be subjected to the IV&V process. The safety class IA systems undergo the most rigorous IV&V, followed by IB, IC, and NINS

Table 3.4 Documents required for carrying out IV&V of computer-based systems of different safety classes

Submittal	1A	1B	1C
1. Requirements			
1.1 System requirements	✓	✓	✓
2. Plans			
2.1 System development plan	✓	✓	✓
2.2 Configuration management plan—development	✓	✓	✓
2.3 System security plan	✓	✓	✓
2.4 QA plan—hardware	✓	✓	✓
2.5 QA plan—software	✓	✓	✓
2.6 Verification plan—hardware	✓	✓	✓
2.7 Verification plan—software	✓	✓	✓
2.8 System validation plan	✓	✓	✓
3. Design outputs			
3.1 System architectural design	✓	✓[a]	–
3.2 Hardware requirements specification	✓	✓[a]	–
3.3 Hardware design	✓	✓[a]	–
3.4 Software requirements specification	✓	✓[a]	–
3.5 Software design	✓	✓[a]	–
3.6 Software programmes (source code)	✓	✓[a]	–
3.7 Programming guidelines	✓	–	–
4. Analysis reports			
4.1 Confirmation of safety function implementation	✓	✓[a]	–
4.2 Failure analysis report (for single failure criterion)	✓	✓[a]	–
4.3 CCF analysis	✓	–	–
4.4 Hardware reliability analysis	✓	✓	–
5. System validation			
5.1 System validation report	✓	✓	✓

[a] Intensity of review of 1B class system can be less than that for 1A and should confirm that system failures cannot have adverse effect on safety functions and will not make frequent demands on class 1A functions

systems. The list of documents required for IV&V of these systems is shown in Table 3.4.

It can be seen from Table 3.4 that for safety class IA system, a complete code walk-through is a must; in addition, all the design outputs shall be subjected to IV&V. For safety class B systems too, documents pertaining to design outputs should be reviewed, but to a lesser degree than that of safety class IA. Code walk-through is not considered essential for safety class IB systems.

3.3.6.3 The IV&V Audit

The review reports generated during the IV&V process are once again reviewed by an audit team, as presented in Table 3.5, before clearing the system for deployment in the field.

3.3.6.4 IV&V for Generic Design Certification

If a generic component or block is used in the design and implementation of CBS, it may happen that the IV&V was carried out on these components, but the same needs to be scrutinized through audits to determine if the IV&V was carried out as per the guidelines. For I&C systems built with commercially available off-the-shelf components, a different set of guidelines is available.

Table 3.5 List of documents which are subject to auditing before clearing the computer-based I&C system for deployment in the field

Documents	1A	1B	1C
1. Plans			
1.1 QA plan—hardware	✓	✓	✓
1.2 QA plan—software	✓	✓	✓
2. Design output			
2.1 Software requirements specification	✓	✓	✓
2.2 Software design	✓	✓	–
2.3 Programming guidelines	✓	✓	–
2.4 System integration and test plan	✓	✓	–
2.5 System integration and test report	✓	✓	–
2.6 System validation report	✓	✓	✓
3. Verification reports			
3.1 Verification report of software requirements specification/application programming requirements	✓	✓	✓
3.2 Verification report of software design	✓	✓	–
3.3 Verification report of code/programs in application programming languages	✓	✓	–
4. Certification reports			
4.1 Certification/regulatory approval reports (report of compliance to general design criteria, safety criteria and quality policy)	✓	✓	✓

3.3.6.5 General Guidelines for Software Development for Safety Systems

For developing software for safety systems, the guidelines specified in the international standard IEC 60880 (Nuclear power plants—Instrumentation and control systems important to safety—Software aspects for computer-based systems performing category A functions) are to be followed. Some of the salient points to be noted are:

- Programs and program parts shall be grouped systematically.
- Modules shall be clear and intelligible.
- Use of the operating system software shall be restricted.
- Use of interrupts shall be avoided/restricted.
- Plausibility checks shall be performed (defensive programming).
- Diagnostic features to check the healthiness of CPU, I/O modules, and memory shall be incorporated. On detecting system failure, fail-safe outputs shall be produced.
- Simple addressing techniques shall be used.
- Dynamic changes to executable code shall be restricted/avoided.

3.3.6.6 Motor Industry Software Reliability Association C Guidelines

For developing safety systems, one of the best choices is to use the C language recommended by the Motor Industry Software Reliability Association (MISRA), which is an adaptation of the ANSI C programming language. This is to ensure the development of safe and reliable software products. The conventional C language has many advantages, like easy access to hardware, low memory requirements, efficient runtime performance, etc. However, certain features of the standard C make it possible to write code that is obscure and whose behavior is unpredictable. This is because standard C programs generally do not provide run-time checking for common problems, which include errors/exceptions like divide by zero, overflow, validity of pointers, or array bound errors, etc.

In order to overcome such issues, MISRA first developed coding guidelines for the C language in 1998. These were specific to the C programming language. Since then, MISRA has added a coding standard for C++ too. The current version, MISRA C:2022, has evolved over several years and includes 182 rules and 18 directives.

For I&C systems built with commercially available off-the-shelf components, a different set of guidelines is available.

Some of the rules of MISRA C are:

- All object and function identifiers shall be declared before use.
- Tests of a value against zero should be made explicit, unless the operand is effectively Boolean.
- Non-constant pointers to functions shall not be used.

Use of MISRA C helps in building safe, secure, portable, and reliable software.

3.3.6.7 Tips for Developing Safety I&C Systems Using PLDs

While MISRA 'C' helps in building robust and reliable safety systems, for developing safety systems using PLDs (including FPGA/CPLD/PLA) using HDL, some tips are available for building safe I&C systems:

- Use of clean power supplies is a must.
- Power sequencing is to be carefully chosen for PLDs using multiple power supplies.
- Power ramp-up rate during switching on and switching off of the PLDs is to be respected.
- Sudden inrush of current due to too many gates switching on can cause a lot of misbehavior of PLDs. To reduce the ground bounce and the switching noise, power pins should be decoupled using decoupling capacitors.
- It is advisable to smoothen noisy input signals with suitable circuitry, like Schmitt triggers. Similarly, the lowest slew rate has to be chosen for outputs.
- Unused pins should be properly tied to either ground or power supply, depending on the requirements.
- Proper reset of all flip-flops across the PLD should be ensured.
- Behavior of a state machine should be defined for both used and unused states.
- Techniques for mitigating single-event upsets (change in the contents of a memory cell due to external radiation) should be considered, including one-time-programmable devices, checking of critical logic blocks using online/offline diagnostics, etc.
- Multiple clock domains should be avoided.
- Meta-stability issues should be avoided.
- Lifetime of the PLDs (around 10 years for programmable PLDs) should be considered where configuration data is stored.

3.3.6.8 Use of Safety Critical Application Development Environment

For safety I&C systems developed on embedded systems, an alternative to using manual coding is to use the Safety Critical Application Development Environment (SCADE) available commercially. The SCADE environment uses model-based specification of system requirements. The requirements are graphically represented, compiled, and the code is automatically generated. The SCADE suite drastically simplifies design and facilitates automatic generation of MISRA-C compliant code so that errors that normally creep in during manual coding are avoided. Use of SCADE helps in reducing the time required for software development. Since the SCADE suite includes tools for simulation and verification and results in certified code generation, it also reduces the product certification costs and helps in managing the product life cycle.

3.3.6.9 Summary

The IV&V process builds up the confidence that a safety I&C system has been specified, designed, implemented, integrated, and tested in the field so that it can be deployed in the field for a safety application. It may be kept in mind that the IV&V process is a time-consuming process and hence must be taken into account while forming the project schedule.

3.3.7 Guidelines for the Design of Safety Systems

The following are the salient points from the various safety guides of AERB as applicable to I&C systems of NPPs, which are applicable as well to test reactors and fast reactors:

- Instrumentation to monitor plant parameters during normal operation, during expected events (like grid power supply failure or loss of flow in heat transfer circuits), and also after any accident.
- Safety instrumentation shall have adequate diversity and redundancy.
- Instrumentation for detecting failed fuel is a must.
- Instrumentation for monitoring the proper cooling of fuel subassemblies.
- Appropriate controls shall be provided to maintain safety related variables within prescribed ranges and annunciate if they exceed their set limits.
- Design and layout of instrumentation system shall be such as to permit periodic testing and preventive maintenance.
- Instrumentation for post-accident shall take cognizance of the harsh environmental conditions during that state (including high radiation, high temperature, etc.).
- Fail-safe design shall be incorporated in all safety related instrumentation and control systems.
- Instrumentation shall be provided to permanently record critical plant signals so as to facilitate analysis of plant events.
- Adequate instrumentation and recording equipment shall be provided for predicting the location and quantities of radioactive materials possibly escaping.
- Instrument power supplies (viz. hydraulic, pneumatic, and electrical) shall be designed to ensure adequate availability and reliability commensurate with their safety functions.
- Main Control Room shall facilitate safe operation of the plant even under accident conditions.
- Layout of I&C Panels shall be such as to provide the operator adequate and comprehensive information on the state and performance of the plant during all operational states and accident conditions.
- Jumpering of any safety condition shall be automatically indicated and identified in the control room.

- Appropriate measures to prevent unauthorized access to plant vital areas shall be ensured.
- Audio/Visual Alarms shall be provided for abnormal process conditions.
- Backup/Emergency control room shall be provided with sufficient instrumentation and control equipment, so that the reactor can be placed and maintained in a safe shutdown state, in case the main control room becomes inaccessible due to any reason like fire, sabotage, etc.
- Reliable air-conditioning and ventilation system shall be provided to the main control room (MCR), backup control room (BCR), and instrumentation rooms housing safety related instrumentation.
- Provision for monitoring of oxygen shall be available to facilitate the safety of personnel required to enter the inserted areas.
- Special instrumentation, including auxiliary neutron sources, shall be provided to ensure safety during initial loading of the core, sub-critical monitoring in the fuel handling state, and during first approach to criticality.
- Plant Protection System shall be designed to protect the plant under all unsafe conditions. It shall be designed for high functional reliability and periodic testability (including online testing) commensurate with the safety functions to be performed. The design shall provide diversity, independence, and physical separation between redundant channels to ensure that no single failure results in loss of protection function (known as the single failure criterion).
- Manual backup: In spite of safety actions initiated automatically, there shall be a manual backup to safely shut down the reactor.
- Effect of Lightning/Flooding shall be taken into account.
- No operator action should be necessary in a time scale of approximately 30 min following an unsafe plant event.

3.4 Instrumentation Specific for SFRs

The instrumentation and control for fast reactors are specifically designed to meet the demanding requirements that are unique to fast reactors. Here, we will deal only with I&C systems specific to SFR. The instrumentation pertaining to systems meant for the utilization of steam and generation of electric power, termed as balance-of-plant (BOP) systems, is considered conventional and hence cannot be covered here.

The following is a list of some main process variables which are measured/detected in a typical fast reactor:

- Temperature of sodium in the sodium systems, at the outlet of fuel subassemblies in the core, of structures like reactor vessel, top-shield, etc.
- Flow of sodium in various sodium circuits, including the primary and secondary sodium systems.
- Level of sodium in various sodium capacities like main vessel/reactor vessel, surge tank, storage tank, sodium pumps, etc.

- Leak of sodium from various sodium capacities like main vessel, surge tank, storage tank, Steam Generator, etc., as well as from sodium pipelines.
- Pressure of argon cover gas in various sodium capacities.
- Speed of the sodium pump, etc.

3.4.1 Factors That Influence Design of I&C Systems for Sodium Systems

There are a number of factors that are taken into account while designing the instrumentation for SFRs:

a. **Chemical incompatibility of sodium with air**. This calls for a rugged design of probes with no scope for sodium leak.
b. **High radioactivity**. Both gamma and neutron activity need to be considered.
c. **High temperature**. High temperature (600 °C) calls for the deployment of suitable material that can withstand the same. For example, mineral-insulated cables are required for such applications which are less flexible.
d. **Low pressure**. Low pressure of sodium is a boon, calling for thinner structures for probes.
e. **Slow loss of reactivity**. This is also a boon so that manual control of reactor power is enough, as the burn-up losses can be easily compensated by manually raising the control rods.
f. **Shorter neutron life time**. This is one order less than thermal reactors, making power multiplication faster by the same order. The neutronic instrument channels need to be fast enough to safely shut down the reactor in case of unexpected reactivity transients.

3.4.2 Sodium Temperature Measurement

For measuring the temperature of sodium, the thermocouple chosen is ANSI type-K (chromel-alumel), due to reasons like the ability to cover the required range, good accuracy and linearity, proven experience to withstand intense gamma-radiation and neutron flux, etc. Ungrounded, metal-sheathed-mineral-insulated 1 mm diameter thermocouples have proven to be very effective in measuring sodium temperature in the presence of neutron- and gamma-radiation, and hence are the preferred choice of sensor for core temperature measurement. It is also used in all other sodium temperature measurement, except that for core temperature measurement, accuracy class 1 is chosen for probe construction, whereas for other sodium temperature measurement, accuracy class 2 is sufficient. Also, the boron content in the MgO used for constructing probes to be used in core temperature measurement should be very less

(<30 ppm) since boron is an effective neutron absorber. Thermocouples are inserted in thermo-wells in all the locations to facilitate ease-of-replacement, except for measuring the temperature of sodium exiting the central fuel sub-assembly, since the neutron flux peaks at this location. This is to quickly detect any temperature change arising from neutron flux changes in that location. Extensive testing on such probes has confirmed that Type-K thermocouples are capable of excellent performance under such conditions, except for a slight increase in the diameter of thermocouples due to swelling and a marginal increase in the response time of individual thermocouples.

3.4.3 Use of Computers in Core Temperature Measurement

The measurement of the temperature of sodium at the outlet of each FSA is a relatively simple task; however, taking safety action against high sodium bulk temperature (due to reactivity transients) and sodium plugging in any FSA is more involved. Deployment of digital computers has helped immensely in carrying out the same. Calculation of the mean of bulk sodium outlet temperature, determination of differential temperature across the core, and estimation of expected temperature rise across each fuel sub-assembly require extensive calculations, and hence, the core-temperature monitoring systems are computer-based safety class IA systems.

3.4.4 Sodium Level Measurement

For measuring sodium levels in various sodium capacities, mutual inductance type level sensors are used. The Mutual Inductance Type Level Probe (MILP) works on the principle that the mutual inductance between two windings changes due to the presence of sodium around the probe. The MILP has two windings wound in bifilar fashion on a non-magnetic stainless steel former. This level probe is inserted in a stainless steel pocket provided in the sodium tank in which the level of sodium is to be measured. The primary winding of the probe is excited with a constant alternating current at a fixed frequency. The choice of the excitation frequency depends on maximizing the output and minimizing the effect of sodium temperature variations; a typical frequency is 1.5–2.5 kHz. When the primary coil is excited, an emf is induced in the secondary coil. Due to this, an eddy current is produced in the sodium that surrounds the probe. The magnetic flux due to the eddy current so produced will oppose the main flux produced by the primary winding. Hence, the net flux linked with the secondary of the probe decreases, and hence the secondary voltage reduces in proportion to the sodium level. Hence, the secondary voltage is an inverse linear function of the sodium level. When the sodium temperature increases, the sodium resistivity increases. This results in a reduction in the eddy current, leading to less opposition to the secondary flux and consequently more secondary voltage. The sodium level will then be indicated erroneously as less than the actual level. In order

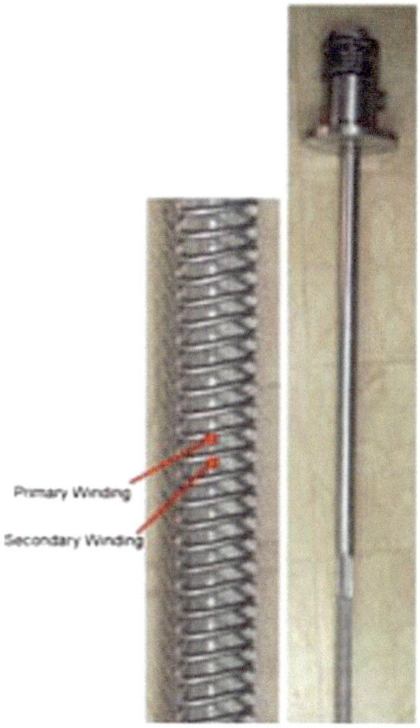

Fig. 3.5 Construction of MILP probe

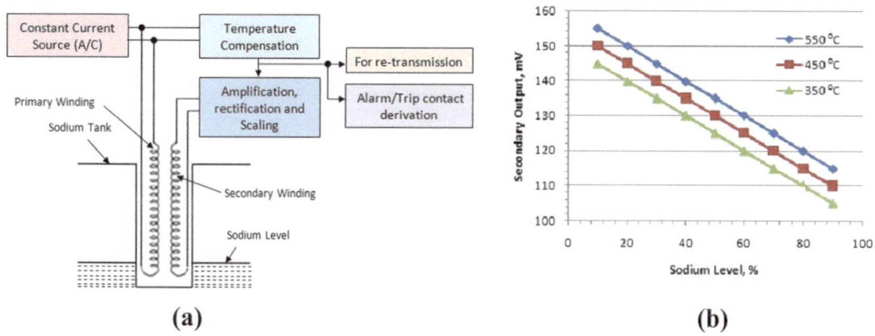

Fig. 3.6 **a** Schematic drawing for MILP. **b** Variation of secondary emf of a MILP at different sodium temperature

to avoid this, a suitable compensating circuit to offset the effect of temperature is required so that the level indicated is independent of sodium temperature but depends only on the actual level of sodium. The construction of the MILP probe, its schematic, and the variation in secondary emf at different sodium temperatures are shown in Figs. 3.5 and 3.6 respectively.

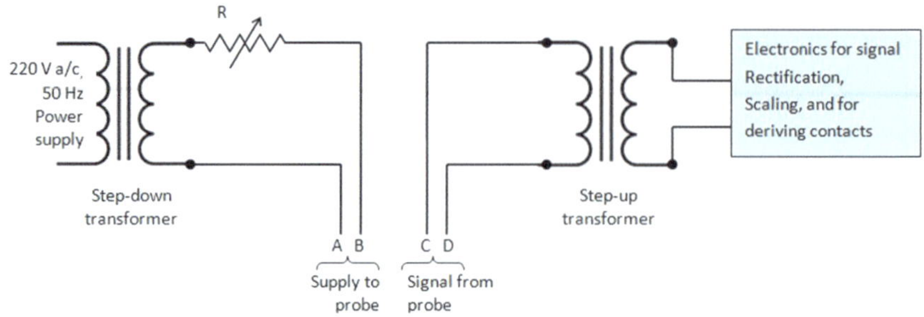

Fig. 3.7 Excitation for resistance type level probes

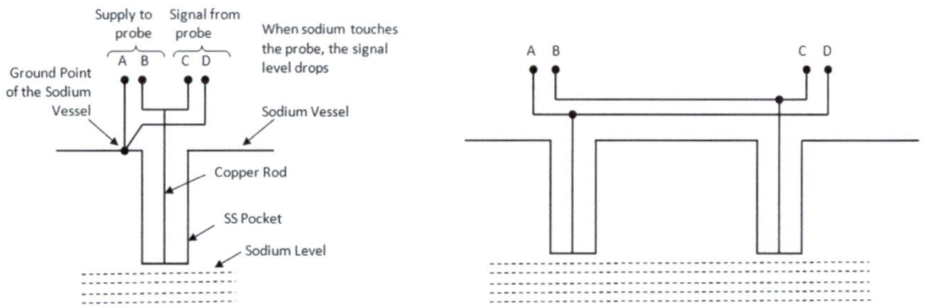

Fig. 3.8 **a** Resistance type level probe for low-level detection. **b** Resistance type level probe for high-level detection

3.4.5 *Sodium Level Switches*

For detecting whether the sodium level is above or below a certain level in the tank/capacity in which the probe is inserted, two types of level detectors can be used: resistance type and mutual inductance type.

3.4.5.1 Resistance Type Sodium Level Switch

In this, the fact that sodium is a good conductor of electricity is made use of. The probe consists of a copper rod kept in the axis of a stainless steel (SS) pocket cylinder and brazed at the bottom. The probe (Figs. 3.7 and 3.8) is inserted in the sodium tank/capacity where the sodium level is to be detected. The probe extends into the tank to such an extent that the presence of sodium above or below that level is required to be detected. For low-level detection, the probe extends into the tank and hence has adequate length of copper to create a voltage drop (Fig. 3.7). For detecting high sodium levels, the probe length is shorter, and hence two probes are connected in series to get a measurable voltage drop (Fig. 3.8). The copper conductor in series with the SS pocket is excited with a constant power line frequency (50 Hz), constant-amplitude alternating current. When the sodium level in the tank is below that of the probe, this results in a measurable voltage drop across the copper conductor-SS cylinder combination. When sodium touches the bottom of the probe, it effectively short-circuits part of the resistance offered by the SS pocket, resulting in a considerable reduction in the voltage drop, which is a clear indication that sodium is near or above the bottom of the probe. The output is rectified, and the resultant DC voltage is compared against a threshold value. The output of the comparator can be used to annunciate an alarm or used in the interlock for the operation of related equipment like pumps.

The resistance type level probes suffer from one disadvantage in that their output also depends on the sodium temperature. Hence, the value of the series resistance and the DC threshold level in the comparator need to be optimized to prevent spurious alarms for the entire range of sodium temperatures. Also, the high-level probes can spuriously actuate due to deposition of sodium vapors at the bottom of the probe, even though the sodium is much below the probe.

3.4.5.2 Mutual Inductance Type Sodium Level Detection

The resistance type level switch requires wetting of the level probe for detecting the presence of sodium. Though SS has good wetting properties by sodium, both the response time and accuracy of detection can be improved by adopting a non-contact method of sodium level detector which does not require sodium wetting the probe. Hence, mutual inductance type level detectors were developed, which employ the same technique as MI level probes described in Sect. 3.4.4. This probe also has the advantage that it can be inserted in an SS pocket welded to the sodium capacity and hence is amenable to easy maintenance, including replacement. In addition, it is possible to have more than one level switch in a single probe. The sensor consists of a bobbin, which houses a primary winding and a secondary winding wound over an SS former. The windings are made of mineral-insulated, metal-sheathed copper cables. The primary coil is excited with 2.5 kHz alternating current. This frequency has been chosen to ensure that the secondary output has much less dependence on sodium temperature. Hence, no separate temperature compensation is required. When the primary is excited, the secondary output will depend on whether sodium surrounds the bobbin or not. When the sodium level is below the bobbin, the secondary output is maximum. When more than one bobbin is there, all the primary coils are connected in series and excited with the same power supply. The MI type sodium level detector probe and construction of the bobbin in MILD are shown in Fig. 3.9a, b.

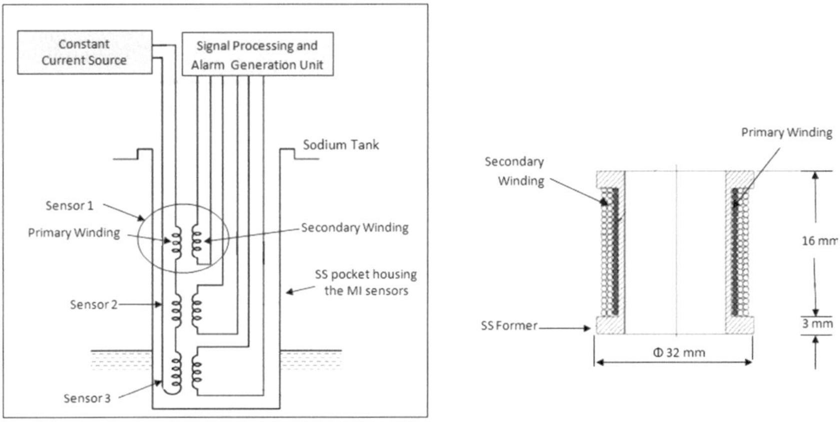

Fig. 3.9 **a** MI type sodium level detector probe and the housing SS pocket. **b** Construction of bobbin in MILD

3.4.6 *Sodium Flow Measurement*

For measuring sodium flow in SFRs, permanent magnet flow meters or eddy current flow meters are employed.

3.4.6.1 Electromagnetic Flowmeters

Faraday's law of electromagnetic induction says that when a electric conductor moves in region in which magnetic flux is present, an emf is induced in the conductor; the magnitude of the emf is given by the equation $E = Blv$, where B is the magnetic flux density, l is the length of the conductor and v is the velocity of the moving conductor. This is the basic principle with which a DC output in proportion to the sodium velocity of sodium is obtained. Using the diameter of the pipeline, the output of the flow meter can be used to provide the volumetric flow of sodium. If the magnetic field is produced by an electromagnetic coil with an external power supply, it is known as an electromagnetic flow meter. When the magnetic field is generated by a permanent magnet, it is known as a permanent magnet flowmeter (PMFM). A PMFM shown in Fig. 3.10 basically consists of a non-magnetic pipe mounted in the transverse magnetic fields between the poles of a permanent magnet assembly. Electrodes are positioned diametrically opposite to each other and welded to the outer surface of the sodium pipe, in such a way that their central axis is perpendicular to that of the magnetic flux. Here, sodium flowing in the non-magnetic SS pipeline is the moving conductor. The flux density is provided by a permanent magnet assembly typically made up of ALNICO.

The reason for providing permanent magnets instead of electromagnets is to minimize the maintenance on the flowmeter, since in primary sodium loops of loop-type

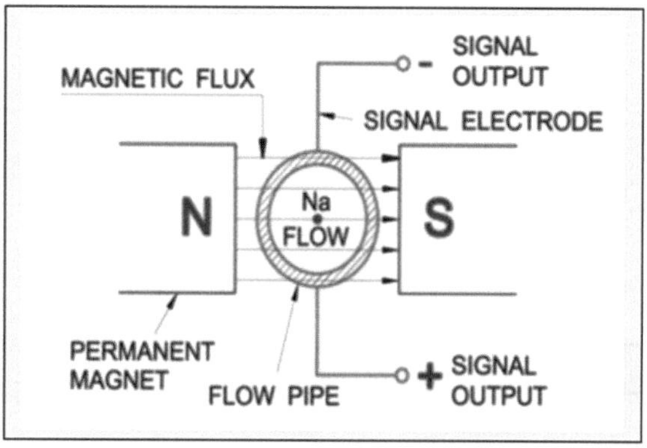

Fig. 3.10 Principle of operation of permanent magnet flow meter

reactors like FBTR, they handle radioactive sodium. One factor that can affect the DC output from the flow meter is the change in flux density generated by the permanent magnet over time. In order to recalibrate the flow meters in such a case, there is a provision to measure the magnetic flux. The ALNICO magnets are proven for service in such high temperature (500 °C) and radioactive applications. The magnetic flux is also affected by the temperature of the magnet, the orientation of the magnet assembly with respect to the sodium pipe, and the presence of magnetic material close by. The output is also affected by the resistivity of the pipe material compared with that of sodium. Straight length requirements need to be respected for electromagnetic flow meters too, akin to those of other flow meters, to ensure a uniform velocity profile.

The PMFM also has another advantage for safety applications. The meter is capable of giving more than one output in a single assembly. Hence, when used for detecting sodium flow in primary and secondary sodium main circuits, a meter with three independent sets of electrodes is used. Even though they share a common permanent magnet, the signals can be considered as independent.

The pure DC output from the PMFM with very low source impedance is normally processed by chopper-stabilized amplifiers and converted into standard signals like 0–10 Vdc or 4–20 mA, which is proportional to the volumetric flow rate.

The pipe size in various circuits of a typical 500 MWe SFR ranges from 15 to 800 NB. PMFMs can be used for pipelines up to 200 mm. For larger sizes, the PMFM becomes too bulky. Saddle-type flow meters have been tried up to 355 mm dia. pipes.

3.4.6.2 Bypass Type Electromagnetic Flow Meters

For sodium pipelines greater than 200 mm diameter, the bypass type flow meter shown in Fig. 3.11 can be used, where a small proportion of the main sodium flow is made to pass through a bypass line of lesser diameter, where a PMFM is installed. The bypass flow measured by the PMFM is proportional to the main sodium flow. The

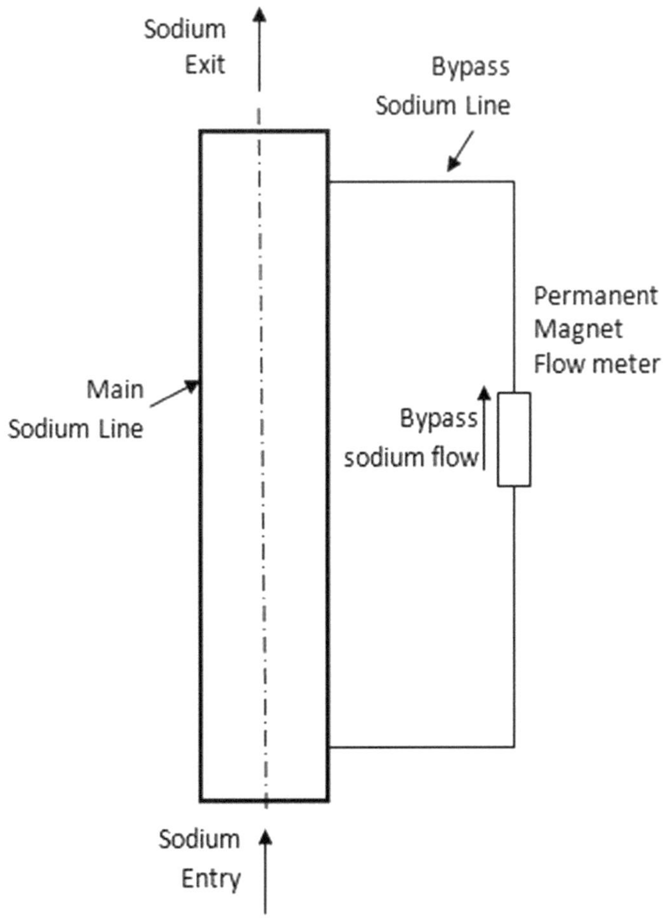

Fig. 3.11 Principle of operation of bypass type permanent magnet flow meter

relation between the velocity V of sodium in the bypass line to that of sodium velocity V in the main pipe line is given by $v/V = \sqrt{k_m/k_b}$, where k_m is overall pressure drop coefficient across the main line and k_b is the overall pressure drop coefficient across the bypass line. The size of the bypass line is optimized to maximize the sensitivity and resolution of the flow measurement.

3.4.6.3 Eddy Current Flow Meters (ECFM)

Another convenient way of measuring the sodium flow in SFRs is to deploy the eddy current flow meter, whose principle of operation is demonstrated in Fig. 3.12.

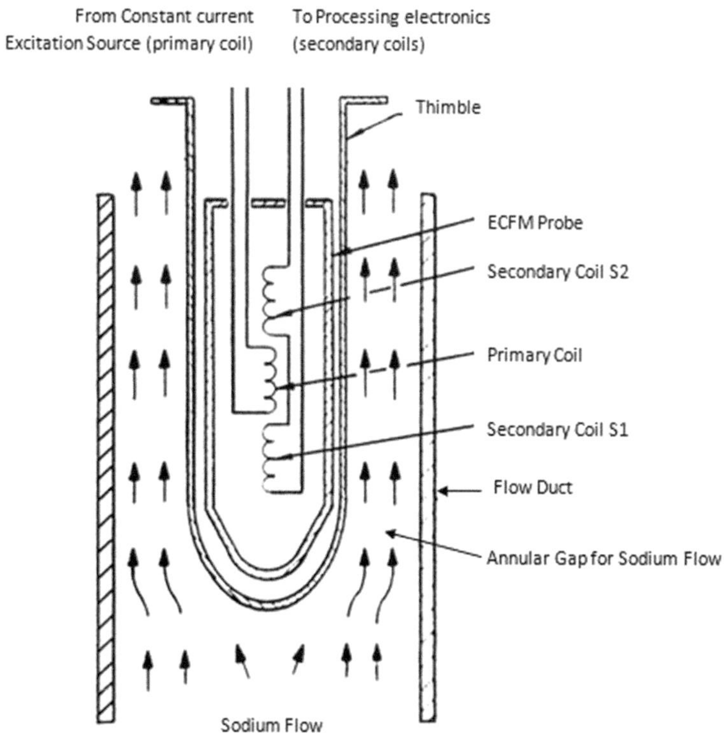

Fig. 3.12 Principle of operation of eddy current flowmeter

In an ECFM, there is a primary coil and two secondary coils S1 and S2. The primary coil is excited with an alternating current of appropriate frequency. The voltages induced in the secondary coils are measured. The secondary voltage is due both to the transformer action (E_{trans}) as well due to motion of sodium (E_{motion}). Under static conditions, ideally, the induced voltage E_{trans} in S1 and S2 are equal, and their difference is zero. When sodium flows, the resulting magnetic field distortion increases the voltage induced in S2 by E_{motion}, while it decreases the voltage induced in S1 by E_{motion}. In order to maximize the sensitivity, the difference (S1 − S2) is taken. This value can be proved to be proportional to the volumetric flow rate of sodium. One of the major factors affecting the output of ECFM is the temperature of sodium. As sodium temperature increases, the resistivity of the sodium increases, resulting in less eddy currents. This affects the net output of the induced secondary voltages. In order to reduce the effect of temperature and make the output depend only on sodium velocity, the ratio (S1 − S2)/(S1 + S2) is recommended. The ECFMs for sodium applications need to withstand high temperatures. Hence, the primary and secondary windings are made of MgO-insulated, SS-sheathed cables with nichrome conductors. Apart from their ability to withstand high temperatures, nichrome cables are also used due to their extremely low temperature coefficient of resistance.

3.4.7 Sodium Leak Detection

Detection of a leak of sodium from any sodium pipeline or tank is essential to alert the operator and or to initiate safety action, since loss of sodium can hamper the heat extraction, and also can lead to reaction of sodium with air, resulting in hazardous smoke and/or fire. Even though enough care (like having double-walled pipelines, 100% radiography testing of all welded joints, avoiding non-welded joints, etc.) is taken to ensure the integrity of sodium-carrying equipment, still, sodium leakage from any part of the sodium-carrying pipelines/tanks should be detected reliably with appropriate instrumentation. Such instrumentation will also help in identifying the location of the leak.

3.4.7.1 Sodium Leak Detection in the Incipient Stage

Sodium in the heat transfer lines/capacities of a SFRs is always kept at a minimum temperature (say 180 °C, as in FBTR), which is more than the sodium freezing temperature of 98.6 °C at atmospheric pressure. If there is a breach in the integrity of the sodium capacity at such high temperatures, sodium leaking will vaporize and lead to the formation of sodium aerosols, which can be detected by the sodium ionization detector (SID). The principle of operation of the SID is shown in Fig. 3.13. The medium (argon, air, or nitrogen) in which sodium aerosols may be present is made to pass through the SID assembly, which houses a platinum filament. The filament is kept at a higher electric potential than the cylindrical SS collector surrounding the filament. The platinum filament is electrically heated, which helps to selectively ionize the sodium present in the medium. The resulting Na^+ ions are attracted to the collector, which leads to the establishment of an ionic current. This current is proportional to the sodium ion concentration present in the detector. The ionization of sodium is selective compared to the presence of other elements like oxygen and hydrogen. Hence, an increase in the ionic current from the background value indicates the presence of sodium in the medium. The leak of sodium can be detected in the incipient stage itself. The ionic current for 100% thermal ionization efficiency is given by the equation $i_0 = ((\mathrm{FAVC})/M)$, where F is the coulomb constant, A is the projected area of the filament, V is the velocity of the sample gas in the detector, C is the weight concentration of sodium, and M is the atomic weight of sodium.

3.4.7.2 Gross Sodium Leak Detection

Detecting heavy sodium leakage is a relatively easy task, since sodium is a very good conductor of electricity. Leaking sodium electrically grounds a suitably designed sensor, and the same is detected using suitable electronic circuits. In order to detect sodium leak from pipelines, wire-type leak detectors are used, while spark-plug type

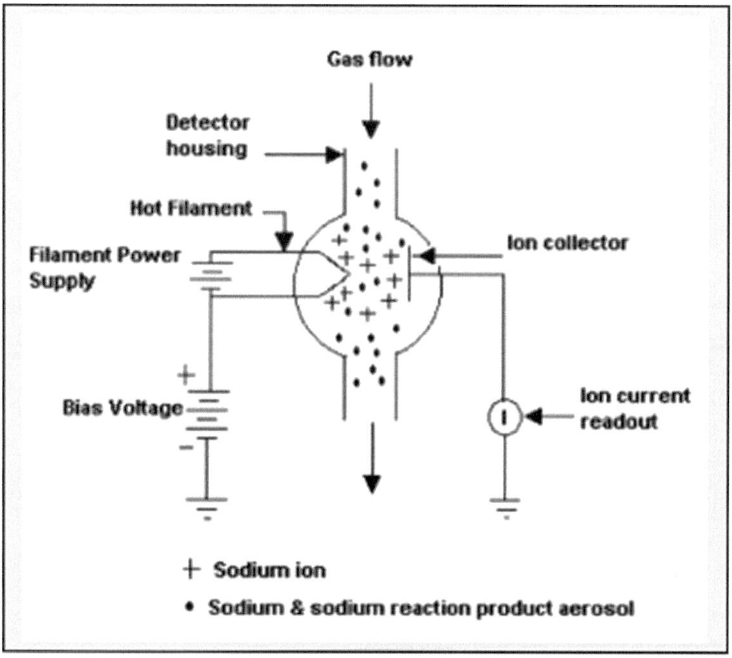

Fig. 3.13 Principle of operation of Sodium Ionization Detector (SID)

leak detectors are used to detect sodium leak from tanks, valves, etc. As described earlier, for detecting sodium leaks in the incipient stage itself, SIDs are used.

The principle of operation of the sodium leak detector (wire type or spark-plug type) is demonstrated in Fig. 3.14. In the absence of sodium, the input to the inverter IC is high; the output of the inverter is hence low, and the relay is in an energized condition. When sodium leaks, it electrically grounds the input of the inverter, making it low, and hence the output of the inverter goes high. The relay de-energizes, and the resulting change in the status of the contact can be used for alarm or in a logic circuit.

There are two types of leak detectors deployed in SFRs: wire type and spark-plug type. The wire-type leak detectors are suitable for detecting sodium leaks from pipelines, which are electrically grounded. Thin nickel wires, passing through suitably designed ceramic insulating beads, are wound helically for vertical pipe sections, while they are kept below the horizontal pipelines up to 250 mm in diameter. For larger horizontal pipelines, wire-type leak detectors are kept in 4 locations, at 0°, 90°, 180°, and 270°, where 0° is considered to be at the top. Even a drop of sodium leaking from the equipment can activate the leak detector. A typical sodium pipeline section showing the details of mounting of leak detectors is shown in Fig. 3.15, which also demonstrates the other requirements of sodium pipelines, like heaters for keeping the sodium in molten condition, thermocouples for surface temperature

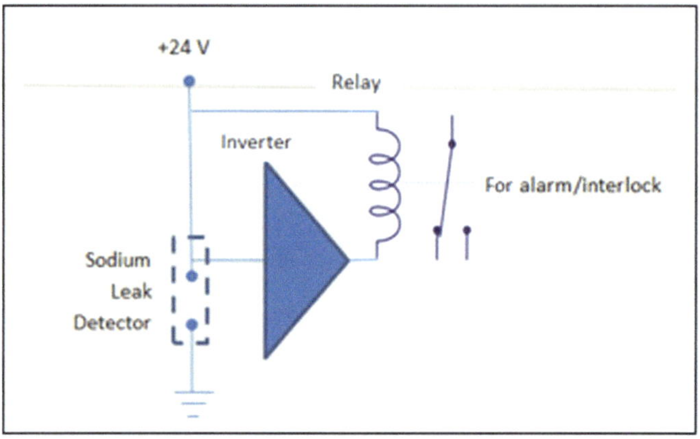

Fig. 3.14 Principle of operation of sodium leak detector

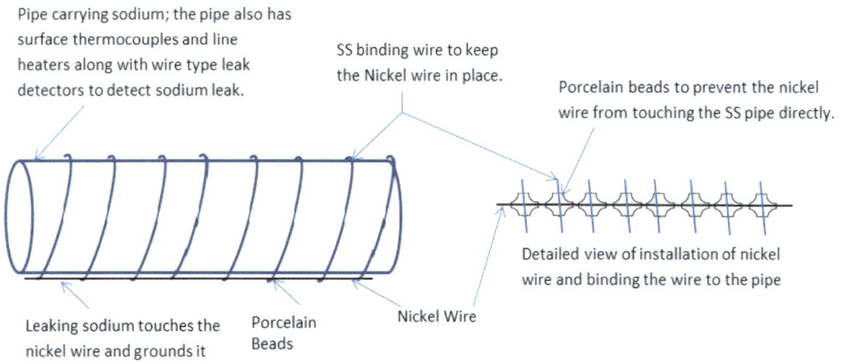

Fig. 3.15 Installation of wire-type sodium leak detector in a section of pipe

measurement, insulation for conserving heat, junction boxes for terminating the leak detectors and electrical heaters, etc. The wire is made of nickel, while the ceramic beads help the leak detector be mechanically supported on the pipelines with the help of metallic binding wires. The ceramic beads are cross shaped so that leaking sodium touches the metallic wire and grounds the electronic circuit. The spark-plug detector resembles the spark-plug of a petrol IC engine. The central electrode is surrounded by a ceramic insulator and is mounted into a metallic housing, such that any sodium leak will bridge the annular gap between the electrode and the housing, resulting in the electrical grounding of the central electrode, which is supplied by a DC voltage. This is processed by the leak detector electronics.

3.4.7.3 Special Instrumentation for Steam Generator Leak Detection

The steam generator is one of the most critical components of an SFR, in which the hot secondary sodium transfers the heat to the feed-water and converts the same into superheated steam in a single stage. The steam generator (SG) is a shell-and-tube heat exchanger in which water flows in the shell side from top to bottom, while feed water flows in the tubes from bottom to top. In the FBTR steam generator, sodium is at a lower pressure (around 3 bars) while the feed water is at a higher pressure (around 120 bars). Typically, the SFR SG is a once-through, counter-flow type heat exchanger, which implies that the feed-water completely gets vaporized and becomes superheated and flows in the outlet steam header. Steam from the other steam generators functioning in parallel is collected in this common steam header. In FBTR, the SG shell side is un-insulated, and the whole SG is housed inside a leak-tight metal casing, which has several leak collection trays at the bottom to collect the hot sodium in case the shell side of the SG leaks. Spark-plug type leak detectors located in the leak collection trays help in detecting the presence of sodium in the leak collection trays. SID-based leak detection system is deployed to provide advanced warning of such a leak, as the air surrounding the SG casing is sampled and tested for sodium aerosols.

The tubes carrying the high-pressure feed-water can also develop failures and become a source of leaks. In that case, water/steam leaking out of the failed tube will react with the surrounding sodium chemically and produce hydrogen as a primary reaction.

$$2\text{Na} + 2\text{H}_2\text{O} \rightarrow 2\text{NaOH} + \text{H}_2 \uparrow$$

The secondary reactions of sodium with sodium hydroxide and sodium with water also produce hydrogen. Hence, detection of this hydrogen will help to detect the tube side failure in the steam generator. The failure of one tube can cause the failure of other tubes too. Hence, it is essential to detect the tube side leakage in SG as quickly as possible, which is achieved with either of the two techniques described in the next section. The type of reactions that occur depends on the sodium temperature, too. If the sodium temperature is more than 400 °C, detection of hydrogen-in-sodium is very helpful. If the sodium temperature is less than 350 °C, detection of hydrogen-in-argon is more reliable, as the hydrogen generated in the failed SG escapes to the argon cover gas space of the sodium capacity in the downstream of the steam generator, known as the *surge tank*. This boils down to detecting the presence of hydrogen-in-sodium or in argon. Accordingly, we have *hydrogen-in-sodium* detectors and *hydrogen-in-argon* detectors. When the hydrogen evolution is very high, the pressure of the blanket cover gas space itself increases and becomes an indicator of SG tube side failure.

Hydrogen-in-Sodium Detectors

Electro-chemical Hydrogen Monitors (ECHM)

ECHM is basically an electrochemical concentration cell that can be used to measure hydrogen concentration in sodium. It uses $CaBr_2$-CaHBr bi-phasic mixture as the electrolyte which conducts only hydrogen ions. A mixture consisting of calcium and magnesium, along with calcium hydride, is used as the reference electrode that fixes the hydrogen partial pressure of the reference side at the meter operating temperature. Hydrogen concentration in sodium fixes the hydrogen partial pressure on the sample side. The operating temperature of the meter has been chosen as 450 °C. Because of the difference in hydrogen partial pressure in these electrodes, an EMF develops across the electrolyte that can be calculated from Nernst's equation:

$$E = (RT/2F)\ \log_e(\mathrm{pH_2}/\mathrm{pH_2Na})$$

where pH_2 is the hydrogen partial pressure in the reference electrode, pH_2Na is the hydrogen partial pressure in sodium, F is the Faraday constant, T is the absolute temperature of sodium, and R is the universal gas constant.

Sample sodium taken from the SG outlet is passed through the ECHM via a re-heater, which is electrically heated to maintain the sodium temperature at 450 °C, to make the ECHM output sensitive only to hydrogen concentration.

Mass Spectrometry

By sending the sampled sodium through a nickel diffuser, the hydrogen present in the sodium can be made to diffuse through the nickel diffuser. The shell side of the nickel diffuser is kept under ultra high vacuum (10^{-8} to 10^{-7} torr) with the help of sputter ion pumps (SIP). By deploying a quadrupole mass spectrometer tuned for hydrogen, the presence of hydrogen can be detected.

Sputter Ion Pump Based Hydrogen Detectors

The shell side of the nickel diffuser is kept under high vacuum with the help of the sputter ion pumps (SIP), which maintain vacuum with the sputtering process. The current drawn by the SIP itself is an indication of the amount of vacuum. In case of a SG tube leakage and subsequent generation of hydrogen through sodium-water reaction, the hydrogen that gets generated diffuses across the nickel diffuser and reaches the shell side, where vacuum is maintained by the SIP. As the SIP tries to maintain a vacuum, the current drawn by it increases, clearly detecting the presence of hydrogen and hence a possible SG failure.

Hydrogen-in-Argon Detectors

There are two methods to detect the presence of hydrogen in the argon cover gas space.

Thermal Conductivity Based Hydrogen Detectors (TCD)

By installing a nickel diffuser in the argon cover gas space of the expansion tank in the upstream of the SG, the presence of hydrogen-in-argon can be detected. The hydrogen diffusing across the nickel detector passes through a thermal conductivity detector, which contains a balanced bridge containing four arms. One arm of the bridge contains a heated filament and is kept cooled by means of a flow of sample gas. Another arm of the bridge contains another heated filament and is kept cooled by means of a flow of reference argon. Normally, both the heated filaments will only see argon, and the bridge will be in equilibrium. During the SG tube leak, the hydrogen so generated escapes to the cover gas space of the surge tank, diffuses across the nickel diffuser, and flows through the sample chamber. Due to the higher thermal conductivity of the hydrogen present in argon, the resistance of the filament present in the sample chamber changes as it now gets cooled faster. Since the resistance of the reference chamber remains the same, the bridge gets unbalanced. The output of the bridge is amplified and is calibrated in terms of hydrogen concentration. This sensor is useful in the range of 50–1000 ppm_v of hydrogen.

Tin Oxide (SnO_2) Based Detectors

The surface electrical conductivity of some semi-conducting oxides, viz., SnO_2, ZnO, etc., gets significantly altered in the presence of trace levels of reducing/oxidizing analyte gas in ambient air. This value process gets restored to the original value when the trace analyte gas is removed from the ambient. This property is used for monitoring trace levels of hydrogen-in-argon cover gas. The sensor consists of a thin film of tin oxide deposited on one side of an alumina substrate, which has a printed platinum heater on the other side. This platinum heater helps to maintain the sensor at 350 °C. The sensor housing has provisions for a gas inlet and outlet. The tin oxide sensor is positioned in such a way as to get a mixture of argon and air. When the argon contains hydrogen as a result of SG tube leakage, the surface conductivity of the sensor changes, and the resulting change in the resistance of the sensor is measured with the help of electronics. The output is calibrated in terms of hydrogen concentration. The SnO_2 sensor is useful for the measurement of hydrogen in the range of 5–100 ppm_v of hydrogen.

3.5 Qualification of I&C Systems for Safety Applications

3.5.1 Environmental Qualification of Electronics Safety I&C Systems

I&C systems deploy a number of electronic components, which, depending on the safety classification of the system and/or environmental conditions in which they will be employed, may use industrial grade/military grade components, or a mixture of both grades. Normally, commercial-grade components are avoided in safety applications. The specifications of I&C systems should clearly state the environmental conditions that of I&C system is likely to face during its transportation/use/storage. For example, the environment may have high/low temperature, high/low pressure, high humidity, high vibration, etc. The I&C system may have to be stored under highly humid conditions. In order to ensure that the I&C system works reliably in the field, the prototype may be subject to a number of tests, as specified in various parts of IS:9000. Some of the tests are listed in Table 3.6.

Table 3.6 List of environmental qualification tests of safety instrumentation

Sl. no.	Name of test	Standard	Purpose
1	Dry heat test	IS 9000 part (III)	To evaluate the ability of the I&C system to operate in high temperature environment
2	Dry cold test	IS 9000 part (II)	To evaluate the ability of the I&C system to operate in low temperature environment
3	Damp heat test (steady state)	IS 9000 part (IV)	To evaluate the ability of the I&C system to operate in high humidity conditions
4	Damp heat test (cyclic)	IS 9000 part (V)	To evaluate the ability of the I&C system to be stored in non-condensing, high humidity conditions
5	Temperature cycling test	IS 9000 part (XIV)	To evaluate the ability of the I&C system to operate in an environment where the temperature alternates between a high and a low value

3.5.2 *Example of Environmental Qualification Test: Dry Heat Test*

An I&C system for a NPP is typically installed in a 19″ rack with forced ventilation. Also, the panel may be located in an area where the temperature is maintained around 25 °C either with forced ventilation systems or with split Air Conditioning units. This may be the ideal normal working environment for the I&C system. However, we need to consider the performance of the system, in the absence of panel ventilation, area ventilation, or both, during which the temperature of the operating environment of the I&C system goes up. Performance under such conditions can be evaluated by the dry-heat test.

I&C systems considered for dry-heat tests are classified into two types: heat-dissipating and non-heat-dissipating. Heat dissipating electronic systems are those whose surface temperature is 5 °C more than the surrounding ambient temperature in the absence of forced ventilation. Also, the change in temperature can be considered as gradual or sudden.

Dry heat tests are conducted in electric ovens, in which the equipment under test is introduced. Cables required to power the Equipment Under Test (EUT) and the signals meant for observation are brought out through special openings.

Normally, such tests are performed at varying severities, including temperatures ranging from 40 to 200 °C for durations ranging from a minimum of 2–96 h. A typical value for testing a class IA I&C system can be a temperature of $55 \pm 2\,°C$ at 50% RH (relative humidity) for a period of 16 h. ("Dry heat" means the absolute humidity shall not exceed 20 g of water vapor per cubic meter of air.) The EUT is first introduced to the testing chamber, and then the temperature is increased at a specified rate ($\leq$1°C/min), if the testing is for gradual change in temperature. The EUT is introduced to the testing chamber, which is already at the desired testing temperature, if the testing is to evaluate performance for a sudden change in temperature. Testing of the EUT is carried out during the designated testing intervals (Fig. 3.16).

3.5.3 *Seismic Qualification of Safety Instrumentation*

In order to ensure that a safety instrument delivers its intended function even under seismic conditions, the instrument is subject to seismic qualification tests. The EUT is fixed on a shake table and is subject to vibration in the three axes. The magnitude of the vibration is as specified for the OBE and SSE level in Table 3.3.

The difference between vibration testing as specified in IS:9000:Part 8 and seismic testing as per the IEEE standard. 344-2013 (IEEE standard for seismic qualification of equipment for nuclear power generating stations) is that the purpose of vibration testing is to evaluate the effect of the vibration during transportation and/or installation at places where there is a lot of vibration induced by pumps, compressors, blowers, etc. On the other hand, the purpose of seismic testing is to evaluate the

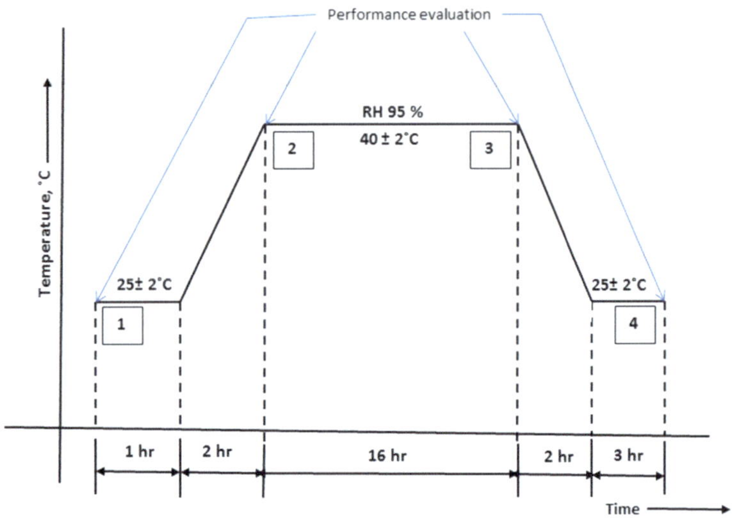

Fig. 3.16 Damp heat test

effect of the seismically induced vibration on the structures in which the instrument is located. Also, in seismic testing, the frequency of vibration is quite less (around 30 Hz), whereas the vibration test involves higher frequencies (<5 kHz). The acceleration amplitudes for vibration tests can go up to 50 g, while the maximum acceleration amplitude for seismic testing is around 0.2 g in the horizontal direction. For vibration testing, for lower frequencies (<10 Hz), the amplitude of displacement is specified rather than the acceleration amplitude.

3.5.3.1 Seismic Qualification Test Procedure

Floor Response Spectrum (FRS):

A safety I&C is very likely to be installed in a panel/rack which is located in a certain civil structure of the NPP at a certain elevation with respect to the ground. With this information and based on the study of seismicity of the site, the characteristics of the very likely seismic event that the I&C system is likely to be exposed to an OBE event or an SSE event are arrived at and are known as the Floor Response Spectrum (FRS). This is basically a plot of the peak acceleration that the safety I&C system may be exposed to during the said seismic event for different frequencies.

Test Response Spectrum (TRS):

For subjecting the safety I&C system to seismic qualification test, a test response spectrum is first obtained from the FRS such that the TRS envelopes the FRS throughout the range of frequencies of interest. This TRS is translated into an acceleration in the time domain, and at each frequency, the acceleration is maintained for 30 s.

Test procedure:

The unit under test is anchored to the seismic shake table and switched on. Parameters to be input and observed are connected to the respective source/monitor. First, the shake table is excited with a low-acceleration value for the full range of frequencies to find out if any resonance is taking place. If any resonance is there, it has to be arrested before proceeding with further testing. The shake-table now excites the EUT with FRS. The response of the instrument is monitored, and if the acceptance criteria are met, the safety instrument is cleared.

3.5.4 EMI/EMC Qualification of Safety Instrumentation

As per AERB safety guide SG-D-20, instruments deployed for safety must be qualified for EMI/EMC as per MIL-STD 421. EMI testing and qualification ensure that the safety instrument can deliver its intended function in the presence of external sources of electromagnetic noise. EMC testing and qualification ensure that the instrument does not interfere with the functioning of other instruments located nearby.

It is well known that electromagnetic noise is generated due to many reasons, including natural causes like lightning, and other transients like switching, relay chattering, fluorescent lamps, automobile spark plugs, leakage currents, by the use of wireless communication equipment, etc. This electromagnetic noise can interfere with a safety instrument by radiation, by conduction, or a combination of both, unless suitable measures are taken against it. Hence, the radiated susceptibility (RS) tests and conducted susceptibility (CS) tests are carried out to ensure that the Equipment Under Test (EUT) is not susceptible to noise passed on to it by external sources of noise. Similarly, radiated emission (RE) tests and conducted emission (CE) tests are carried out to ensure that the Equipment Under Test (EUT) does not generate noise. In addition, there is an effect of the susceptibility to noise passed on to it by external sources of noise. Figure 3.17 shows the source and receiver for electromagnetic noise and the means of coupling between the two.

Apart from the four main tests listed above, there are other tests too, which are conducted to ensure that the safety instruments can perform their function in a typical nuclear installation, as listed in the Table 3.7.

The EMI/EMC tests have to be conducted under specially built facilities, basically consisting of an anechoic chamber, antenna, and RF measurement devices. The Society for Applied Microwave Electronics Engineering and Research (SAMEER), under the Ministry of Electronics Information Technology, Government of India has

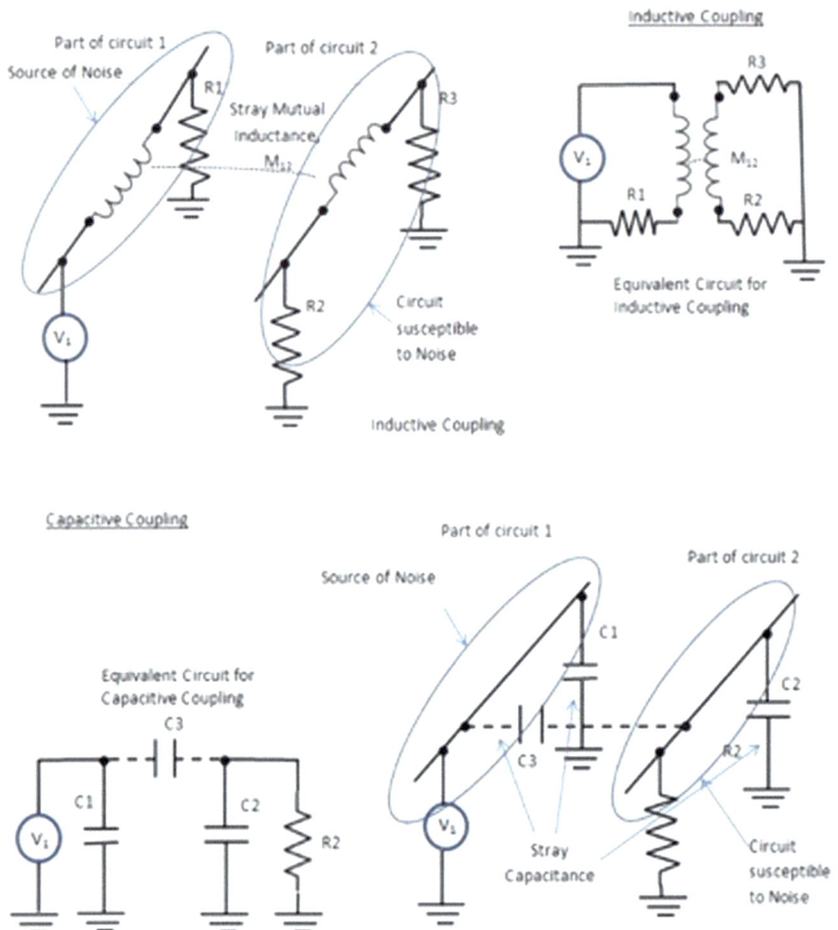

Fig. 3.17 Inductive and capacitive coupling of electromagnetic noise

such facilities at different places in India, which can be used for evaluating, qualifying and certifying that safety instrumentation meant for deployment in NPPs meet the specified EMI/EMC requirements.

The EMI/EMC performance of safety instrumentation is normally improved by adopting several techniques, some of which are: use of decoupling capacitors, reduced frequency of operation, use of transient suppression devices at inputs, power supply power factor correction, use of clock dithering to reduce emissions, proper PCB layout, use of ferrite beads, use of EMI gaskets, use of metal doors, use of line filters, use of optical isolation, use of EMIC/EMC MCBs, closing all open portions with metallic ones and grounding them, improved grounding, use of shielded cables and twisted-pair cables and physical isolation of power and signal cables.

Table 3.7 List of EMI/EMC tests for qualifying safety instrumentation

Sl. no.	Type of test	Name of the test	Ref. standard
1	Immunity	Electrostatic discharge (up to ±15kV)	IEC61000-4-2
2	Immunity	Radiated susceptibility (80 MHz–1 GHz)	IEC61000-4-3
3	Immunity	Electrical fast transient	IEC61000-4-4
4	Immunity	Surge protection	IEC61000-4-5
5	Immunity	Conducted RF	IEC61000-4-6
6	Immunity	Power frequency magnetic field	IEC61000-4-8
7	Immunity	Pulse magnetic field	IEC61000-4-9
8	Immunity	Voltage dips and interruption	IEC61000-4-11
9	Immunity	Damped oscillatory (1 MHz or 100 kHz)	IEC61000-4-12
10	Immunity	Variation of power frequency	IEC61000-4-28
11	Emission	Conducted emission	CISPR-11
12	Emission	Radiated emission	CISPR-11
13	Emission	Harmonic emission	IEC61000-3-2
14	Immunity	Distortion of power supply voltage harmonicity	IEC61000-4-13

3.6 Conclusions

In this chapter, the basic requirements of Instrumentation and control systems required for safety applications in sodium fast reactors, in particular, and in NPPs in general, have been brought out. The basic guidelines for designing such systems, as detailed in the relevant AERB safety guides, have been indicated. The qualification tests required for deploying safety I&C systems for such critical applications have been introduced. The specific instrumentation systems deployed in SFRs have also been brought out. This can give the reader a fair insight into the kind of requirements for I&C systems deployed in SFRs.

Chapter 4
Core Monitoring Aspects in Thermal Reactor Systems

Umasankari Kannan and Sudipta Samanta

4.1 Introduction

The Pressurized Heavy Water Reactors (PHWRs) are the pillars of the Indian nuclear power generation [1]. There are 20 operating PHWRs out of which 15 are rated for 220 MW(e), two are of 540 MW(e) and three are of the higher capcity 700 MW(e) reactors. Apart from these, two Boiling Water Reactors (BWRs) at Tarapur and two VVERs of 1000 MW(e) capacity at Kudankulam. The process systems in all these types of reactors are different and thc phenomenon-based detection is also varied in all these thermal reactors. The reactor instrumentation in all these types of reactors is different. The detection mechanism is classified into major categories that cater to core monitoring during the first approach to criticality, during normal operation at rated power and during shutdown state. Special instrumentation is also provided in certain prolonged shutdown scenarios.

4.1.1 Neutron Detection Principles

The neutron density in a nuclear reactor is dependent on space, energy, angle, and time, represented as $n(r, E, \Omega, t)$ which is required to be measured in the reactor. The detectors measure the neutrons that reach it at any instant of time and energy-sensitive material is used to detect them. For example, Boron-10 or He-3 has good neutron

U. Kannan (✉)
Retired, Reactor Physics Design Division, Bhabha Atomic Research Centre, Mumbai, India
e-mail: umadishanka@gmail.com

Former Senior Professor, Homi Bhabha National Institute, Mumbai, Maharashtra, India

S. Samanta
Reactor Physics Design Division, Bhabha Atomic Research Centre, Mumbai, India
e-mail: sudiptas@barc.gov.in

U. Mangalanathan et al. (eds.), *Nuclear Instrumentation and Control*,
https://doi.org/10.1007/978-981-97-1283-0_4

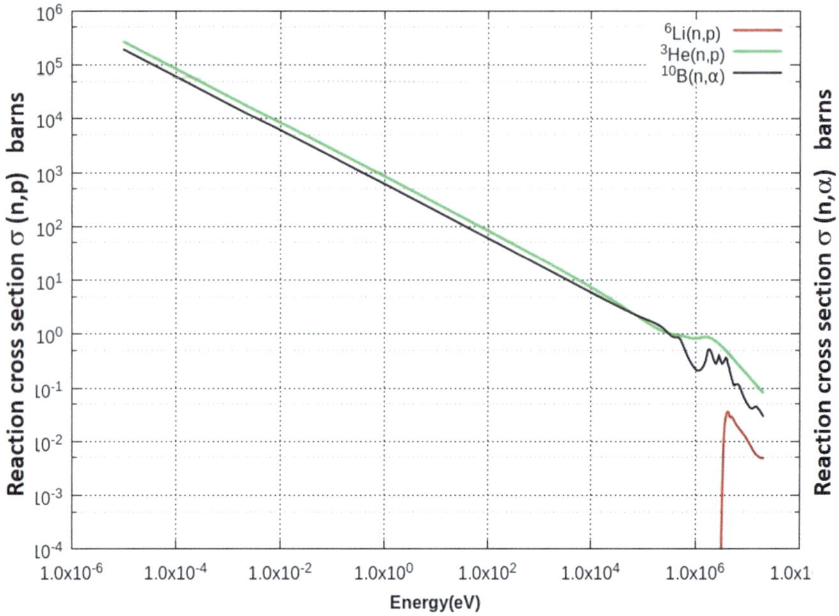

Fig. 4.1 Energy-dependent cross section from $^6\mathrm{Li}(n, p)$, $^3\mathrm{He}(n, p)$, and $^{10}\mathrm{B}(n, \alpha)$

absorption cross section for thermal energies and is used as the medium in gas-filled detectors. The neutron detection is done when the incident neutron interacts with a nucleus and produces a secondary charged particle such as a beta or gamma which ionizes the medium. The governing reactions for ^{3}He and ^{10}B are given below:

$$^3\mathrm{He} + {}^1n \longrightarrow {}^1\mathrm{H} + {}^3\mathrm{H} + 764\,\mathrm{keV}$$

$$^{10}\mathrm{B} + {}^1n \longrightarrow {}^7\mathrm{Li} + {}^4\mathrm{He} + 2.39\,\mathrm{MeV}$$

The energy-dependent cross section of typical nuclear reaction used for neutron detection is shown in Fig. 4.1. The main methods of detection involve either capture of the neutron where a charged particle is produced, activation reaction where the activated nucleus emits a radiation with sufficient half-life and energy to aid detection or use of fission reaction whereby the fission products move creating ion pairs in the medium [2].

When a neutron enters inside the detector volume, it will create certain amount of charge. This charge in turn will induce some amount of current at the respective electrode. Depending on the number of incident particles, detectors use pulse mode or current code to record the interactions. Reactors are designed to use different types of detectors based on the application required. The widely used detectors for neutronic instrumentation consist of ion chambers (IC), fission chambers (FC), and self-powered neutron detectors (SPNDs). In an ion chamber used for PHWRs, the detection is by the $^{10}\mathrm{B}(n, \alpha)$ mechanism. These detectors have a thin coating of ^{10}B on its walls and when the neutron interacts with ^{10}B, the resulting alpha particle or lithium ion causes ionization of the gas in the detector, and the resulting charges from this direct ionization process are then collected.

Fission counters work according to the principle of neutron-induced fission (n, f) reaction. The fission fragments generated are positively charged and heavy. Fission fragments carry about 20–22 units of positive charge. They interact while moving inside the active detector volume toward the cathode and creates electron and ion pairs on their path. Ion chambers or ionization chambers work in the averaging mode (DC mode) and fission counters work in both averaging and pulse modes.

Detectors in the proportional region are either ^{10}B lined (natural Boron or enriched in ^{10}B) proportional counters or BF_3 counters, which work in the pulse mode. When the applied voltage is so adjusted that the production of electron and ion pairs is in the proportional region, the detection of incident radiation is realized by gas multiplication.

Another class of neutron detectors is the self-powered neutron detectors (SPNDs) which do not require any external voltage or power. These detectors consist of emitters having large activation cross section and result in the formation of β or γ after interaction with the incident neutron. The emitter is surrounded by an outer electrode called the collector. A current signal is produced when a β-particle is collected by collector. It is measured between the emitter and the collector. This current signal is proportional to the number of incident neutrons on the emitter.

4.1.2 Reactor Instrumentation

The reactor core is effectively instrumented to cater to all phases of reactor operation and safe shutdown. The important measurements can be broadly defined as neutron flux measurements, thermal power measurement, and the spatial distribution of neutron flux [3]. These measurements form a complete neutron instrumentation for any reactor core. A brief of these aspects will be given here and the details of the specific systems will be elaborated in subsequent sections.

4.1.2.1 Neutron Flux Measurements

Neutron flux is defined as the number of neutrons across a particular cross-sectional area per unit time and expressed as neutrons/cm^2/s. In a critical reactor the neutrons are in a dynamic equilibrium. The rate of change in the neutron population will be a measure of deviation from this equilibrium and required to be monitored. Neutrons are produced directly from a fission reaction and neutron flux is the basic measurable quantity. Based on the detection principles mentioned in Sect. 4.1.1, the reactor core is heavily instrumented with neutron monitors. Care is taken to place the detectors in proper locations so as to sense the neutron multiplication in the core. Also, detectors are chosen with respect to their energy sensitivity for the neutron required to be monitored.

4.1.2.2 Thermal Power Measurements

The next most significant measurement parameter is the thermal power which is the measure of the amount of energy produced. The fission energy is transferred to the coolant which is in the form of heat and carries this heat to the secondary side. The thermal power measurement is dictated by other physical conditions such as the system pressure, the change in the temperature across the heated surface, coolant flow, and the coolant conditions. This is instrumented by sensors for measuring this change in various regions in the primary and secondary sides of the reactor core.

4.1.2.3 Flux Mapping

Spatial distribution of neutron flux is another important measurement required for reactor control. Reactor cores are designed to derive maximum power from an optimized flux distribution in the core. Any deviation from this distribution leads to local perturbation which will affect in turn the power production. Reactor cores are thus provided with a flux mapping system which gives a complete distribution of the neutron flux in the entire core through a set of neutron monitors placed at as many locations as possible.

4.1.3 Classification of Neutron Instrumentation

The neutron detectors are also classified with respect to their location. Reactor instrumentation mainly consists of ex-core and in-core detectors. The neutron flux which is leaking out of the core is measured by the detectors placed external to the core or ex-core detectors. These are mostly ionization chambers which detect the neutrons based on their interaction with the filled gas medium. They are also called power range monitors and the neutrons are usually attenuated by about 4–5 decades with respect to flux in the core [3]. The advantage of ex-core detectors is that they can be conveniently located with minimum penetrations into the pressure boundary of a reactor core and by virtue of their placement, the life of the detectors also could be extended. The typical saturation values of fluence in these detectors are about 10^{18} $n/\text{cm}^2/\text{s}$. These detectors are also supposed to provide monitoring during shutdown or refueling outages. The measurement of neutron flux will be effective if the ex-core detectors are placed as close to the core as possible. In most reactors, the B-10 lined counters are used for ex-core monitoring. In a strong gamma environment, fission counters are used.

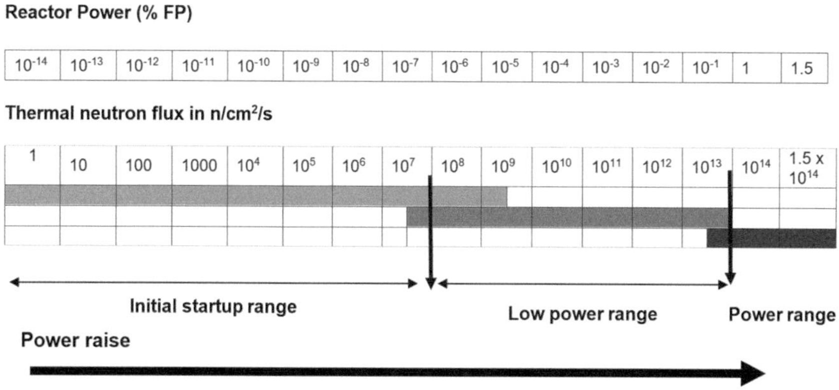

Fig. 4.2 Neutron detection overlap regimes

In-core instrumentation consists of a set of detectors for flux mapping and sensors for measurement of physical conditions of coolant or moderator. SPNDs are used for in-core flux monitoring. Since these detectors operate at ambient temperatures, the SPND emitters should have a high melting point. The output current of SPNDs seems to be heavily dependent on energy spectrum of incident neutrons. Some types of SPNDs have an added disadvantage of having slow response time.

4.1.4 Neutron Detection Ranges

The range of neutron detection to be covered is from a sub-critical regime to critical and transient situations [4]. With regard to the magnitude of the neutron flux, the range to be covered is several decades or order from about 10^{-5} full power (FP) to ~150% FP where the neutron flux density will be about almost 0–10^{14} n/cm^2/s [5]. The detectors are generally classified as source or start-up range, intermediate range, and power range monitors. The detection is so designed as to have sufficient overlap between these regions which is explicitly shown in Fig. 4.2.

4.2 Neutron Monitoring During Different Operational Regimes

Monitoring of neutron population in a nuclear reactor at any operational stage is utmost important for the safe operation of the reactor. Nuclear instrumentations for a nuclear reactor are designed to monitor and control change of neutron flux level of several decades. Neutron flux level changes from few neutrons at reactor start-up to about 10^{14} at full power operation. Appropriate instruments or detectors are required for different operating ranges of the reactor. A single of detector type or channel is not suitable to capture these large changes of neutron flux. The sensitivity or response of a detector is also dependent upon energy spectrum. Nuclear detector with the associated electronics is called channels. According the neutron flux level in a nuclear reactor, the operational regime, i.e., from sub-critical to FP, has been divided into three parts. It has also been discussed in Sect. 4.1.4. The detectors are operated in pulse mode or as a counter in source range. As the count rate is low at source range, pulse mode is suitable for background gamma discrimination. The choice of the neutron detector at this regime should be detector of higher sensitivity. At intermediate and power ranges, detectors are operated at current mode. Boron-coated ion chambers with lower sensitivity will be a good choice. As the rate of incident particle is high an average current will be produced which is proportional to the neutron flux. There should always be some overlapping region between each regime for smooth and efficient transfer of monitoring channels.

4.2.1 Principles of Monitoring During First Approach to Criticality (FAC)

There are different operational stages of a reactor, namely, when the fuel is fresh and reactor is to be started for the first time, operation at different power levels, shutdown state, and refueling state. The neutron monitoring aspects are different in each of these conditions. Criticality is defined as achieving a steady level of neutron population, i.e., a dynamic equilibrium between the production and loss of neutrons. The first approach to criticality is achieved in different manner in different reactors. The methods used in thermal reactors are increase in moderator height, decrease in boron concentration or boron dilution, removal of control absorbers, etc. The neutron counts are due to spontaneous fission neutrons in the beginning. As the fission is enhanced by moderation or removal of absorber, the neutron counts increase. This increase is monitored by the in-core and ex-core detectors. The criticality is measured by multiplication of neutron and is represented by a factor K called the multiplication factor [3]. The neutron multiplication in the sub-critical regime is given by

$$S = \frac{S_o}{1 - K} \cong \frac{S_o}{\rho} \cong S_o C \tag{4.1}$$

Criticality is attained when $K = 1.0$. If K_1 and K_2 correspond to sub-critical states, then

$$S_1 \cong \frac{S_o}{1 - K_1} \cong \frac{S_o}{\rho_1} \cong S_o C_1 \tag{4.2a}$$

and

$$S_2 \cong \frac{S_o}{1 - K_2} \cong \frac{S_o}{\rho_2} \cong S_o C_2 \tag{4.2b}$$

where S is the neutron density at any instant and S_o is that at the beginning of the chain reaction. If S_1 corresponds to sub-critical state K_1 and S_2 that of state K_2, the factor $1/\rho_1$ can be defined as a measure of multiplication C_1 and similarly for C_2. Equation 4.1 is valid only if $K < 1$.

The quantity $(1 - K)$ is measure of departure from criticality[1] and denoted usually by ρ. Here ρ_1 and ρ_2 are the measure of sub-critical multiplication or reactivity of the core at that instant. Thus, any sub-critical state can be easily related to the counts provided an initial state is known.

$$\frac{\rho_2}{\rho_1} \cong \frac{S_1}{S_2} \tag{4.3}$$

Equation 4.3 shows that the inverse of count rate is a direct measure of the sub-criticality [3]. The observables are the neutron counts and its reciprocal is continuously monitored.

During approach to criticality the multiplication is closely monitored with in-core detectors. In PHWRs, B-10-coated counters are used if the fuel is fresh. In case the fuel has been irradiated, then there is significant amount of gamma source strength background and the B-10 counters would not be able to discriminate the neutron counts. In such cases fission counters are being used [6]. In VVERs, only ex-core ion chambers are available for neutron detection and the leakage neutrons are monitored.

The approach to criticality is done carefully where it is usually seen that the step increase in multiplication is >20 to 30% higher than the previous step. For example, in PHWR the core has a large amount of boron in moderator which renders it highly sub-critical. The positive reactivity is introduced by removing boron in small steps so that the ratio of count rates is well within calculational bounds. In general, the deviation in these sub-critical stages from prediction is fixed as per previously estimated safe limit in the range of 20–30% [7]. The graph of $1/C$ with boron content is then plotted. The intersection of this plot with the X-axis will then be the point of criticality where the counts are infinite and the reactor is on a stable reactor period.[2]

In some reactors, the control rods are completely inside the core to make it sub-critical and they are gradually withdrawn in small steps where the neutron count rates are monitored. Then the inverse count rates are plotted with the observable parameter, i.e., the control rod position. The rate of reactivity insertion is so controlled that the extrapolation to criticality is estimated and the added reactivity in the next step is one-third of the extrapolated difference. This is known as the one-third rule. In many

[1] Reactivity is defined as a fractional departure from criticality, which is represented as $(k - 1)/k$. The units of reactivity are usually expressed as *mk* (*miili-k*) which is one-thousandth of the multiplication factor k.

[2] Reactor period is defined as the time over which the neutron population increases e-fold, i.e., the neutron population at any instant is given by $n(t) = n_0 e^{-t/T}$, where T is defined as the reactor period.

situations the curve may not be linear. This gradual approach ensures safety and observable change in neutron population. A typical approach to criticality curve for a PHWR is shown in Fig. 4.3a and another approach where criticality is achieved by increasing the moderator height is shown in Fig. 4.3b [8].

Another aspect of core monitoring in deep sub-critical regimes is the availability of spontaneous neutrons in cores having U-238 or Pu-239 in the fuel material. The detectors will pick these neutrons first and when fission reactions are enhanced the source multiplication will be significant. After a prolonged shutdown, the neutron counts decrease below the detectable limit and also the core will have a significant gamma source. In these situations, fission counters are used as neutron detectors which can detect neutrons in the background of about 10^5 mR/h. Examples of such situations are prolonged shutdown for *enmasse* coolant channel replacement (EMCCR)[3] in PHWR or refueling in a BWR or PWR which is done in shutdown state [1].

4.2.2 Commissioning and Start-Up Physics Experiments

Reactor physics experiments are required before commissioning of a reactor core to validate the design parameters and operational characteristics with the physics design calculations. The start-up physics experiments are done usually at low power to check the safety and regulation channels, verify worth of control rods and functioning of shutdown systems. The measured neutronic safety parameters of the reactor core during operation and transients give an estimate of design safety margins. As an example, the procedures for PHWR are described below. The commissioning of a reactor core consists of the following three phases:

Phase A: Pre-operational commissioning tests (functioning of instrumentation channels, leakage tests, fuel handling machine, etc.).

Phase B: It consists of initial fuel loading, addition of moderator/coolant, initial or first approach to criticality (FAC), and low power tests (i.e. at 0.1% of Full Power (FP)).

Phase C: Power tests (from 0.1 to 100% FP) are conducted in this phase (measurement of Xenon load is also done in BWRs and PWRs).

The importance of low-power physics experiment during commissioning is to measure and estimate uncertainties in design parameters. A series of low-power physics experiments are carried out after initial approach to criticality. Measurements of few important design safety parameters include the following:

- Worth of reactivity devices: Calibration of control rod, measurement of shut-off rod (SOR).
- Measurement of critical boron concentration.

[3] En-masse coolant channel replacement (EMCCR) is carried out in PHWR as the life in coolant channel is limited due to irradiation.

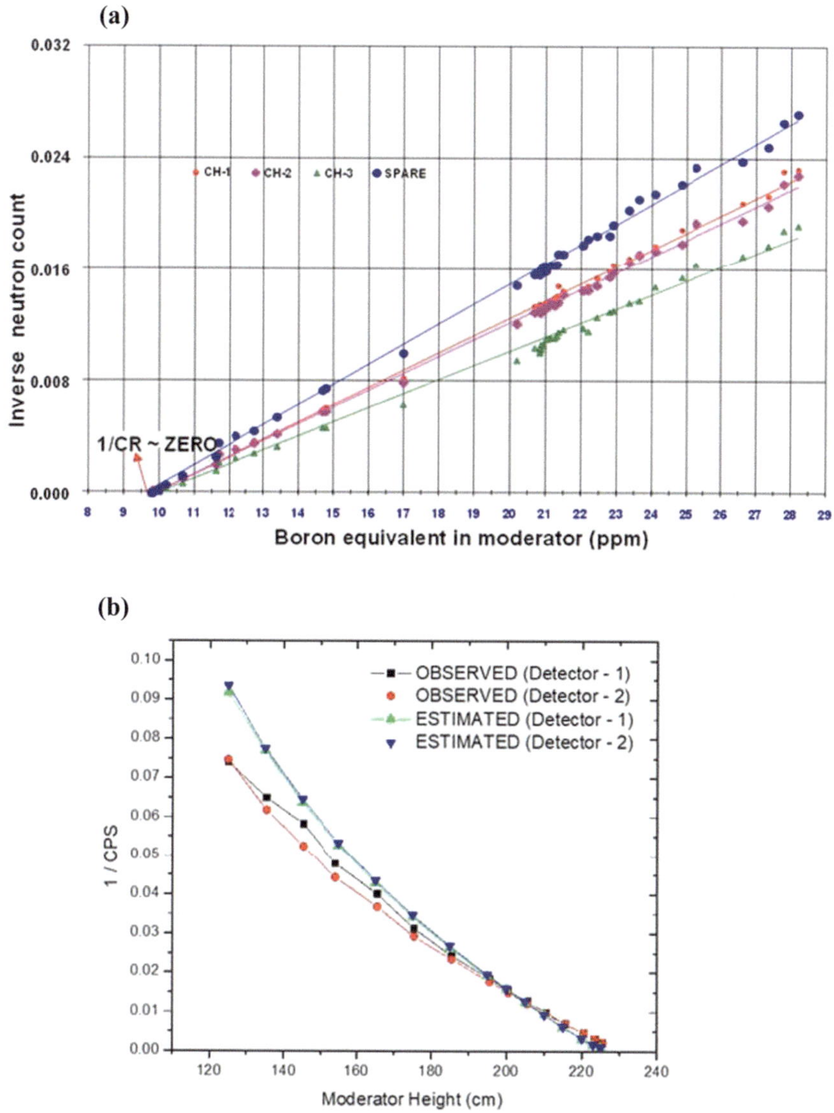

Fig. 4.3 **a** Neutron counts versus boron in moderator in a typical PHWR (criticality achieved by removal of Boron in moderator). **b** Inverse count rate with increase in moderator height for typical criticality in AHWR critical facility

- Reactivity coefficient: Measurement of moderator temperature coefficient, boron coefficient etc.
- Measurement of Xenon load.
- Neutron flux measurement and characterization of the core.

Measurement of core power distribution and power calibration is done at higher power levels (30–50% of rated power) after low power tests. The principles of reactivity measurements are presented in the following sections.

4.2.3 Principles of Reactivity Worth Measurements

The two main experimental methods for measurements of reactivity are static and dynamic. Sub-critical and substitution methods are called as static methods. The neutron count rate observed to determine reactivity worth in static method is time-independent behavior of neutron population at two different sub-critical states. Dynamic methods like asymptotic period method, rod drop method, rod oscillator method, and inverse kinetic method are used to measure reactivity worth by measuring dynamic characteristic of neutron population. Sub-critical methods are useful to determine worth for sub-critical systems. Dynamic methods are useful to study neutron kinetic parameters like effective delayed neutron fraction, prompt neutron lifetime.

4.2.3.1 Sub-criticality Method/Static Method

With a neutron source strength of S, the sub-critical system will attain a steady state in which the neutron population is a constant, though at a very low power level. After addition of negative reactivity in the form of movement of a control rod (CR) or shut-off rod (SOR) in sub-critical system the neutron population will attain another steady state. The worth of individual rods can be estimated using sub-critical count methods. The stable neutron count rate at known amount of sub-critical state (K_1) is S_1. Now, after insertion of portion of the CR/SOR the system reaches to higher sub-critical state K_2. If the stable count rate after insertion of a CR/SOR is S_2, the individual worth can be evaluated using the sub-critical multiplication formula [3].

Similar to the arguments in Sect. 4.2.1, if S_1 and S_2 are count rates in two different reactivity states, then

$$S_1 = \frac{S}{1 - K_1} \tag{4.4a}$$

and

$$S_2 = \frac{S}{1 - K_2} \tag{4.4b}$$

then

$$\frac{S_1}{S_2} = \frac{1 - K_2}{1 - K_1} \tag{4.4c}$$

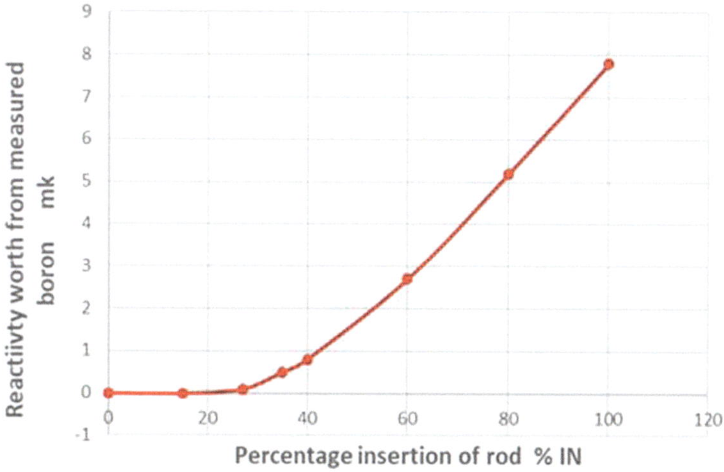

Fig. 4.4 Calibration of adjuster rod in a typical PHWR: reactivity worth from measured boron

For the measurement of reactivity of control rods here:

S_1 Count rate when all rods are up (Corresponding to a sub-critical state K_1).
S_2 Count rate when a particular rod is inserted (fully IN) (Corresponding to another sub-critical state K_2).
K_1 Known sub-criticality corresponding to moderator level.
K_2 is the estimated sub-criticality after rod insertion.
$K_2 - K_1$ will then be the worth of the particular rod.

Accuracy of these measurements improves as reactor is near to criticality. For example, the worths of the control rods in a PHWR are calibrated using a known amount of boron in moderator. The rods are made to move out with the addition of known quantity of boron. The reactivity is measured in steps where each step is a critical state [3]. A graph showing the typical change in reactivity measurements for calibrating the control rods in a PHWR is shown in Fig. 4.4.

4.2.3.2 Dynamic Methods

The worth of reactivity device by static methods is a measure of reactivity difference between two known steady states where the neutron counts are stabilized. During reactor operation and shutdown the control rods/shut-off rods are dropped instantly and the dynamic worth is realized in the core. The reactor instrumentation is required to have a fast response to record the change in reactivity in these dynamic conditions. In dynamic methods, reactivity worth of portion of control device is inferred by measuring the rate of change of reactor power with time. There are three main dynamic methods for experimental determination of neutron kinetics parameters and are presented below.

Measurement of Differential Worth by Asymptotic Period Method

The stable nuclear reactor period or asymptotic period, T, is defined as the e-folding time of the neutron density. The measurement of reactivity worth by period method is based on the determining the rate of change of exponential rise of neutron flux during step change of reactivity [3]. In this method, reactor is made critical and the portion of the reactivity device is inserted. The rise of reactor power is recorded. The rate of change of power is governed by the amount of reactivity inserted in the core. It can be described from the solution of point kinetics equation (Refer Chaps. 8 and 9 in this book). Assuming six groups of delayed neutrons solution for neutron density can be written as

$$n(t) = \sum_{i=0}^{6} A_i \exp(\omega_i) \tag{4.5a}$$

which reduces to Eq. 4.5b time t $\gg$ 0, i.e., when the transient part (negative roots) has died

$$n(t) = A_0 \exp(\omega_0 t) \tag{4.5b}$$

where $1/\omega_0 = T$ is the stable reactor period or asymptotic period of reactor. In general, by measurement of asymptotic period the change in reactivity in the system can be estimated directly from in-hour equation:

$$\rho_0 = \omega_0 \left(\Lambda + \sum_{i=1}^{6} \frac{\beta_i}{\omega_0 + \lambda_i} \right) \quad \text{and} \quad \omega_0 = \frac{1}{T} \quad \text{and} \quad T = \frac{\beta_{\text{eff}} - \rho}{\lambda_{\text{eff}} \rho} \tag{4.6}$$

β_{eff} = effective delayed neutron fraction; λ_{eff} = effective decay constant of delayed neutron precursors.

The worth of portion of the reactivity devices can be measured by making the reactor slightly super-critical and the stable reactor period ω_0 can be measured. As the reactor is super-critical again the other rod is moved inside the core to make it critical. Again, the control device is withdrawn in small steps and the stable reactor period ω_0 is measured to estimate the differential worth of the reactivity device which is a signature of the movement of the reactivity device.

Rod Drop Method

The total worth of shutdown devices is measured by this method. Usually shut-off rods fall under gravity and a large amount of negative reactivity is inserted in a short period of time. This makes it suitable to estimate the reactivity worth by prompt drop approximation. From prompt jump approximation, the neutron count rate S_1 before trip and neutron count rate S_2 immediately after rod dropped in a critical reactor ($t \sim 0$) is related by

$$\frac{S_2}{S_1} = \frac{\beta - \rho_1}{\beta - \rho_2} \text{ and } \rho_2 = \beta \left(1 - \frac{S_1}{S_2}\right) \tag{4.7}$$

here, $\rho_1 = 0$

ρ_2 is reactivity worth in mk.
β is delayed neutron fraction.
S_1 is count rate when reactor was tripped.
S_2 is count rate after prompt drop.

The count rates are then recorded by the neutron detector channels and the worth of shut-off rods can be estimated. Rod drop method is usually used to measure large amount of reactivity, i.e., integral worth of all control rods. The traveling time of rods limits the prompt drop approximation hence the accuracy. Also, this method is associated with reactor shutdown and subsequent start-up. This method can be employed in both PHWRs and PWRs. For a known amount reactivity change the kinetic parameters of delayed neutrons can also be estimated.

The inverse kinetic method is more accurate method to measure reactivity worth. In this method inverse of point kinetic model is solved. The power variation after reactivity insertion is noted and the inverse point kinetics equation is solved to find out the reactivity which caused the variation.

Source Jerk Method

In principle the source jerk or source perturbation method is same as rod drop method. The neutron source is removed rapidly from a sub-critical state of a reactor and the transient response of neutron count is recorded. From the prompt drop of neutron count rate, the sub-criticality of the reactor can be determined. Following the sudden removal of external source, the neutron population will undergo a sharp drop. The neutron count rate immediately after a source removal from a sub-critical reactor is related by

$$\rho = \beta \left(1 - \frac{S_0}{S_1}\right) \tag{4.8}$$

where S_0 is the neutron count rate with presence of external source and S_1 is the neutron count rate immediately after the source removal/jerk. This method is not suitable for power reactors with high burn-up fuel which masks the external source and the external source cannot be simply removed from the core. The variation of neutron flux in rod drop method and source jerk method using the prompt jump approximation is illustrated in Fig. 4.5.

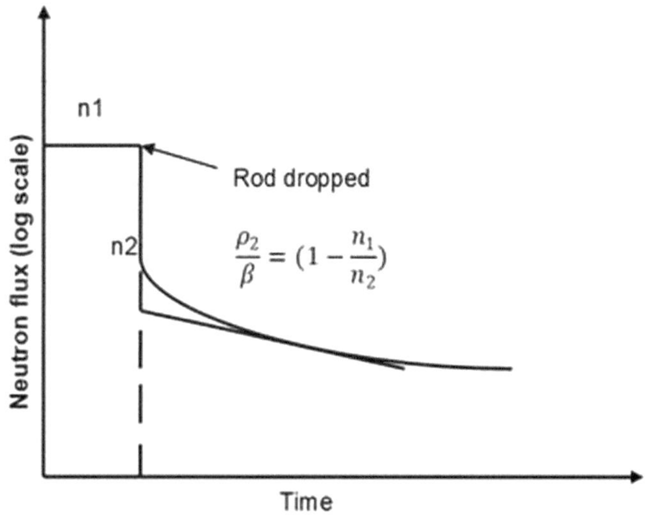

Fig. 4.5 Variation of neutron flux in rod drop method and source jerk method

4.2.4 Neutron Flux Measurement (SPNDs)

The ion chambers are bulky and not practical to be used for in-core flux measurements. The neutron flux inside the core is measured by self-powered neutron detectors (SPNDs). The neutron when incident on the emitter directly gets captured and generates a current due to the emanating beta or gamma rays [7]. Since these detectors do not require any external power, they are called self-powered. Typical emitter materials are Vanadium, Cobalt, Platinum, Platinum clad Inconel, and Inconel. A typical layout of the SPNDs in a reactor core is shown by a schematic in Fig. 4.6. For example, Vanadium is used as detector in many power reactors for achieving a fast response and the decay product has a small half-life.

$$^{51}V + {}^{1}n \rightarrow {}^{52}V \xrightarrow[3.8\,\mathrm{m}]{\beta} {}^{52}\mathrm{Cr} \tag{4.9}$$

One interaction mechanism is $(n,\ \beta)$ where the beta-decay current is proportional to the rate of capture of the neutrons. Since this depends on the beta decay, the response will be dictated by the half-life of the reaction. Another reaction is (n,γ) where a Compton or photoelectron is generated by the capture gammas and this is instantaneous. Gammas from fission products also can directly interact with the emitter and produce electrons.

In a steady-state operation or during a perturbation, the local reactivity changes are measured by the SPNDs. The flux measured at each detector location is then integrated in a flux mapping algorithm.

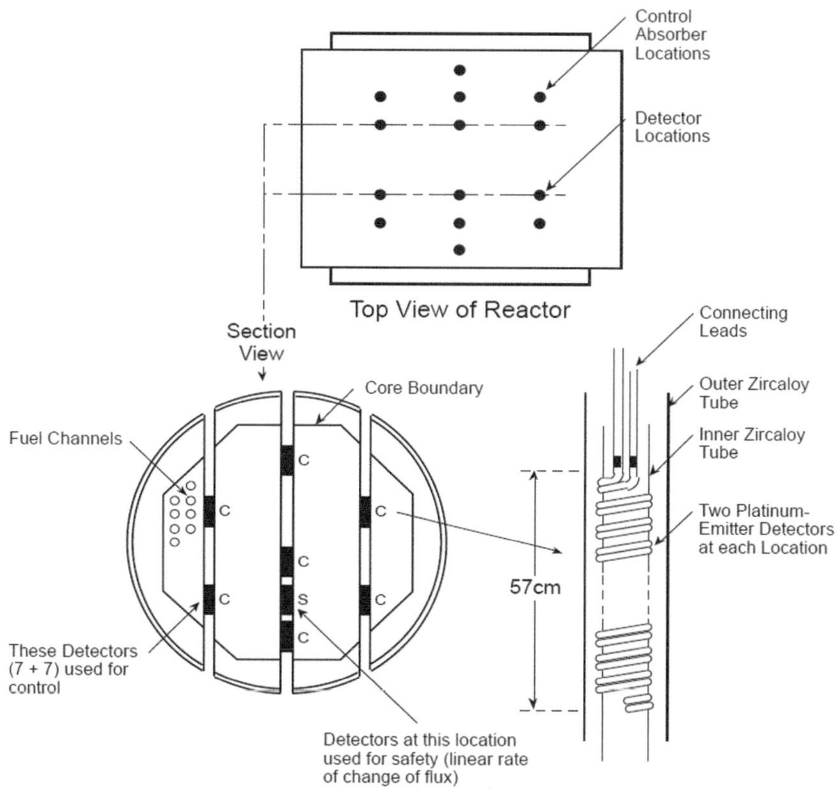

Fig. 4.6 Schematic showing in-core detectors in a PHWR. *Courtesy* Ref [5]

When the reactor is in steady state, there is a balance between production and loss of neutrons and the detectors will generally read a constant value with respect to the reactivity configuration of the core and the fluxes will converge asymptotically to the fundamental mode. During a perturbation, the higher modes of the fluxes also get manifested and die down with respect to the magnitude of the reactivity perturbation. Then the scalar flux at any instant t and location r is given by combination of all possible modes of the flux as shown below [9]:

$$\Phi(r,t) = \sum_{n=1}^{N_m} a_n(t)\Phi_n(r) \tag{4.10a}$$

where n is the number of detectors and m is the eigenmode of the neutron flux. The measured flux at the detector location will be

$$\Phi_{dk} = \Phi(r_k) = \sum_{n=1}^{N_m} a_n \Phi_n(r_k) \tag{4.10b}$$

Here k denotes a specific point k in the core. The SPND measurements are very useful inputs to the flux mapping system and help in controlling the zonal or regional flux or power distribution.

4.2.5 *Principles of a Reactivity Meter*

Reactivity meter is a device to display the amount of sub-criticality during first approach to criticality (FAC) or the sub-critical reactor operation of a nuclear reactor. The principle of a reactivity meter is based on the solution of inverse point kinetic equations [10]. A robust inverse point kinetics solver can be used to design an online digital reactivity meter. This method works for both critical and sub-critical states of a reactor. The advantage of inverse kinetic method is that the reactor need not to be shut down and as long as power profile is available this method works. The accuracy of this method depends on estimation of the source term. When the reactor is operating at high power level and core has no external neutron source, the neutron source term may be neglected. But in lower power and sub-critical state the source term is quite significant, which is difficult to estimate. Also, low neutron count rate poses challenge in using this principle in low flux region.

Other method employed in online reactivity measurement is based on "neutron noise", which a stochastic approach [3]. In this approach, the fluctuations in observed count rate are analyzed by using various models like "Feynman alpha" and "Rossi-alpha". One can measure only the relative sub-criticality by using the deterministic method like inverse kinetics. But neutron noise-based approach can be used to measure the absolute value of reactivity. However low neutron count rate is a bigger challenge for noise-based method. At higher count rate noise method will give the absolute value of reactivity with good accuracy. Both approaches are complementary to each other. Online reactivity measurement by noise method was demonstrated during the refueling of a PWR.

4.2.6 *Thermal Power Measurements*

The fission power is transferred to the coolant and the coolant from the primary side carries the heat to the secondary side. In power reactors, the total thermal power is estimated by calculating heat transferred to the different components of the reactor using associated temperature changes in them. The heat energy from core will be dissipated and result in heating of coolant, reactor structural materials, core structural materials, pipelines, etc. The heat loss takes place in the pipelines of primary coolant system, due to water evaporation, heat loss across the heat exchanger, and leakage through various components.

(a) *Thermal power based on primary side parameters* [5]
This method is used to calculate thermal power based on the following reactor parameters:

(i) Average change in temperature in coolant channel (ΔT).
(ii) Gross design coolant flow.

If the average boiler ΔT corresponding to operating power = $(\Delta T)_{av}$, and m in the number of coolant channels, then the operating P is given by

$$P = \text{Coolant flow} \times (\Delta T)_{\text{av}} \times m \tag{4.11a}$$

(b) *Thermal power based on secondary side parameters* [5]
Enthalpy balance method is used to calculate thermal power based on secondary side parameter. These parameters are listed below:

(i) Feed water flow at secondary side.
(ii) Temperature of feed water.
(iii) The reheat drain flow.
(iv) Reheat drain temperature.
(v) Boiler steam pressure.

Then thermal power is estimated using the following formula:

$$P = (h_s - h_f) \times F_f + (h_s - h_r) \times F_r \tag{4.11b}$$

where

h_s is enthalpy of steam as a function of pressure.
h_f is enthalpy of feed water as a function of temperature.
h_r is enthalpy of reheat water as a function of temperature.
F_f is feed water flow.
F_r is reheat water flow.

Temperature sensors are provided in each coolant channel in a PHWR to measure the temperature difference across the pressurized channel. In advanced PHWRs where boiling is allowed, the channel outlet temperature is monitored up to non-boiling regime. In the non-boiling regime, the thermal power is estimated from a few representative instrumented channels which measure flow, pressure, and other parameters.

In advanced PHWR designs, Thermal Power Measurement System (TPMS) is used to measure the reactor power based on the primary and secondary side parameters [5]. Pre-boiling bulk reactor thermal power is computed based on primary parameters (i.e., differential temperature and flow values obtained from a few selected channels). These instrumented channels are equipped with temperature and flow sensors. These measurements are also used to compute post-boiling reactor bulk thermal power based on secondary parameters (such as feed water flow, feed water temperature, steam flow, steam generator blow down flow, steam generator pressure).

4.2.7 Temperature and Flow Measurement

The temperature of various components is required for monitoring the process parameters in a reactor core. Reactor instrumentation consists of several thermocouples at several regions of the core. With respect to pressure of reactors, the coolant temperature is monitored at the exit phase of the coolant in each channel. In a PHWR, the temperature difference across the fuel from inlet to outlet is measured by series of thermocouples fixed at the channel outlet. In PWRs, each fuel assembly has provision for such instruments as shown later in Sect. 4.3.1. As mentioned above the assembly power is determined by the temperature difference and flow measured by these sensors.

The coolant flow is also an important input to process monitoring. The inlet to each channel is provided with a venturi where the inlet coolant flow can be measured. In vessel-type reactors, since the coolant mixes in an inlet header, the flow is measured at the outlet of this header.

In boiling water reactors, coolant enters from the bottom of the core in single phase and exits the fuel assembly in two phase of water and steam. In such reactors, the exit steam quality is an important parameter and sensors are provided to measure the steam fraction and hence enthalpy of the exiting coolant is estimated.The temperature measurements require exhaustive network of sensors to determine the temperature at every region/assembly of the core and requires detailed designing. In some modern reactors, acoustic sensors are used for determining the steam-void fraction but these instruments cannot be placed inside the core and therefore require detailed experimentation before their implementation.

4.2.8 Failed Fuel Detection Principles in Thermal Reactors

In order to detect the fuel failure, the activities from fission products are the major signatures and can be detected online. The typical radio-isotopes used for detection should be able to produce betas or gammas within a detectable time limit. The isotopes of noble gases, i.e., Kr and Xe are ideal for this purpose. Soluble isotopes having long half-lives such I, Np, Cs, and Sr are also used in the online gamma spectrometry measurements. These measurements include activities of ^{131}I, ^{132}I, ^{133}I, ^{134}I, and ^{135}I, activities from ^{133}Xe and ^{135}Xe in the off-gas system and activities from other radioisotopes in the coolant. These activities are indicators of failed fuel, contamination in the fuel rod surface and also fuel burnup. In typical CANDU reactors, this is performed by the delayed neutron monitoring systems which is a complex network of signals from detectors from the individual channels where the individual channel also can be identified online and reactor can be shutdown to remove the leaking fuel. In VVERs and BWRs, the leakers cannot be removed immediately. The total iodine activity in the coolant is monitored continuously. However, for individual fuel assemblies the coolant activity is monitored during shutdown by a sipping method and the leakers are identified.

4.3 Core Neutron Monitoring in PHWRs

PHWRs are designed to have several detection mechanisms for each of these ranges, namely, start-up, intermediate, and power ranges. The reactor is gradually taken from a sub-critical state to criticality where neutron balance is achieved. The reactor is enabled with start-up detectors, ex-core ion chambers, and in-core detectors (SPNDs). PHWR is a horizontal reactor with all reactivity mechanisms entering from the top. These detectors are housed in vertical and horizontal chambers which are penetrations in the calandria and running perpendicular to the fuel channels (Fig. 4.7a) [7].

In deep sub-critical regimes, it is always difficult to get sufficient counts in the detectors. During the first approach to criticality, the reactor core is equipped with start-up counters, namely, boron-coated counters or fission counters as the case may be to monitor the filling of heavy water, removal of boron from moderator, etc. When the in-core counters saturate, the neutron monitoring is transferred to the ex-core counters in a gradual manner at about 10^{-3}% FP. A typical layout of the ex-core ion chambers is shown in Fig. 4.7b [7]. Criticality is achieved by removal of boron. During normal operation, the core is monitored by a network of self-powered neutron detectors (SPNDs) for spatial distribution of flux and the ex-core ion chambers for monitoring of neutron power. In PHWRs, there is also provision of measuring the coolant outlet temperature in each fuel channel which is a measure of the power produced by that fuel in that pressurized channel. Since PHWR uses on-power refueling, the fuel loading is carefully done so as not to alter the core flux distribution. During shutdown conditions, due to the large gamma background, fission counters are used. In the following sections, certain basic principles of neutron monitoring are given.

4.3.1 In-Core and Ex-Core Monitoring in BWRs and PWRs

In Boiling Water reactors (BWRs), the in-core instrumentation consists of Low Power Range Monitors (LPRMs) and Traveling in-core Probes (TIPs) which are located in a detector housing throughout the core in between four fuel assemblies. A number of detectors are also present axially along the fuel channel. The LPRM signals are

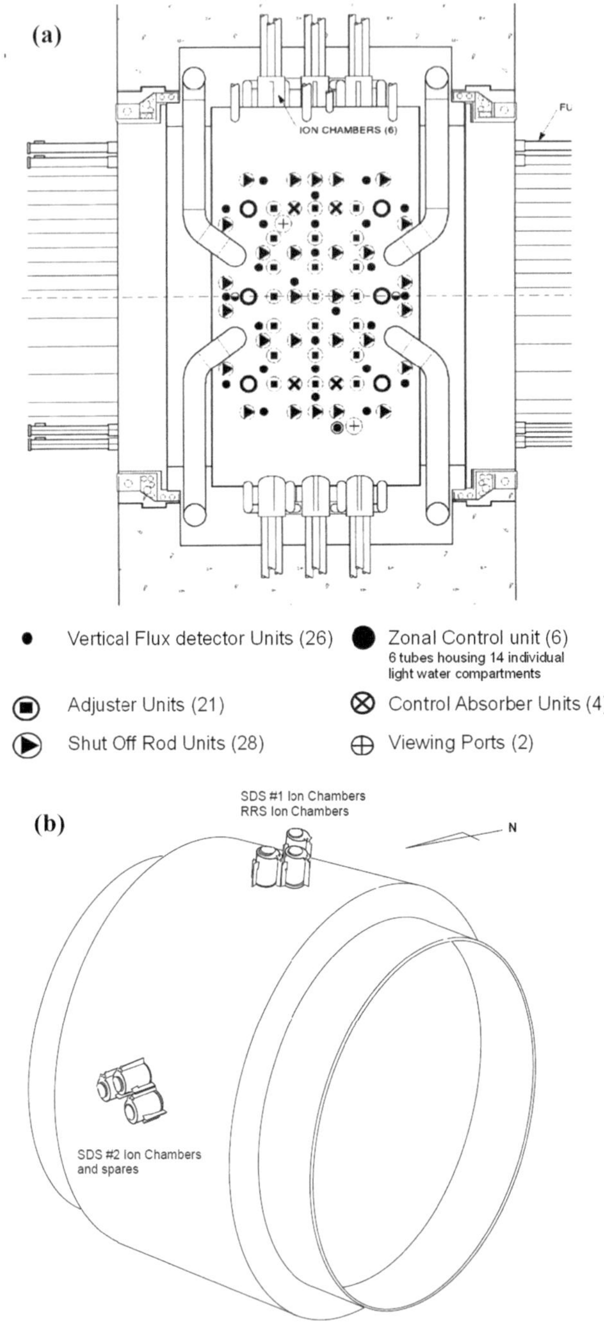

Fig. 4.7 **a** Placement of reactivity devices and vertical flux units in a typical PHWR (CANDU-6) [5]. **b** Typical schematic of location of ion chambers in a PHWR. *Courtesy* Ref [5]

calibrated to the 3D power distribution in the core. Since BWRs have a two-phase flow, the flux distribution is heterogeneous and so the detailed detector readings are compared with theoretical estimations to obtain the fuel limits and thermal margins. Both the LPRMs and TIP readings are used to quantify the thermal margins in the fuel assembly [11]. The LPRM signals are also integrated for larger regions in the core to estimate the average power and affect protection of the core during normal operation. The other detectors functioning in start-up and intermediate power range are used during shutdown and low power ranges.

In typical PWRs, the neutron flux measured by in-core SPNDs and ex-core ion chambers. The SPND currents are directly calibrated to the local flux and hence the local power, from which the parameters like linear heat generation rate can be determined [11]. The margin to nucleate boiling which is a safety parameter is then determined online from these measurements. A typical layout of the ex-core detector in PWRs is given in Fig. 4.8a, b. The reactivity management during the operating cycle for burnup is through soluble boron in the moderator. The control rods are mostly out of the core during normal operations. The in-core instrumentation coverage is given by the Rhodium-based SPNDs which are present in about one-third of the fuel assemblies in the core [12]. The instrumentation channel in each assembly has provision for loading seven SPNDs axially and temperature monitoring thermocouples. The protection feature is achieved by redundant set of ion chambers in the out-of-core locations. During refueling outage, fission counters placed in ex-core locations are used in the high gamma environment for monitoring the core.

4.3.2 Instrumentation in a PWR Fuel Assembly

In all PWRs, each fuel assembly has a location called instrumentation channel where SPNDs, thermocouples for coolant temperature measurements and flow sensors are loaded. The instrumentation thimble contains 6–7 SPNDs placed at regular intervals axially and thermocouples to measure the temperature at inlet and outlet of each fuel assembly. These data are used as inputs for estimating the rod peaking factors and volumetric heat generation rates. The cross section of 17×17 PWR fuel assembly and layout of the flux monitors and thermocouples are illustrated in Fig. 4.9a, b. The spatial distribution of neutron flux is estimated by this network of flux monitors and the detector housing locations are optimized with respect to operational scenarios for effective flux mapping.

4.3.3 Movable Probes for Neutron Flux Measurements

The traveling in-core probes which are based on fission counter were discussed earlier. In some advanced PWRs, a special in-core monitoring system called Aeroball monitoring system (AMS) is provided, which uses the principles of neutron activation

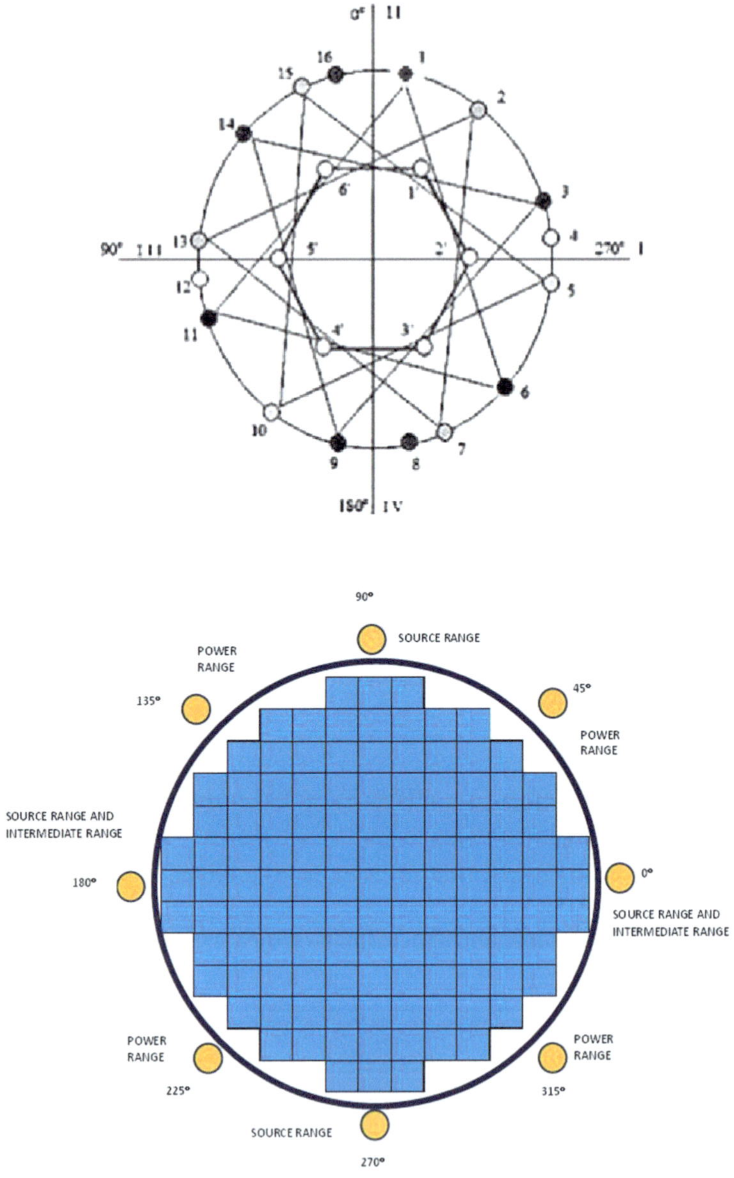

Fig. 4.8 Layout of ex-core detectors in **a** VVER and **b** PWR

to estimate the in-core reaction rate and hence assess the neutron flux or irradiation level in the core [13].

The AMS is used as a neutron flux mapping system (FMS) based on movable activation probes (or balls) which can operate on demand. It serves as a reference

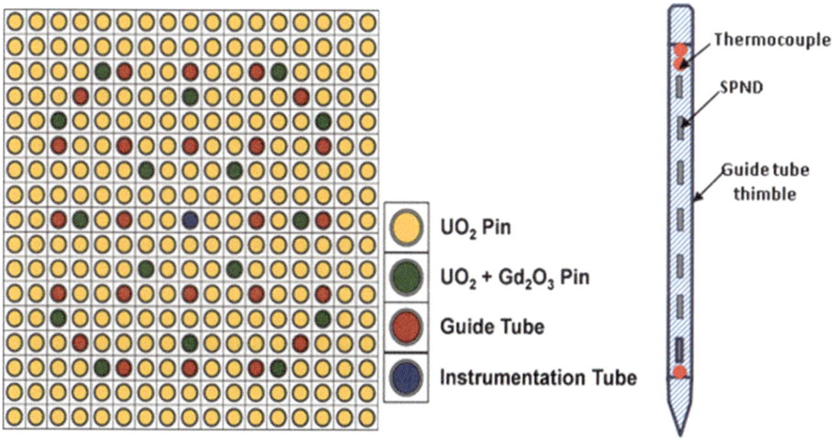

Fig. 4.9 **a** Cross section of a 17 × 17 PWR fuel assembly and **b** axial arrangement of flux monitors and thermocouples in instrumentation tube

instrumentation system to determine relative neutron flux density in the core and is used to calibrate the SPND-based in-core flux mapping system. Moveable probes are used in the Aeroball stacks, in a column length that spans the active core height. It is used to determine the relative neutron flux density in the reactor core. The Aeroballs are typically steel balls containing vanadium. A mechanism with nitrogen gas as driving medium is used to transport the Aeroball stacks to the core where they become irradiated. The balls are driven into the core pneumatically and are irradiated for a specific duration and drawn out and the activation acquired is counted out of core. The frequency of introduction of these balls is decided by the level of accuracy in measurements required.

4.4 Core Monitoring in Advanced Reactors

Reactor instrumentation essentially gives the signature of the phenomena happening in the reactor with measurable quantities of the process. However, these would require extensive instrumentation and also their healthiness will have to be assessed from time to time. The current reactors are designed with new types of sensing techniques such as acoustic based or noise based and with the advent of better diagnostics, the signals from such instruments are effectively monitored. An example of the neutron noise-based diagnostics is the CORTEX project being developed in Sweden as part of Euratom Research and Training programme [14]. In some modern approaches, predictive behavior is also assessed from the previous operational history using Artificial Neural Networks (ANNs). With the advent of faster computers and better computational methods, several new algorithms are being developed to understand the core behavior.

4.4.1 *Nuclear Reactor Simulators*

The dynamic response to any plant operation is studied using simulators which is a computer-based system programmed for various scenarios of the operating core [15]. They require a large amount of inputs in the form of core power distribution, reactivity coefficients, operational and accident sequences, instrumentation responses, and process system parameters. Traditionally, nuclear power plant simulators are widely used for operator training and licensing. With the requirements of physical and functional fidelity these Full Scope Simulators (FSS) are dependent on the finalized process design and operational data of reference plant, and are generally made available during the final stages of plant development cycle. Another category of simulators known as Nuclear Power Plant Analyzers (NPPAs) typically consisting of detailed models of only nuclear and steam supply systems is used for design and safety studies. However, in the past few decades, availability of inexpensive hardware with advanced computational capabilities and improved knowledge on process modeling has led to paradigm shift in the way in which simulators are built and used. Especially, use of another category of simulators called engineering simulators in various design phases of the plant has proved to have huge impact on the quality of control and instrumentation design and plant commissioning time. Unlike FSS, these simulators employ models of only major process, instrumentation, control and protection systems, and have no stringent requirement on control room exactness. With inclusion of auxiliary plant models and integration with replica control room, engineering simulators can further be extended to FSS for training and licensing purposes.

The main objective of an engineering simulator includes simulating the dynamic response of plant under normal and transient situations. These can be used for several purposes such as verification and validation (V&V) of control algorithms and safety interlocks, hardware-in-loop (HIL) response testing of prototype control and instrumentation systems, validation of performance and interfaces of control room human factor engineering (HFE), and virtual plant commissioning.

4.4.2 *Detection of Antineutrino and Monitoring Safety of Nuclear Reactors*

Antineutrinos are produced in nuclear reactor from nuclear fission process. They are weakly interacting sub-atomic particles which comes from beta decay of fission products. Because of the low interaction cross section antineutrinos are capable of pass-through shielding of nuclear reactor core. So, antineutrino detection can be very useful tool for nuclear safety and security. About $10^{18}\bar{\upsilon}_e/s$ antineutrinos are produced in a nuclear reactor producing 1MW of power. Antineutrinos can be detected through inverse beta-decay process:

$$\bar{\upsilon}_e + p \rightarrow e^+ + n$$

Special liquid scintillator detectors with gadolinium (Gd) doping are used for detection of antineutrinos [16]. The antineutrino spectrum is distinct for different fissionable isotopes. Each fissionable isotope has different yield of fission product and each fission product has different branching ratio which makes the antineutrino spectrum unique for a fuel inventory in a nuclear reactor. So, the total antineutrino spectra give an estimate of the inventory of fissionable content in a given nuclear reactor throughout the operating cycle. The measurement of antineutrino count rate gives insight into the reactor power level. The difference in antineutrino spectra from different fissile isotopes (^{233}U and ^{235}U) allow us for diversion detection of irradiated nuclear materials. The Indian Scintillator Matrix for Reactor Anti-Neutrino (ISMRAN) is being actively pursued at BARC Dhruva reactor facility [17].

Acknowledgements The authors are very thankful to Dr. A.P. Tiwari, Former Director, Knowledge Management Group, BARC for reviewing this manuscript and giving several useful suggestions. The authors also would like to thank Dr. V. Balagi and Shri. Amit Mishra, BARC; Dr. Sherley Ray, EX-NPCIL; Shri. Susheel Kumar, AERB for several inputs in preparing this manuscript.

References

1. S.S. Bajaj, A.R. Gore, Nucl. Eng. Des. (The Indian PHWR) (236) (7), 701–722
2. G.F. Knoll, *Radiation Detection and Measurement*, 3rd edn. (Wiley, 2000)
3. P. Mohanakrishnan, O.P. Singh, K. Umasankari (eds.), *Physics of Nuclear Reactors* (Academic Press, San Diego, 2021)
4. Neutron Fluence Measurements, IAEA Technical Reports Series No.107, IAEA, Vienna (1970)
5. Fundamentals of Power Reactors Module One Science, https://canteach.candu.org
6. G.P. Srivastava, Electronics in nuclear power programme of India—An overview. Sadhana **38**(5), 897–924 (2013)
7. Science and Reactor fundamentals: instrumentation and control CNSC Technical Group, https://pdf4pro.com/view/science-amp-reactor-fundamentals-a8849.html
8. A.K. Mallick, R. Kumar, S. Samanta, Y. Upreti, K. Deo, D. Bhandari, U. Kannan, V. Shivakumar, S.K. Verma, S.K. De, A. Sharma, Experimental study of voiding effects with thorium based MOX fuel cluster in critical facility. Nucl. Eng. Des. **398**, 111973 (2022)
9. B. Anupreethi, A. Gupta, U. Kannan, A.P. Tiwari, Optimization of flux mapping in-core detector locations in AHWR using clustering approach. Nucl. Eng. Des. **366**, 110756 (IF: 1.869) (2020)
10. E. Suzuki, T. Tsunoda, A reactivity meter and its application. J. Nucl. Sci. Technol. **1**(6), 210–218 (1964)
11. Core monitoring from commercial reactors: improvements in systems and methods, in *OECD Workshop Proceedings* (Stockholm, 1999), https://www.oecd-nea.org/upload/docs/application/pdf/2019-12/2208-core-monitoring.pdf
12. S.K. Agrawal, A. Chauhan, A. Mishra, The VVERs at KudanKulam. Nucl. Eng.Des. **236**, 812–835 (2006)

13. U.S. EPR Nuclear In core Instrumentation Systems Report, ANP-10271N (2006)
14. C. Demazière, P. Vinai, M. Hursin, S. Kollias, J. Herb, Noise-based core monitoring and diagnostics, in *Overview of the CORTEX project, Proceedings of Advances in Reactor Physics (ARP-2017) IT-04* (Mumbai, India, 2017)
15. K.N.V. Sairam, A.K. Mishra, S. Subudhi, T.U. Bhatt, M. Naskar, S.R. Shimjith, C.C. Verghese, D. Das, A.P. Tiwari, Engineering simulator for advanced heavy water reactor. BARC Newsletter (2018)
16. Nuclear reactor safeguards and monitoring with antineutrino detectors. J. Appl. Phys. **91**, 4672 (2002). https://doi.org/10.1063/1.1452775
17. P.K. Netrakanti et al., Indian Scintillator matrix for reactor anti-neutrino detection. https://doi.org/10.1088/1742-6596/1216/1/012023

Chapter 5
Developments in Gamma Detectors Leading to Advanced Characterization Tools in Materials Science and Medical Diagnosis

R. Govindaraj

5.1 Introduction

Radiations are extensively used for advanced material characterization, medical diagnosis, and for specialized medical treatments. Radiations include both uncharged and charged particles. Uncharged radiations involve X-rays, gamma rays, and neutrons. With respect to applications in medical field X-ray and γ-rays are widely used. X-rays are used for imaging in medical field. X-ray diffraction analysis are employed for finding out the structure of molecules. Gamma rays are used in cancer treatments and also in advanced imaging techniques including positron emission tomography (PET) [1–8]. Gamma camera is also used widely in medical imaging [9–11]. All these types of radiations find extensive applications in advanced materials characterization. On the other hand X-rays provide a very powerful tool for materials characterization studies including X-ray diffraction for crystal structure analysis. Most of the other techniques exploit the absorption of X-rays by core electrons of the solid of interactions resulting in X-ray photoelectron spectroscopy for elemental identification, chemical state of elements, and valence band structure [12, 13]. While neutrons similar to X-rays are used for material characterization widely right from imaging, radiography, and crystal structure [14–17]. Though chargeless similar to X-ray photon, neutron has spin and a non-zero magnetic moment. Hence neutron diffraction provides information on long-range magnetic ordering in addition to structural ordering of solids. Both X-ray- and neutron-based small angle scattering such as small angle X-ray scattering (SAXS) and small angle neutron scattering (SANS) are used to deduce structural, compositional information of precipitates present in matrices of interest which might play a crucial role in controlling defects movements leading to varying mechanical properties of the materials. Gamma ray-based

R. Govindaraj (✉)
Materials Science Group, Indira Gandhi Centre for Atomic Research, Kalpakkam, Tamil Nadu, India
e-mail: govindaraj_ramanujam@yahoo.com; govind@igcar.gov.in

U. Mangalanathan et al. (eds.), *Nuclear Instrumentation and Control*,
https://doi.org/10.1007/978-981-97-1283-0_5

techniques such as Compton scattering, positron annihilation, and time differential perturbed angular correlation-based spectroscopic techniques [1–5, 18–32] emerge as very powerful techniques for characterization of defects in materials. In particular, positron annihilation spectroscopy has emerged as a powerful technique for studying open volume defects in materials. Various advancement of techniques of positron annihilation spectroscopy for defect studies has also led to the application of positron annihilation as a powerful medical diagnosis tool as will be discussed in this chapter.

Charged particulate radiations include right from energetic electrons, protons, helium ions, and heavy ions. Electron beams could be used for the studies of crystal structure, microscopic studies. High energetic charged particle beams of light and heavy ions could be obtained using accelerators. High energetic proton beams can be obtained using accelerators and are widely used for controlled doping of hydrogen in semiconductors, eye cancer therapy, based on the fact that the high energetic protons loose their energy at their range defined as Bragg peak. Hence high-energy proton beam-based therapy is quite powerful due to the fact that the cancerous cells located at specific locations can be killed selectively by means of fine-tuning their energy so that the range of the particles match with the location of interest.

As this chapter extensively discusses the applications of gamma rays for different advanced materials characterization studies and also for the studies of biological problems, which depend crucially on the measurements of the energy of a gamma ray, time difference between the detection of two gamma rays, and position of interaction of gamma ray with a detector with respect to the desired application.

Gamma rays basically interact with the electrons leading to ionization processes in the medium of interaction. Interaction cross section of gamma rays is strongly dependent upon Z, the atomic number of the detecting medium besides the incident energy of the gamma ray. Basic processes of interactions of gamma rays mainly with the electrons are photoelectric absorption and Compton scattering which are dominant while the energy of gamma rays is less than 1.02 MeV. While the incident energy of gamma rays exceeds 1.02 MeV, the gamma rays interact mainly with the nuclei of target atoms leading to the emission of electron and positron, each emitted with some value of kinetic energy the sum of which is given as the excess value of the energy of the incident gamma rays from 1.02 MeV, the threshold energy for the production of electron–positron pairs. This process is known as the pair production. Table 5.1 shows the energy and atomic number dependence of different processes of the interaction of gamma rays with the detecting medium. When the energy of gamma rays exceed well beyond 1.02 MeV, the threshold energy for the production of electron–positron pairs, the gamma rays interact more strongly with the nuclei of the target leading to gamma-induced nuclear reactions. This article is primarily focused toward the efficient detection of gamma rays of energy less than or close to 1 MeV leading to various spectroscopic applications.

As this article is focused toward exploitation of the advancements in the detector technology for various applications with respect to advanced materials characterization techniques and medical applications, it is important to discuss as to how these detectors are selectively used for measurements of different parameters such

Table 5.1 Various processes of interaction of gamma rays with matter and the dependence of the cross section of these processes on atomic number of the detector and the incident energy of the gamma rays

Process of interaction	Mechanism of interaction	Dependence of cross section in terms of atomic number Z and energy of incident gamma rays E
Photoelectric effect	Absorption of the energy of gamma photons by bound electrons of atoms leading to the emergence of photoelectrons with kinetic energy (h$\gamma - E_B$)	Z^m/E^n ($m \approx 4-5$ and $n \approx 3-4$)
Compton scattering	Partial absorption of energy of photons by electrons with the scattered photon carrying the remaining energy	Z with complex inverse E dependence having strong tendency for forward scattering with enhanced incident energy
Pair production	While the energy of gamma ray is much larger than 1.02 MeV, the threshold energy for its interaction with nucleus of atoms leading to electron–positron pair production. e^- and e^+ emerge with finite kinetic energies $E_{e+} + E_{e-} = (h\gamma - 1.02\,\text{MeV})$	Z and increases significantly with increasing E

as energy with high precision. The following are some of the important parameters that can be measured efficiently by means of proper choice of detectors.

i. Measurement of energy of gamma rays.
ii Measurement of time interval between gamma photons of two distinct energies of the order of hundreds of picoseconds.
iii. Measurement of energy of gamma rays of mean energy of hundreds of keV accurately so that even small changes of the order of few keV in the spectral line shape of gamma rays, i.e., changes in full width at half maximum.
iv. Position-sensitive measurements using nuclear detectors.

Detectors displaying good energy and time resolution find extensive applications in nuclear spectroscopic techniques many research related to studying materials and biomedical problems.

Probability of the above interaction processes is strongly dependent upon the incident energy of the gamma rays and the atomic number of the detector material which is shown schematically in Fig. 5.1.

Thus the gamma rays result in ionization such as the excitation of electrons from valence band to conduction band leading to formation of electron–hole pairs in

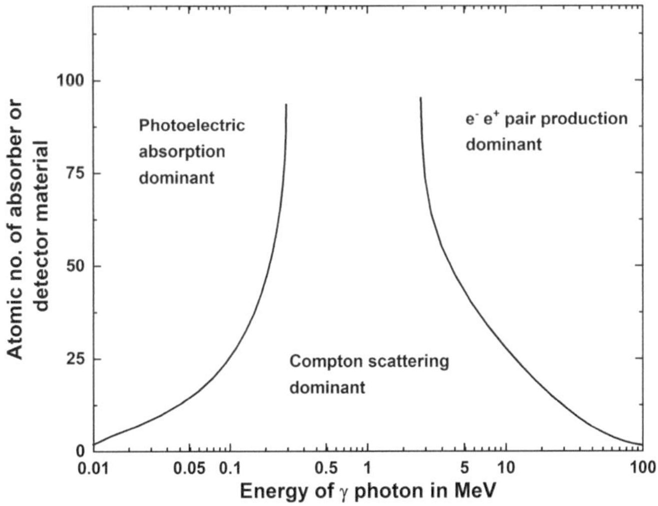

Fig. 5.1 Dependence of various processes of gamma interaction on the incident energy of the gamma rays and the atomic number of the detector material

the case of semiconductors while the gamma interaction leads to ionization of gas molecules resulting in electron-ion pairs in gas-based detectors (Table 5.2).

5.2 Measurements of Different Parameters of Gamma Rays Using Radiation Detectors

For any discussion on the measurement of different parameters of gamma rays, it is first of all important to deduce the energy of the gamma rays. Even while discussing other applications involving the measurement of the interval of time between the emission of gamma rays of different energies by a radioactive source, it is first of all important to deduce energies of gamma rays. Hence energy deduction is a part of the timing spectroscopy [33].

5.2.1 Measurement of Energy of a Gamma Ray

The nature of signal pulse due to an interaction of a single photon with the detector depends on the input characteristics of the circuit to which the detector is connected. While R and C represent input resistance and equivalent capacitance of both the detector and circuit [33] (Fig. 5.2).

Table 5.2 Summary of some of the very important parameters of different types of most common detectors that are employed in detecting low-energy gamma rays

Type of detector	Mechanism/threshold energy for production of information carriers or charges	Advantages	Disadvantages
Gas-based detectors	Ionization of gas; detector has to be used in proportional counter mode for spectroscopy. Tens of eV	Detection of low-energy gamma rays efficiently; does not suffer from radiation damage	High-energy gamma rays cannot be detected
Semiconductor detector	Excitation of electrons from valence band to conduction band	Much better energy resolution than gas- and scintillation-based detectors	Suffers from radiation damage, limitation with respect to keeping the detector at low temperatures while the detector is biased or unbiased
Scintillation detectors	Scintillation-induced light and electron multiplication by photomultiplier tube; hundreds of eV	Scintillation-based detectors provide very good energy/time resolution	Indirect process of production of charge carriers, scintillation photons are made to fall on photocathode with high quantum efficiency resulting in emission of photoelectrons. These photoelectrons are accelerated and multiplied through dynodes
Microchannel plate	Scintillation-induced light, electron multiplication; 100 eV	Very good spatial and time resolution	MCP contains number of pores with each pore containing channel coated with high quantum efficient material
Hybrid detectors (gas-based + scintillation-based detectors)	Ionization and scintillation leading to enhanced signal/noise ratio; Tens of eV	Very good energy resolution	Cathode of the gas detector is coated with high quantum efficient material leading to significantly high signal/noise
SQUID-based transition edge sensor (TES)	Excitation of cooper pairs to normal electrons	Provides the best energy resolution	Detector has to be kept at very low temperature requiring adiabatic demagnetization-induced cooling (mK)

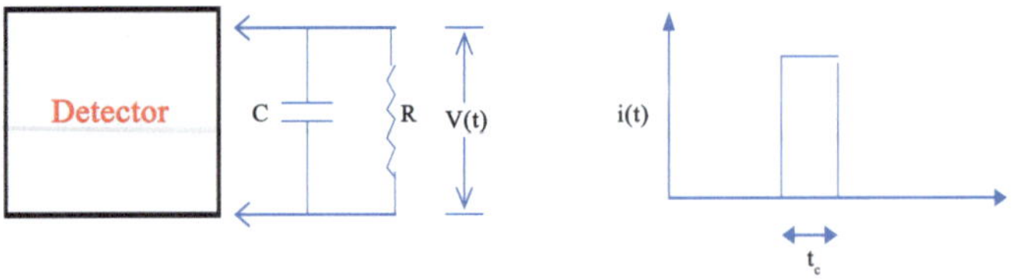

Fig. 5.2 Schematic of generalized detector

Whereas the current integrated over a period of charge collection time is nothing but the total charge due to a single interaction

$$\int_0^{t_c} idt = Q$$

Q is in turn proportional to the energy of incident quantum of radiation.

Representation of signal voltage for two extremes of time constant as compared to the charge collection time is made in the following figure. When the time constant RC is much smaller than the charge collection time, it results in the signal $V(t)$ which would at any instant of time represents the instantaneous current flowing in the detector. While for large value of the time constant of the detector as compared to charge collection time RC $>>$ t_c, very little current flows through the load resistance during the charge collection time with the detector current which is integrated in the capacitance.

$$V(t) = Q/C = 1/C \int_0^{t_c} idt$$

With V_{max} = maximum amplitude of the signal = Q/C. Preamplifier of the detector is used in charge-sensitive configuration to preserve the magnitude of Q. Thus when operated in pulsed mode the individual pulse amplitude carries information regarding the charge generated by that particular radiation interaction in the detector. While analyzing a large number of pulses small differences in the radiation energy, manifested as small changes in the pulse amplitude therefore represents small differences in energy or fluctuations in the inherent response of detector for a monochromatic radiation. While a radioactive source emits gamma rays of different energy signals of different amplitudes can be observed in an oscilloscope. Pulses of different amplitudes and the associated small variation in the amplitude of each of these pulses can be represented as a differential pulse height distribution function. In a differential pulse height distribution spectrum coordinate refers to a differential number dN of pulses observed within an amplitude increment of dH given as dN/dH. Horizontal scale either refers to pulse amplitude. Number of pulses with amplitude between H_1 and H_2 is given as $\int(dN/dH)$ dH. A recorded distribution of pulse amplitudes reflects the corresponding distribution in energy of the incident quantum of radiation [33–38] (Figs. 5.3 and 5.4).

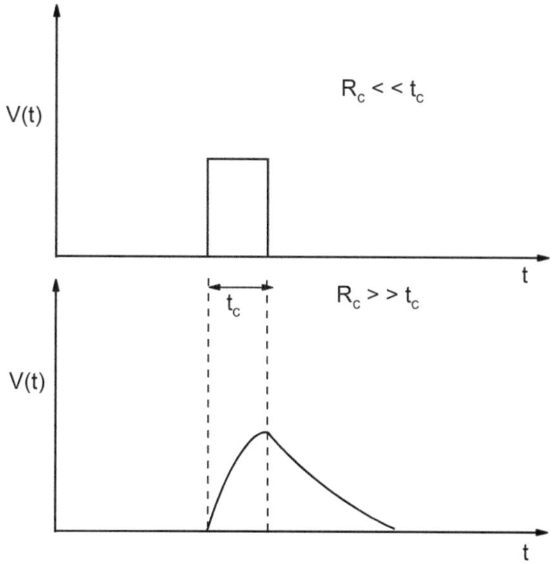

Fig. 5.3 Time dependence of the output signal of the generalized detector for the cases of the time constant RC $<< t_c$, the charge collection time, and the case for which RC $>> t_c$

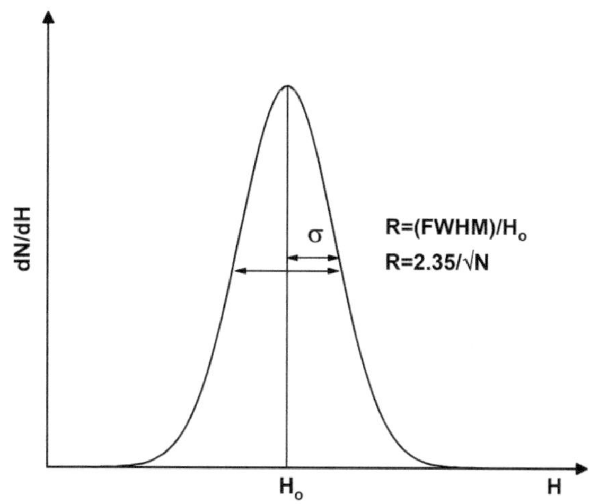

Fig. 5.4 Illustration of obtaining the energy resolution of the detector at the value of pulse height H_o from the pulse height spectrum

Pulse height distribution as shown above with mean value of H_o is given as

$$G(H) = A/(\sigma(2\pi)^{\frac{1}{2}}) \exp(-(H - H_o)^2/2\sigma^2)$$

While Full Width at Half Maximum (FWHM) of the Gaussian function is given as 2.35 σ. As the response of the detector should be linear with respect to increasing energy, i.e., $N \alpha E$.
This means that $H_o = kN$ and $\sigma = k(N)^{1/2}$.
Hence the energy resolution of the detector at H_o is given as

$$R = 2.35\sigma/N = 1/(N)^{1/2}$$

Overall variation in FWHM is thus given as

$$(\mathrm{FWHM})^2_{\mathrm{overall}} = (\mathrm{FWHM})^2_{\mathrm{stat}} + (\mathrm{FWHM})^2_{\mathrm{noise}} + (\mathrm{FWHM})^2_{\mathrm{drift}}$$

The block diagram of scintillation-based detector system for detecting gamma rays of different energies as emitted by a radioactive source is shown in Fig. 5.5. $(\mathrm{FWHM})^2_{\mathrm{drift}}$ can be minimized by choosing a proper bias voltage of the detector so that there is no significant variation in the gain of the detector for any small variation in the bias supply. $(\mathrm{FWHM})^2_{\mathrm{noise}}$ is mainly reduced by a proper processing of preamplifier output signal which in general is sharply rising and with a long tail. Preamplifer signal is shaped by CR-RC-based shaping circuits for taking care of base line shift, pulse pile-up effects. Linear amplifier is used for shaping the preamplifier output signal and for amplification of the signal which in turn can be used for spectroscopy-based studies involving energy detection of single photon.

Pulse height analysis system for measurement of energies of gamma rays emitted by a radioactive sample is shown in Fig. 5.5. It primarily consists of a detector which in this case refers to scintillation crystal optically well coupled with suitable PMT based on the florescence spectrum of the crystal. Preamplifier is used for efficient impedance matching and transfer of signal for pulse shaping and processing by a linear amplifier. Besides linear amplifier plays an important role in pulse shaping providing a shaped and amplified output signal for processing in any spectroscopic studies. In certain studies, differential energy discriminator (SCA) is used for selecting gamma ray of specific energy for counting and processing. For a very good energy resolution of a detecting system, the signal-to-noise ratio should be high with well-reduced pulse pile-up effects and optimize settings of the linear amplifier. Resultant pulse height spectra corresponding to a radioactive source emitting gamma rays of energy well below and above 1.02 MeV, the threshold energy for pair production, as deduced using small detector are shown in Fig. 5.6 (Figs. 5.7 and 5.8).

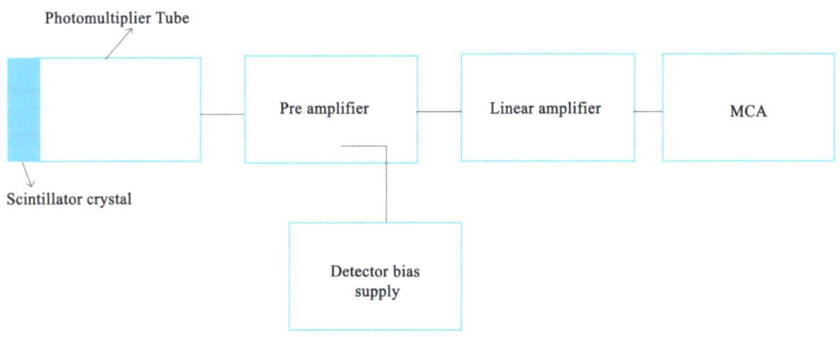

Fig. 5.5 Block diagram of scintillation detector-based pulse height analysis system in spectroscopy studies

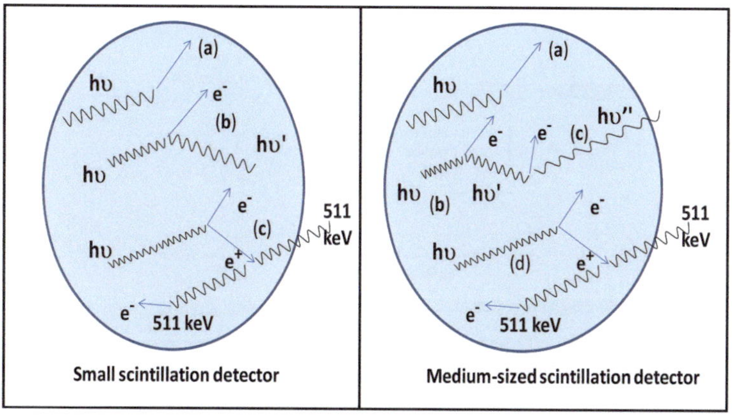

Fig. 5.6 Interaction of gamma rays with small scintillation detector in left panel **a** photoelectric absorption **b** single Compton scattering and escaping of photon **c** pair production and escaping of annihilation photons; medium sized detector in right panel **a** photoelectric absorption **b** and **c** single and multiple Compton scattering and escaping of photons **d** single escaping of annihilation photons

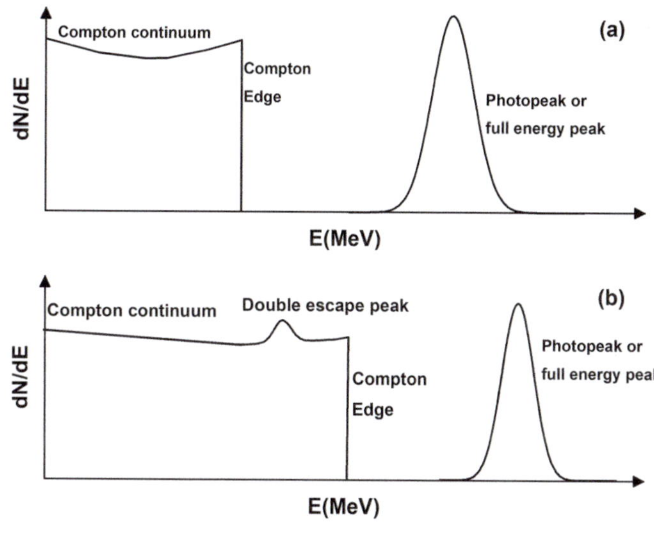

Fig. 5.7 Pulse height spectra as obtained due to interaction of gamma rays with small scintillation detector **a** the case of energy <1.02 MeV and **b** the case of energy >1.02 MeV

5.2.1.1 High Pure Ge Detector for a Very Good Energy Resolution

From the expression for the energy resolution, it can be seen that for the case of high pure Ge detector, due to the low value of energy gap of Ge, the number of electrons excited from valence band to conduction band is enormously large leading to a significantly enhanced energy resolution of the detector [39–41]. As there is likely to

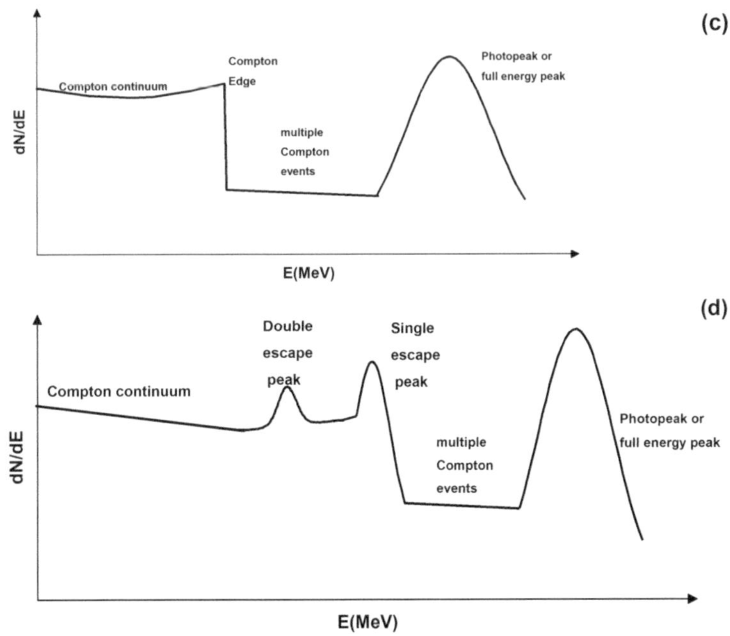

Fig. 5.8 Pulse height spectra as obtained due to interaction of gamma rays with medium-sized scintillation detector **a** the case of energy <1.02 MeV and **b** the case of >1.02 MeV

be thermally induced excitation of electrons at room temperature, it is important to keep the crystal at 77 K before biasing of the detector to achieve a very good energy resolution.

$$R = \frac{2.35\sigma}{N}\alpha\frac{1}{\sqrt{N}}$$

For example, typical value of FWHM as obtained in NaI(Tl)-based scintillator detector is close to 46 keV against a value close to 2 keV as obtained in the case of high pure Ge detector at 662 keV gamma energy of ^{133}Cs. This implies that a significantly large value of the energy resolution of high pure Ge detector as compared to that of NaI(Tl)-based scintillator detector. Hence high pure Ge detectors are used where there is a strong requirement for a good energy resolution to resolve two gamma rays of energies quite close by varying by a few keV. Figure 5.9 clearly illustrates the significantly enhanced energy resolution of high pure Ge detector [22].

5.2.1.2 Gamma Ray Spectrometer Based on Transition Edge Sensor for Analysis of Nuclear Materials

Cryogenic gamma ray spectrometers are based on superconducting thermistors and these provide more than an order of magnitude improvement in energy resolution

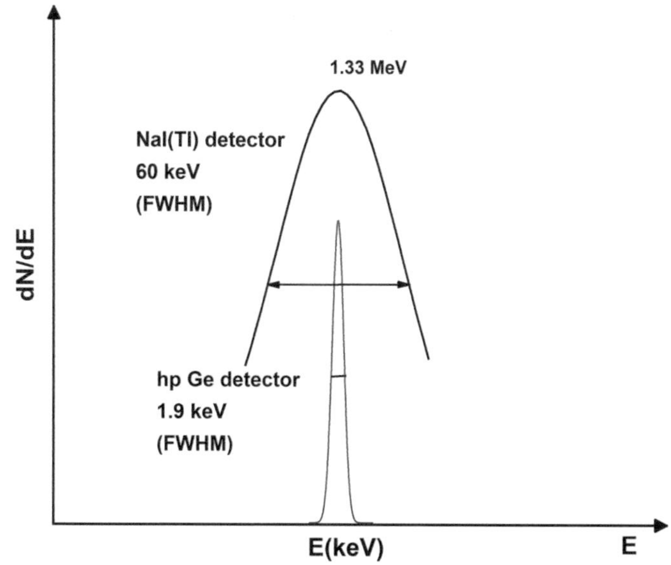

Fig. 5.9 Pulse height spectra as obtained using NaI(Tl)-based scintillation detector is compared with that obtained using high pure Ge detector

over conventional high-purity germanium detectors [39, 40]. Their operation is based on the measurement of the temperature increase upon gamma ray absorption with a sensor operated at the transition between its superconducting and normal states.

Transition Edge sensor (TES) works on the principle of measuring the temperature increase precisely due to the absorption of gamma rays [40–45]. This detector is operated at a temperature very close to and below the superconducting transition temperature to normal state. Hence any small increase in the temperature due to absorption of gamma rays would lead to the superconducting to normal phase transition which occurs very sharply. Energy of excitation is around milli-eV. Hence TES provides the best energy resolution of all the detectors. TES is able to measure precisely the energy of gamma ray by means of measuring the increase in temperature due to the absorption of gamma ray. Schematic shows the operating principle of the detector. TES consists of radiation absorber, thermometer, and the thermal sink so that it is quickly reset to detect the next gamma ray. Thermometer is made by a superconducting thin film that is biased in a constant voltage. Schematic of TES sensor is shown in Fig. 5.10 operated in the region of superconducting phase transition. Joule heating is inversely proportional to the resistance R of the sensor. Thus TES is cooled down due to removal of the deposited energy, by negative electrothermal feedback mechanism. There is a constant flow of heat from the sensor to the heat bath through the weak thermal link and the power reduction of the Joule heating compensates the energy of the deposited gamma ray photon. Based on this negative feedback mechanism, the constant voltage bias stabilizes the operating point on the

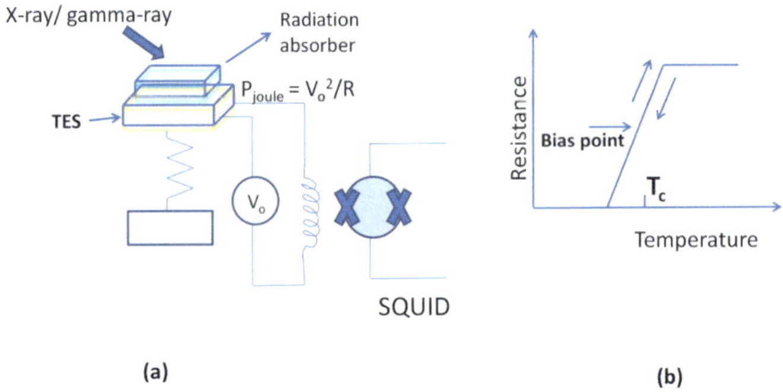

Fig. 5.10 **a** Schematic representation of the coupled electrical and thermal circuits of transition edge sensor, **b** a typical measurement of the resistance of a TES across the superconducting to normal transition. The resistance was measured from the response to a small AC excitation current

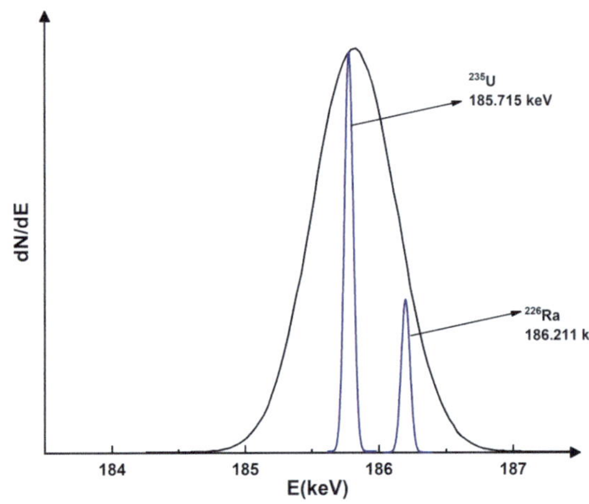

Fig. 5.11 Measurement of the ^{235}U 185.715 keV gamma-ray line and the interfering ^{226}Ra 186.211 keV line. Results shown from a HPGe detector and a Transition Edge Sensor (TES) array are marked as blue line and filled black, respectively. The two gamma ray peaks are easily separated by the TES

transition curve where heat flow to heat bath is equal to Joule heating so that very high count rate is allowed.

These spectrometers have been reported to have achieved an energy resolution between ~50 and 90 eV FWHM for energies below 100 keV, and are thus ideally suited for precise non-destructive analysis of nuclear samples [42–45]. TES exhibits the best energy resolution, as illustrated in Fig. 5.11, which shows the measurement of the ^{235}U 185.715 keV gamma ray line and the interfering ^{226}Ra 186.211 keV line. Results shown from HPGe detector and a TES array are marked as blue line and filled black, respectively. The two gamma ray peaks are easily resolved by the TES while HPGe could not resolve these closely lying energy peaks.

5.2.2 *Measurement of Time Interval Between the Emission of Two Gamma Rays of Distinct Energies Using Scintillation-Based Detectors*

For detecting time elapse between the detection of two gamma rays which are distinct with respect to their energies and time of emission the detectors should exhibit a good time resolution. Deduction of time difference between two gamma rays which are of different energy also basically involves the discrimination of gamma rays by energies. Hence this measurement first involves deducing energy of a gamma ray by a detector. The time clock is set on as triggered by the deduction of the first gamma ray of energy h υ_1 by the first detector and the termination of the clock following the deduction of gamma ray of energy h υ_2 by the second detector. Figure 5.12 shows the schematic of block diagram of a systems to record multichannel time spectrum from a radioactive source emitting coincident radiation and that to record coincident delay curves from a radioactive source emitting coincident radiation in cascade separated by an intermediate state with measurable lifetime typically of the order of hundreds of picoseconds and above. The time resolution of a scintillation detector depends mainly on the scintillation decay time of the scintillator crystal, efficient optical coupling between the crystal and the photomultiplier tube, and the optimal number of stages of a photomultiplier without compromising too much on the gain of the signal. Timing spectroscopy mainly involves measurement system which generates a logical pulse having the leading edge indicating the time of occurrence of an input linear pulse.

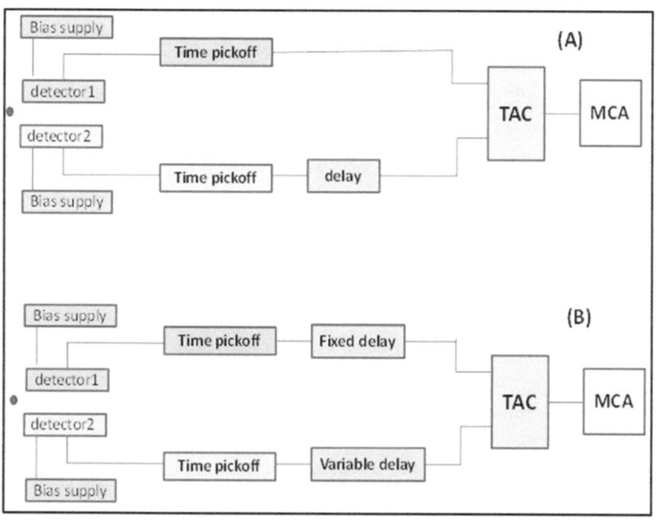

Fig. 5.12 **a** Schematic of block diagram of a system to record multichannel time spectrum from a radioactive source emitting coincident radiation while **b** shows the simplified system to record coincident delay curves from a radioactive source emitting coincident radiation in cascade separated by an intermediate state with measurable lifetime

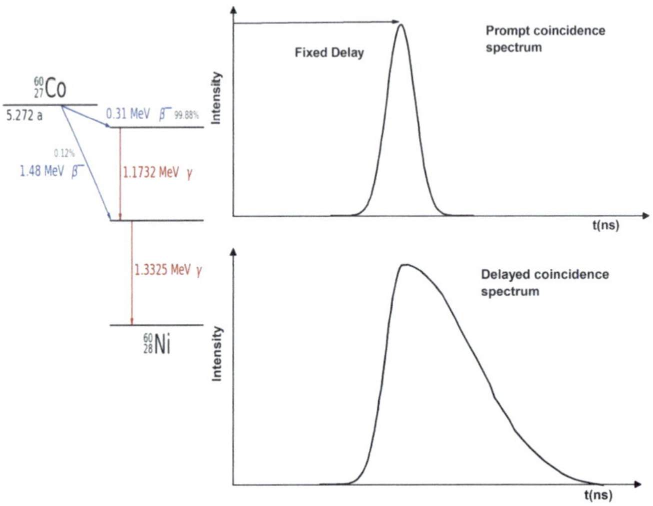

Fig. 5.13 Shown on the left panel is the decay scheme corresponding to ^{60}Co while the right panel shows the prompt coincidence spectrum as obtained using ^{60}Co source in which 1.17 and 1.33 MeV gamma rays are emitted in cascade with a delay of only 3 ps. Typical delayed coincidence spectrum is displayed on the right panel

Electronic devices that carry out this function are called time pickoff units or triggers. Accurate measurement of an occurrence of a gamma ray in terms of the timings in nuclear spectroscopy is affected mainly due to amplitude walk, time walk, and jitter. This is circumvented by means of using constant fraction discriminator [46–49].

Shown in Fig. 5.13 are the time spectra which is the output of the TAC (time-to-amplitude convertor) as obtained in a MCA (multichannel analyzer).

For example, consider ^{60}Co source which emits gamma rays of energies 1.17 MeV and 1.33 MeV having the intermediate state which has only a lifetime close to 3 ps. Scintillation detectors like BaF_2 can be used for proper energy discrimination, for example, detectors 1 and 2 can be fine-tuned by means of SCA to detect 1.17 MeV and 1.33 MeV gamma rays, respectively. But, on the other hand, the time difference between the emission of 1.17 and 1.33 MeV which is dictated by the intermediate state lifetime is only of the order of 3 ps which is too small for any detectors to resolve in time. Hence ^{60}Co is a familiar prompt source with the resulting time spectrum is called as a prompt spectrum whose FWHM is known as the time resolution of the detector. On the other hand, if each of the detector is set for gamma rays of different energies as emitted in cascade by radioactive nuclei, with the isomeric state having lifetime typically of the order of ns which is much higher than that of the time resolution of the detector the coincidence spectrum as observed in the detecting system will be typically called as delayed coincidence spectrum which can be fitted as $I(t) = I_o \ \exp(-t/\tau)$, where τ is the mean lifetime of the intermediate state.

5.2.3 *Spatial or Position Determination of the Emission of Gamma Rays*

Spatial determination of the emission of gamma rays is also important in several applications including nuclear physics, material characterization techniques, and in medical field. Principle of using a scintillator detector for one-dimensional position sensing is discussed here. Figure 5.14 shows a schematic of a long scintillator crystal optically coupled with photomultiplier tubes PM1 and PM2 on both the sides. This is followed by detector bias supply and the respective modules for pulse height analysis as was discussed in the previous section. As light is generated in a scintillator due to interaction of any ionizing radiation it is possible to sense the position of incidence of the ionizing ray [33].

Assuming that well-collimated gamma ray is incident at the position x from the center of the long scintillator of length L. The point x is therefore located at a distance of $(x + \frac{L}{2})$ from the PMT1 and $(\frac{L}{2} - x)$ from the PMT2.

Taking α as the attenuation coefficient of light in the scintillator then the signal E_1 at the PMT1 is given as

$E_1 = \frac{E\gamma P}{E_o}\exp\left(-\alpha\left(\frac{L}{2} + x\right)\right)$ and similarly

$E_2 = \frac{E\gamma P}{E_o}\exp\left(-\alpha\left(\frac{L}{2} - x\right)\right)$

From above equations we get the value of (E_1/E_2) as equal to exp(2 αx)

Hence $x = 1/(2\alpha)\ \ln\ (E_1/E_2)$.

Thus the position of the incident of the gamma ray is proportional to the ratio of the signal as received in photomultiplier tubes located at both the ends of the PMTs. Photomultiplier tube would have photocathode having very high quantum efficiency to convert photons falling on to it into a much larger number of photoelectrons. Followed by that are a number of dynodes which are kept at increasing positive potential in order to enable acceleration of photoelectrons and multiplying into large number of secondary electrons to achieve a large gain in the signal.

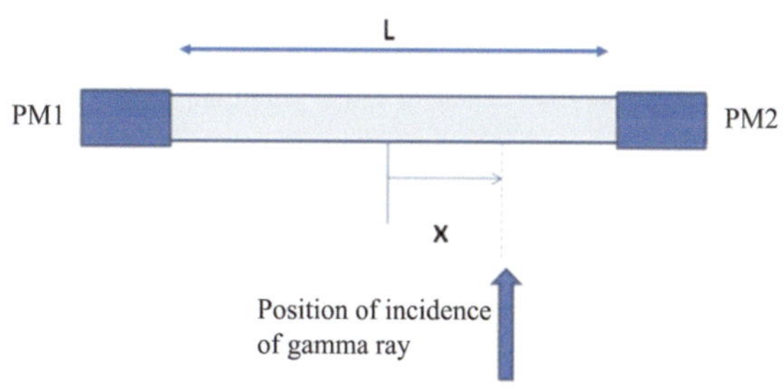

Fig. 5.14 Schematic shows the one-dimensional position sensing using scintillation detector. At both the ends of the scintillator crystal, photomultiplier tube and detector assembly are attached

5.2.4 Microchannel Plate Detector for Position Determination and for Energy, Timing Spectroscopy

Another type of detector which is closely related to scintillation detector with photomultiplier tube with several dynodes is microchannel plate-based detectors which can be used to detect protons, ions, neutrons, and gamma ray spectroscopy. Microchannel plate (MCP) [49] has many separate channel for which the magnified view of single channel is shown in Fig. 5.15. Analogous to PMT, a single particle incident on the inside channel leads to secondary electron multiplication thus providing a powerful tool for position sensing.

Plate, photocathode converts photon in to electron with high quantum efficiency. MCPs amplify electrons as illustrated above in terms of multiplication of secondary electrons accompanied by acceleration of electrons by the rear field of the anode. Patterned anode is used to measure charges as caused by the incidence and interaction of charged particle at specific pores of MCP.

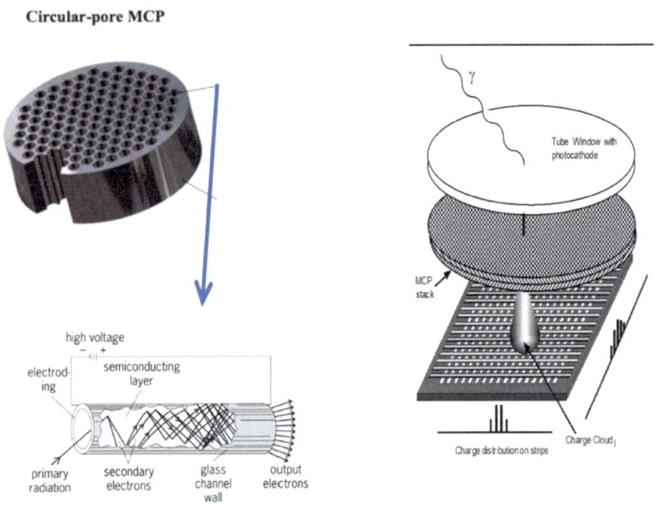

Fig. 5.15 Circular pore microchannel plate is shown on the left top. Enlarged view of each vertical pore which resembles as that of a photomultiplier tube is shown below. Shown on the right half is the schematic of detection of an interaction of a charged particle or a gamma ray with MCP [49]

5.3 Applications of Advanced Gamma Detectors for Material Characterization Techniques

Gamma ray-based imaging techniques are mostly used in different fields such as nuclear medicine, industries, nuclear and elementary particle physics and in high-energy applications [9–11, 35–38]. Positron annihilation spectroscopy (PAS) [18–22] has been established as a very powerful tool for studying open volume defects such as vacancy, vacancy clusters, and voids in materials and basic electronic properties of materials. PAS operates on the principle that a positron which is the antimatter of electron, annihilates with electron resulting with high probability the emission of two gamma rays of energy close to 511 keV, equal to the value of the rest mass energy of positron or electron. The time duration between the emission of a positron from a radioactive source by beta decay and the detection of annihilated photon of energy 511 keV corresponds to the lifetime of positron or positronium.

While the thermalized positrons with kinetic energy close to 0 keV annihilate with electrons exhibiting finite momenta and energy, due to which there will be a Doppler shift in the energy of annihilated photons, i.e., $(511 \pm \Delta E)$ keV. The small variation in the energy would lead to variation in the width of 511 keV gamma photons which can be deduced using detectors with very good energy resolution. Analogously there will be a variation in the angle between annihilated photons from 180°, i.e., $(180 \pm \Delta\theta)$. By means of deducing the variation in θ, the momentum distribution of electrons in the material of interest could be deduced. This in turn can be achieved by position sensitivity of the detector. Therefore for positron annihilation-based techniques such as positron lifetime measurement, Doppler broadening measurement and angular correlation study detectors are schematically explained in Fig. 5.16. The detectors meant for these studies should be having good time resolution, energy resolution, and position (angular) resolution, respectively.

Positrons emitted by beta decay in ^{22}Na into any material loses energy by ionization in tens of picoseconds and gets thermalized and subsequently annihilates with electrons of the constituent atoms following diffusion. Decay schemes of ^{22}Na along with that of ^{60}Co are shown in Fig. 5.17. Thus the positron lifetime is inversely proportional to the electron density at the site of annihilation. As vacancy sites are characterized by much lower electron density as compared to that of defect-free sites in the analogous well-annealed system, lifetime of positrons annihilating at open volume defects such as vacancies will be much higher than that of lifetime of positrons in the well-annealed analogous system. Positron lifetime sensitively increases with increasing open volume of defects such as monovacancy, divacancy, vacancy clusters, and voids. As the positron is positively charged and the effective charge of a vacant site in a metal is negative, positrons exhibit a strong affinity to get annihilated with valence electrons associated with open volume defects hence establishing that the PAS is a very powerful tool for the studies of open volume defects. In metals and/or semiconductors, the implanted positrons annihilate rapidly unless voids such as vacancy defects are present. For insulators such as polymers or zeolites, implanted

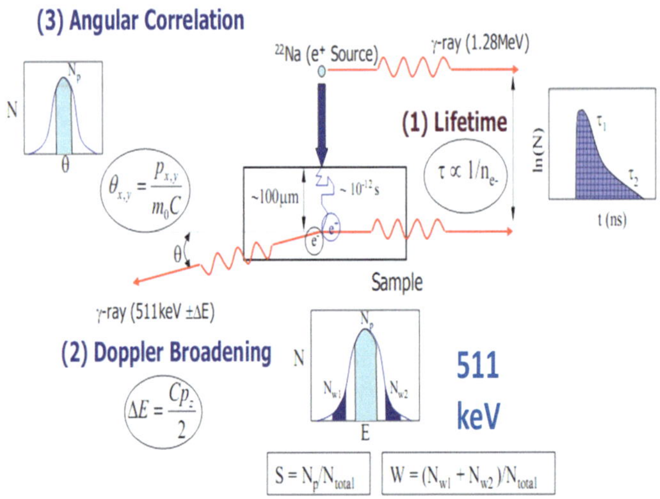

Fig. 5.16 Positron annihilation spectroscopic studies comprise of the measurements of positron lifetime, Doppler broadening of 511 keV annihilated photons, and angular correlation in terms of the measurement of deviation of the annihilated 511 keV photons from 180°. Techniques (2) and (3) enable us to measure the transverse and longitudinal components of momentum of electrons [20]

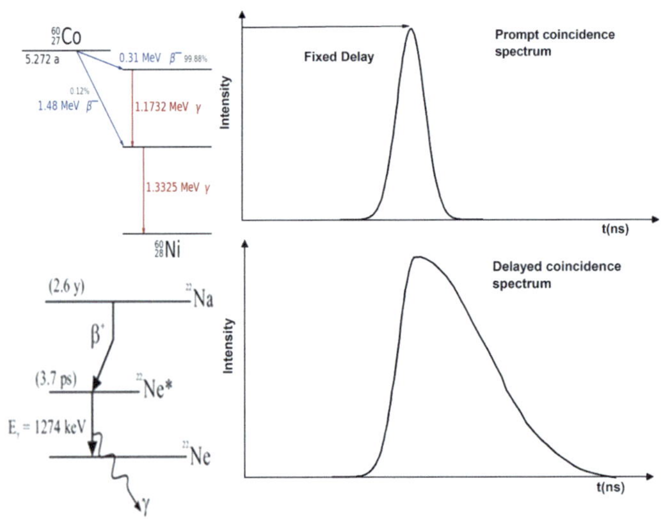

Fig. 5.17 Decay schemes of the radioactive ^{60}Co and ^{22}Na sources which are used for obtaining prompt coincident spectrum and positron lifetime studies. FWHM as obtained using the lifetime spectrometer with prompt source such as ^{60}Co indicates the time resolution of the spectrometer

positrons interact with electrons in the material to form a momentarily stable bound state of electron–positron pair called positronium [25, 26].

Positronium (Ps) is a metastable bound state of hydrogen-like atom having an electron and a positron exiting in two spin states which are *para-positronium* and *ortho-positronium*. *Para-positronium*, p-Ps, is a singlet state having the positron and electron spins in anti-parallel configuration with a characteristic self-annihilation lifetime of 125 ps in vacuum. Ortho-positronium, o-Ps, is a triplet state (with the parallel spin configurations of positron and electron) with a characteristic self-annihilation lifetime of 142 ns in vacuum. In any molecular materials, the lifetime of o-Ps is environment dependent and it delivers information pertaining to the size of the void in which it resides. Ps can pick up a molecular electron with an opposite spin to that of the positron, leading to a reduction of the o-Ps lifetime from 142 to 1–4 ns (depending on the size of the free volume in which it resides). It is known that the human body consists of free molecular spaces. Hence in the process of the PET imaging, the positrons which are emitted by the positron source can pick up electron leading to the formation of positronium marked by a high value of probability. Positron annihilation properties vary with respect to size of the inter- and intra-molecular voids and concentration of molecules such as molecular oxygen. Advanced techniques have been devised recently for positronium imaging by means of registering simultaneously annihilated gamma photons and de-excited photons from pharmaceutical which are labeled with radionuclides.

5.3.1 Positron Lifetime Studies in Materials Which Involve Measurements of Hundreds of Picoseconds

Positron lifetime studies in materials using ^{22}Na are carried out by means of determining the time difference between the detection of 1.28 MeV gamma rays and the annihilated gamma ray of energy 511 keV. BaF_2-based scintillator detectors with a very short scintillation decay time of 0.6 ns are used for lifetime studies. Two such BaF_2-based scintillator detectors which are, respectively, set for 1.28 MeV and 511 keV are used for positron lifetime studies [50, 51]. The schematic of a positron lifetime spectrometer for measurement of lifetimes typically of the order of hundreds of picoseconds is shown in Fig. 5.18.

Such spectrometers are very important to minimize the noise contribution in the timing signal. Possible sources of inaccurate time determination are time walk, amplitude walk, and jitter in timing signal, the details of which are referred to in the literature [50, 51]. Constant fraction differential discriminator can be used for removing time walk, amplitude walk, and jitter in timing signal which would result in accurate time determination thus leading to a very large signal-to-noise ratio in timing signal. Output of the CFDD which processes the detection of 1.28 MeV is fed as the 'START' signal to a nuclear module called as Time-to-amplitude converter (TAC).

Once the "start" of the TAC receives the output of CFDD, it initiates the charging of a capacitor with a constant current. Once the "stop" gamma ray is received by the TAC, the charging of the capacitor is terminated. The resultant voltage developed

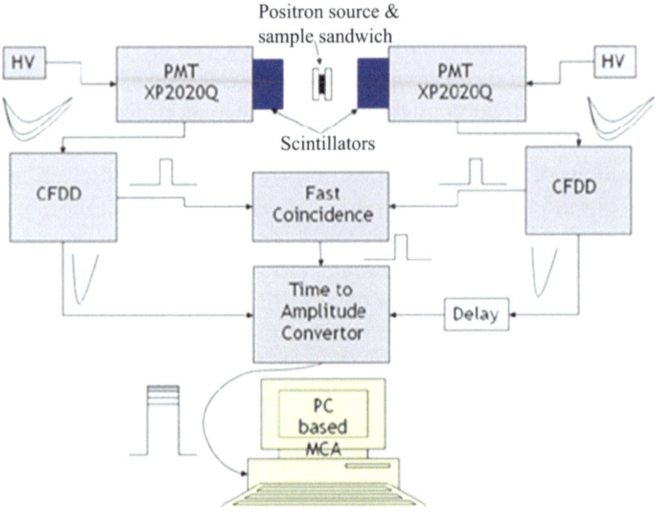

Fig. 5.18 Schematic of positron lifetime spectrometer

across the capacitor measures the time interval between the detection of two gamma rays in terms of voltage leading to time detection in multichannel mode.

By means of de-convolution of the lifetime spectrum into exponential decaying spectra with different values of respective relative intensity and exponential decay as $I(t) = I_1\exp(-t/\tau_1) + I_2\exp(-t/\tau_2)$ and so on the values of relative fractions of positrons (I_1, I_2) annihilating with lifetime such as τ_1 and τ_2, respectively, could be deduced.

5.3.2 *Deduction of Time Dependence of Angular Correlation Due to Nuclear Spin Precession*

Another important timing-based spectroscopy used in materials characterization is called as time differential perturbed angular correlation (TDPAC) which is similar to Nuclear Magnetic Resonance (NMR) spectroscopy [27–32]. This technique is basically dependent upon the measurement of time dependence of angular correlation of gamma rays emitted by selective radioactive nuclei, called as TDPAC probes, for studying material parameters related to local structure and magnetic properties. TDPAC is based on the hyperfine interaction (i.e., interaction between nuclear moments of the probe atoms with solid-state parameters such as local structure and magnetic field as experienced by the probe atoms defined in terms of electric quadrupole interaction and magnetic hyperfine interactions. Magnetic hyperfine interaction is defined in terms of interaction between magnetic dipole moment at the intermediate state of the probe nucleus with magnetic field as experienced by the probe nuclei. This term will be present only in samples which are magnetically ordered. Even in a magnetic sample this term would tend to become zero if the TDPAC measurement is carried above magnetic ordering temperature such as Curie temperature in the case of ferromagnetic sample. The quadrupole interaction is due to the interaction between nuclear quadrupole moment with the electric field gradient (EFG) at the sites of radioactive nuclei. These interactions primarily result in the precession of nuclear spin with the frequency of the nuclear precession is dictated by the magnitude of the hyperfine interaction. By means of deducing the time dependence of the anisotropy of the emission of the gamma rays, solid-state parameters such as electric field gradient (EFG) at the sites of probe atoms indicating the local structure and magnetic hyperfine field implying the strength of magnetic interactions could be deduced. Thus TDPAC is a powerful technique for the studies of defect–impurity interactions and in addition for studying the structural and magnetic phase transition in materials [28, 29].

Hyperfine interaction-induced time dependence of angular correlation is experimentally measured leading to the determination of solid-state parameters. Decay schemes of some of commonly used TDPAC sources are shown in Fig. 5.19 along with that of ^{60}Co emitting prompt gamma rays which are typically used for the measurement of the time resolution of the spectrometer. The schematic of a TDPAC spectrometer along with the photograph of the spectrometer used at IGCAR is shown in Fig. 5.20.

TDPAC and NMR

Nuclear Magnetic Resonance (NMR) is widely used in different applications ranging from nuclear physics, solid-state physics, and to recently introduced medical imaging of cancerous cells. This technique is also used as a powerful characterization tool for addressing various problems in the field of chemistry, biological, and condensed matter. NMR measures the properties of nuclear ground state directly. The energy needed for the absorption of photon to excited state or emission of photon due to de-excitation occurs in the low radio frequency range of 1–10 kHz which in turn has several important consequences.

Basic schematic of an NMR experiment is shown in Fig. 5.21. The sample is subjected to an external magnetic field H_o so that due to the interaction of nuclear

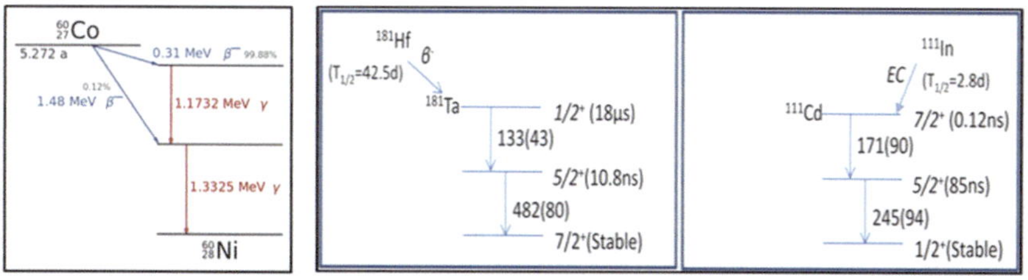

Fig. 5.19 Decay scheme of ^{60}Co prompt source along with the $\gamma 1$–$\gamma 2$ cascade of TDPAC probe nuclei ^{181}Ta and ^{111}Cd [28]

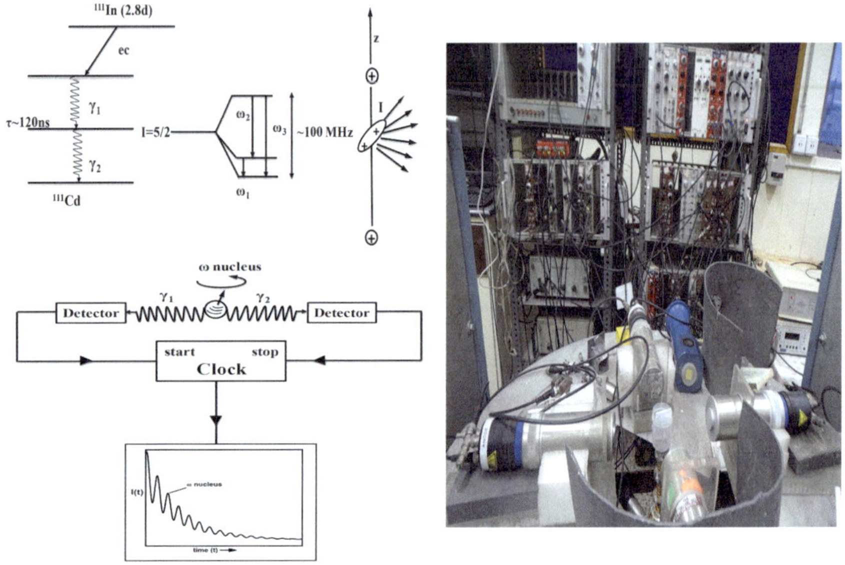

Fig. 5.20 Schematic showing the principle of TDPAC [28, 29] along with the photograph of the in-house developed TDPAC spectrometer

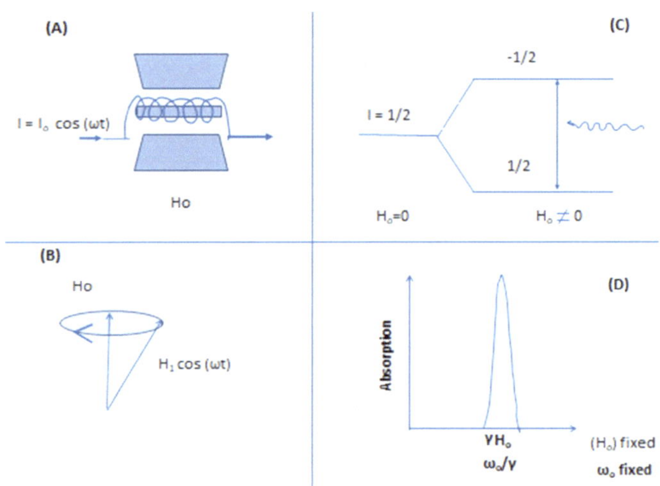

Fig. 5.21 Schematic of a basic NMR experiment. **a** Laboratory scale; **b** microscopic picture; **c** nuclear energy levels ($I = 1/2$); **d** basic NMR spectrum (frequency or field scan) [19]

moments with the external magnetic field, the $(2I + 1)$-fold degeneracy of the ground state is removed. Magnetic resonance refers to the resonant absorption of photons by the nuclear spin system through the nuclear dipolar interactions between energy levels.

There are two ways of absorption of RF energy which can be measured by many means (e.g., directly through the supplementary loss of the energy of the coil or indirectly through the subsequently induced emission of photons). Thus an NMR spectrum consists of a plot of some quantity proportional to the energy absorption (e.g., the quality factor of the coil or the electromotive force created in the coil by the re-emitted RF field) as a function of either the RF field frequency at a constant Ho or as a function of Ho at a constant frequency (ω_o). Since the energy absorption is proportional to the weak magnetic polarization of the ground state and to the small photon energy, the NMR sensitivity is much lower than that of other related nuclear techniques using triggered detection techniques without the need of polarization such as time differential perturbed angular correlation (TDPAC) [30–32].

Advancement of imaging method using NMR which is simply known as Magnetic Resonance Imaging (MRI) has emerged quite powerful mainly based on proton-based NMR studies. In MRI, a powerful magnet which produces a strong magnetic field that forces protons in the body to align with that field is used. The protons are stimulated, and spin out of equilibrium, straining against the pull of the magnetic field when a radiofrequency current is then pulsed through the patient. The MRI sensors are able to detect the energy released as the protons realign with the magnetic field. The time that takes for the protons to realign with the magnetic field, as well as the amount of energy released, is dependent on the environment and the chemical nature of the molecules, once the radiofrequency field is switched off. Hence it is possible to deduce the difference between various types of tissues based on the changes in the magnetic properties. By means of utilizing a large number of imaging data and the detailed software-based imaging analysis, MRI has emerged as a very powerful tool for the studies of biological systems including cancerous cells, tumors in terms of their location, growth, and so on.

5.3.3 Recent Advancement of Digitization of Spectrometers

Conventional TDPAC spectrometer consists of many energy modules such as spectroscopic amplifier, single channel analyzer, and counter and timing modules such as constant fraction discriminator, time-to-amplitude converter (TAC) besides coincidence unit, linear gate corresponding to each of the four BaF_2-based detectors. In a conventional TDPAC spectrometer, time- and energy-related signals are separately converted into analog electrical signals which are subsequently shaped and amplified, then digitized to select energies and measure the time elapsed between two distinct gamma rays of interest. Hence the operation becomes tedious as large number of calibrations are needed. On the other hand, in the digitized version of the spectrometer Fig. 5.22 there is no need for amplifying electronics which require different energy modules and with no need for advanced timing modules such as constant fraction discriminator and TAC. In the digitized version of the spectrometer the photomultiplier anode signal is fed directly to the Digital Signal Processing (DSP) input Figs. 5.22 and 5.23. While the DSP device when triggered by a transient

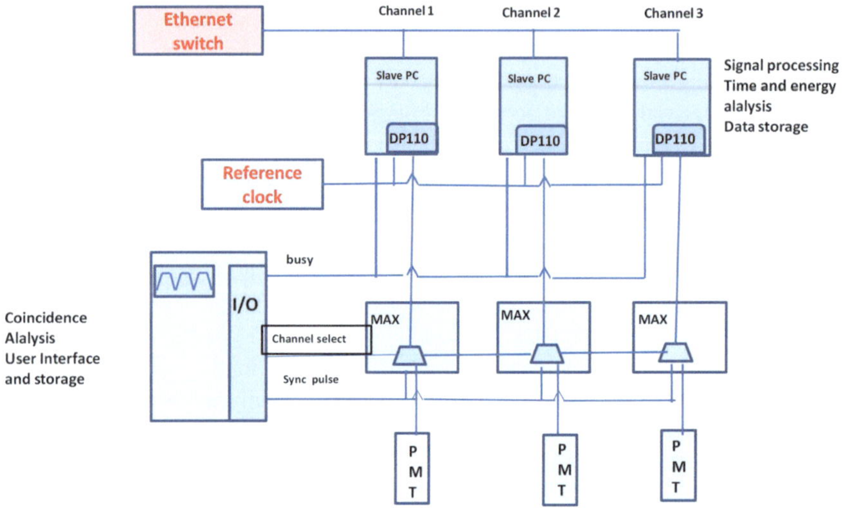

Fig. 5.22 Block diagram of a digital PAC spectrometer [52]

Fig. 5.23 Shown are raw data as array of segments with each segment having 512 data points [51, 52]

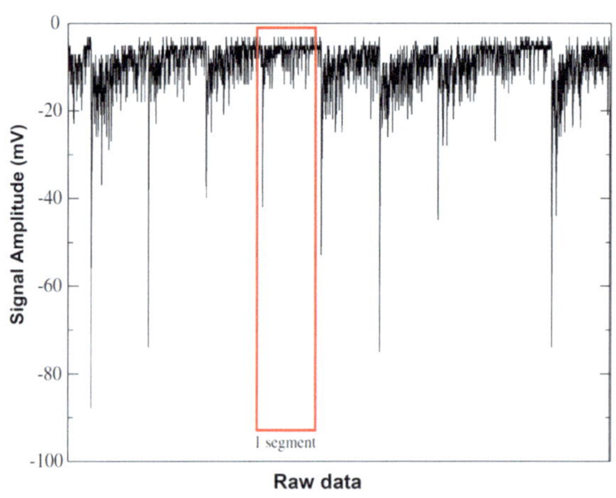

signal, the time of arrival is directly available from the digital recorder. By means of suitable software programs it is possible to digitally filter the signal to perform the function equivalent to that carried by constant fraction discriminator. On the other hand, the energy-related information is derived directly from the transient waveform after single step of calibration measurement. It is important to note that there is no start/ stop asymmetry. The number of detectors can be expanded by simply adding more modules with no additional complications. Hence the complexity of the spectrometer is significantly reduced in digitized version of TDPAC spectrometer which also holds good for the digitized version of positron lifetime spectrometer [50–52].

5.4 Advanced Material Characterization Techniques Involving Deduction of Small Changes in Energies of the Order of Few keV

5.4.1 Measurement of Doppler Broadening of Annihilated 511 keV Photons Providing a Powerful Tool for Material Characterization

Positrons emitted by radioactive source by beta decay gets thermalized and diffuses and get annihilated with conduction electrons with appreciable probability. As the positrons are at rest while annihilating with electrons of finite momenta and energy, depending upon the momenta/energy of electrons there will be a shift in the energy of annihilated photons by ΔE keV, i.e., $(511 \pm \Delta E)$ keV. Such a small energy shift of the order of a few keV could be deduced by means of using a detector with high energy resolution such as high pure Ge (HPGe) detector. Schematic of Ge detector for pulse height spectroscopy is shown in Fig. 5.24. The depletion width is expressed as follows:

$$d = (\frac{2\epsilon V}{eN})^{1/2}$$

For large value of depletion width which dictates the active volume of the detector, besides high value of biasing, the number of impurities should be very low which is achievable in the case of Ge crystal using float zone refinement.

The Ge semiconductor may be ν type or π type or lithium drifted. As discussed earlier the detector should be kept at 77 K while biased so that thermal excitation of electrons from valence band to conduction band is decreased leading to an enhanced signal-to-noise ratio.

Coincidence Doppler spectrometer based on the deduction of both of the 511 keV annihilated photons.

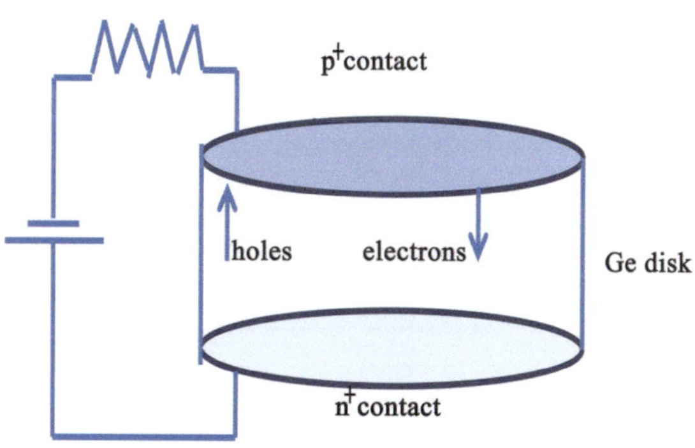

Fig. 5.24 Schematic of a p^+-i-n^+ configuration illustrating the movement of charge carriers in planar HPGe detector

Left panel of Fig. 5.24 shows the block diagram of coincident Doppler broadening setup meant for measuring the annihilation of positrons with core electrons enabling the identification of the chemical environment of the defects in binary or multicomponent alloys. The top portion of the right panel shows the comparison of the Doppler broadening of 511 keV annihilated photons as compared with defect-free and defective samples. This spectrum is simply obtained by means of deducing the broadening with the deduction of single photons of 511 keV. By means of deducing both the 511 keV annihilated photons emitted close to 180° results in a drastic increase in S/N ratio at higher values of moments contributed due to annihilation of positrons with core electrons [19–26]. Conventional Doppler broadening spectroscopy is based on the deduction of the broadening of single out of the two 511 keV annihilated photons by means of using one high pure Ge detector. Based on this technique annihilation of positrons with valence electrons is deduced.

The defect-related signature is obtained using Doppler broadening studies based on the concept that a positron annihilating at a vacancy site in a metallic matrix would result in an enhancement annihilation of positrons at valence electrons as compared to annihilation of positrons with core electrons [21–26].

By means of detecting both the annihilated 511 keV photons in coincidence, it is possible to enhance drastically the signal-to-noise ratio of the Doppler broadening spectrum which enables the deduction of the part of annihilation of positrons with

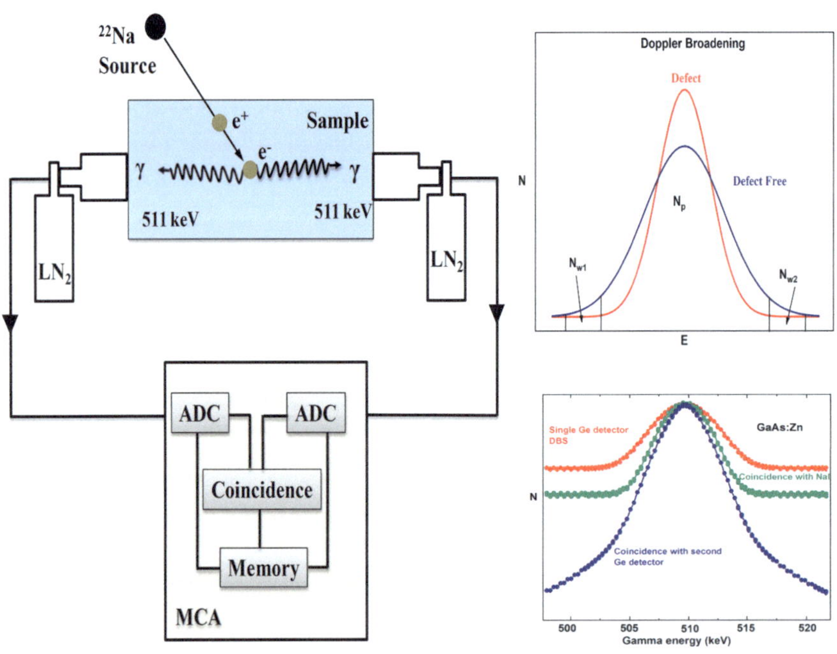

Fig. 5.25 Block diagram of coincident Doppler broadening setup and the respective defect-induced signatures as obtained using this setup is elucidated

core electrons which are characterized by high momenta in the Doppler broadening curve. Shown in Fig. 5.25 is the schematic of the coincidence Doppler broadening spectrometer. Deduction of coincidence Doppler broadening enables obtaining site-specific information of defects which positrons preferentially annihilate in a matrix. Considering that the problem of interest, for example, is to understand the defects in a binary alloy subjected to defect treatments such as quenching and heavy ion irradiation. By means of analyzing the wing portion of coincident Doppler broadening curves, implying the annihilation of positrons at core electrons, it is possible to find out the relative fraction of positrons annihilating at the vacancies due to that of A and B atoms in a binary alloy AB.

5.4.2 *Positron Emission Tomography*

Annihilation of positrons with electrons leads to the emission of two 511 keV gamma photons at nearly 180° is the basic principle being utilized in positron emission tomography (PET) for studying cancer growth in any part of human body. Different radioactive sources of positrons have affinity to get dominantly absorbed at different parts of the body. Therefore depending upon the organs to be scanned, different isotopes of positrons are produced using cyclotron and tagged to a natural body compound. Different isotopes get selectively deposited in different parts of the body. Subsequent to administration of a suitable source of positrons into the patient, the radioactivity as localized in the appropriate parts of the body is detected by the PET scanner. Based on the selective uptake of the source of positrons which is crucially dependent upon the functioning of the concerned portion of the body, the brightness of the PET image varies. A healthy tissue, for example, uses glucose for energy and accumulates to some extent the source tagged glucose which in turn would show up as PET images. But if the tissue is cancerous it accumulates a much larger concentration of glucose than that of the normal tissue and hence the PET image corresponding to cancerous tissues would appear more brighter than the normal tissue. Schematic of PET imaging system is shown in Fig. 5.26. It consists of scintillator detectors arranged in circular manner. As the annihilation of positron with electron dominantly results in the emission of two annihilated photons each of energy 511 keV close to 180°. Hence these 511 keV gamma rays are detected in coincidence by detectors almost close to 180, using the coincidence detecting system shown as the block diagram in the figure. Detector assembly in plane can be rotated at different ways for imaging of the sample of interest. Obtained data are subjected to different corrections with respect to possible sources of errors as explained in a detailed manner in the literature. Also many possible ways for improving the signal-to-noise ratio in these measurements have been recently proposed in the literature by scientists working in these areas of research [2–7]. One such method is to carry out PET under magnetic field so that it is possible to confine the positrons at the location of interest dominantly and result in the emission of annihilation photons. Nowadays PET imaging systems are widely used as an advanced imaging system.

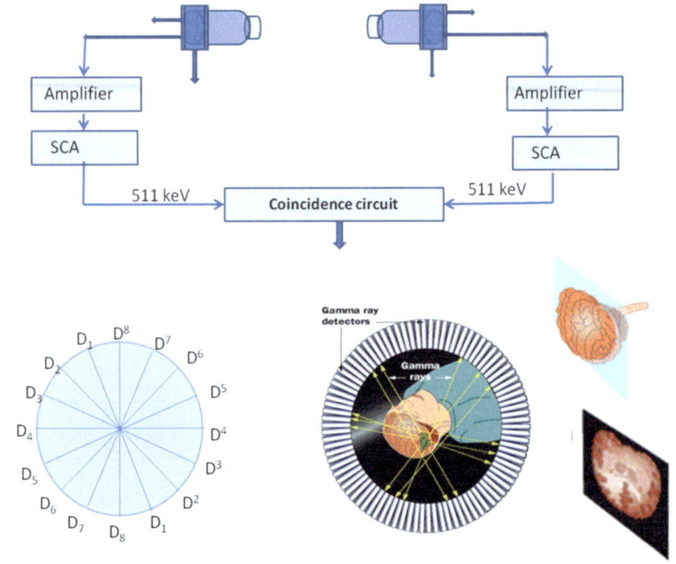

Fig. 5.26 Schematic of a PET camera-based coincidence setup made based on BaF2 scintillators coupled with XP2020Q-based photomultiplier tube assembly is shown at the top. PET image corresponding to cancerous tissues will appear brighter than that of the normal tissues

Other important problems for which PET [2–7] is utilized extensively, viz., patients with conditions affecting the brain, certain types of cancer, Alzheimer's disease, and some neurological disorders. Important radioactive nuclei which are quite often used as source of positrons include ^{11}C, ^{13}N, ^{15}O, and ^{18}F. These are natural elements which are passed through body and can be detected by the scanner. Various drugs and other chemicals can be labeled with these isotopes.

Positron while interacting with electrons associated with the tissues and/or body parts which are zones of interest for scanning, there is also a finite probability for the positron to pick up an electron leading to the formation of positronium which is a stable e^-e^+ pair. As we have seen earlier, Positronium decays are sensitive to local oxygen saturation. After some time the positron and electron of this metastable Ps would annihilate with each other leading to the emission of two or three of the annihilated photons depending upon the spin directions of electron and positron. Ortho-Ps which have the spins of electron and positron in parallel configuration decay into three photons with energies ranging from 0 to 511 keV after a mean lifetime of 142 ns, which is very much larger than para-positronium having spins of electron and positron anti-parallel. Long lifetime of ortho-Ps makes it possible for an appreciable interaction of O-Ps with the biological organs to be studied. Thus PET imaging has emerged as a powerful tool for the studies of medical imaging.

Conclusion

In this chapter, recent advancements in different detectors of gamma rays have been highlighted with respect to energy resolution, time resolution, and spatial resolution. These developments have resulted in remarkable progress leading to advanced characterization techniques in various fields ranging from materials science to biological science. It may be noted by the gentle readers that while this article might serve to provide an overall view of the developments, it does not exhaustively discuss all such progress.

Acknowledgements Author would like to thank Mr. C. G. Karhadkar, Director, Indira Gandhi Centre for Atomic Research (IGCAR) and Dr. B. Venkatraman, former Director, IGCAR for their kind encouragement and full support. Author express his sincere thanks to all the past and present members of the then metal physics section, MSG for several scientific discussions which have helped a lot in writing this article. Author also thanks Mr. Ravi Kumar Yadav for his help in formatting figures and text.

References

1. J.M. Zaucha, S. Chauvie, R. Zaucha, A. Biggii, A. Gallamini, The role of PET/CT in the modern treatment of Hodgkin lymphoma. Cancer Treat. Rev. **77**, 44–56 (2019). https://doi.org/10.1016/j.ctrv.2019.06.002
2. K.M. McCarten, H.R. Nadel, B.L. Shulkin et al., Imaging for diagnosis, staging and response assessment of Hodgkin lymphoma and non-Hodgkin lymphoma. Pediatr. Radiol. **49**, 1545–1564 (2019). https://doi.org/10.1007/s00247-019-04529-8
3. S. Pauls, A.K. Buck, K. Hohl, G. Halter, M. Hetzel, N.M. Blumstein et al., Improved non-invasive T-staging in non-small cell lung cancer by integrated 18F-FDG PET/CT. Nuklearmedizin **46**(1), 09–14 (2007). https://doi.org/10.1055/s-0037-1616618
4. F. Chao, H. Zhang, PET/CT in the staging of the non-small-cell lung cancer. J. Biomed. Biotechnol. **2012**, 783739 (2012). https://doi.org/10.1155/2012/783739
5. H.C. Steinert, PET and PET-CT of lung cancer, in *Positron Emission Tomography. Methods in Molecular Biology*, vol. 727 (Humana Press, 2011), pp. 33–51. https://doi.org/10.1007/978-1-61779-062-1-3
6. A. Aldin, L. Umlauff, L.J. Estcourt, G. Collins, K.G. Moons, A. Engert et al., (Cochrane Haematology Group), Interim PET-results for prognosis in adults with Hodgkin lymphoma: a systematic review and meta-analysis of prognostic factor studies. Cochrane Database Syst. Rev. **1**(1), CD012643. https://doi.org/10.1002/14651858.CD012643.pub2
7. T.S. Khan, A. Sundin, C. Juhlin, B. Langstrom, M. Bergstrom, B. Eriksson, 11C-metomidate PET imaging of adrenocortical cancer. Euro. J. Nucl. Med. Mol. Imaging **30**(3), 403–10 (2003). https://doi.org/10.1007/s00259-002-1025-9
8. H. Minn, A. Salonen, J. Friberg, A. Roivainen, T. Viljanen, J. Langsjo et al., Imaging of adrenal incidentalomas with PET using (11)C-metomidate and (18)F-FDG. J. Nucl. Med. **45**(6), 972–9 (2004)
9. L. Verger et al., Performance and perspectives of a CdZnTe-based gamma camera for medical imaging. IEEE Trans. Nucl. Sci. **51**(6), 3111–3117 (2004). https://doi.org/10.1109/TNS.2004.839070
10. D. Gagnon, G.L. Zeng, J.M. Links, J.J. Griesmer, F.C. Valentino, Design considerations for a new solid-state gamma-camera: Soltice, in *IEEE Nuclear Science Symposium Confer-*

ence Record (Cat. No.01CH37310), vol. 2 (2001), pp. 1156–1160. https://doi.org/10.1109/NSSMIC.2001.1009755

11. M. Bocher, A. Balan, Y. Krausz et al., Gamma camera-mounted anatomical X-ray tomography: technology, system characteristics and first images. Eur. J. Nucl. Med. **27**, 619–627. https://doi.org/10.1007/s002590050555
12. S.L. McArthur, Applications of XPS in bioengineering. Surf. Interface Anal. **38**, 1380–1385 (2006). https://doi.org/10.1002/sia.2498
13. J. Lefebvre, F. Galli, C.L. Bianchi, G.S. Patience, D.C. Boffito, Experimental methods in chemical engineering: X-ray photoelectron spectroscopy-XPS **97**, 2588–2593 (2019). https://doi.org/10.1002/cjce.23530
14. C.G. Shull, M.K. Wilkinson, Neutron diffraction studies on various transition elements. Rev. Mod. Phys. **25**, 100 (1953). https://doi.org/10.1103/RevModPhys.25.100
15. M. Balasko, E. Svab, G. Endrczy, Comparison of neutron radiography with other nondestructive methods. IEEE Trans. Nucl. Sci. **52**(1), 330–333 (2005). https://doi.org/10.1109/TNS.2005.843648
16. B. Winkler, Applications of neutron radiography and neutron tomography, **63**, 459–471 (2006). https://doi.org/10.2138/rmg.2006.63.17
17. N. Kardjilov, I. Manke, A. Hilger, M. Strobl, J. Banhart, Neutron imaging in materials science. Mater. Today **14**, 248–256 (2011)
18. A.E. Dupasquier, A. Dupasquier, P. Hautojarvi, P. Hautojarvi, *Positrons in Solids* (Springer, Berlin, 1979)
19. C.S. Sundar, B. Viswanathan, Positron annihilation spectroscopy, metals, materials and processes **8**, 1 (1996); R.W. Siegel, Positron annihilation spectroscopy. Ann. Rev. Mater. Sci. **10**, 393–425 (1980). https://doi.org/10.1146/annurev.ms.10.080180.002141
20. F. Tuomisto, I. Makkonen, Defect identification in semiconductors with positron annihilation: Experiment and theory (PDF). Rev. Modern Phys. **85**(4), 1583–1631 (2013). https://doi.org/10.1103/RevModPhys.85.1583
21. Y.C. Jean, D.M. Schrader, P.E. Mallon, Principles and applications of positron and positronium chemistry (2002)
22. M. Eldrup, D. Lightbody, J.N. Sherwood, The temperature dependence of positron lifetimes in solid pivalic acid. Chem. Phys. **63**(1–2), 51 (1981). https://doi.org/10.1016/0301-0104(81)80307-2
23. S. Capponi, F. Alvarez, D. Racko, Free Volume in a PVME Polymer—Water solution. Macromolecules **53**(12), 4770–4782 (2020). https://doi.org/10.1021/acs.macromol.0c00472,hdl:10261/218380
24. S. Kahana, Positron annihilation in metals. Phys. Rev. **129**(4), 1622–1628. https://doi.org/10.1103/PhysRev.129.1622
25. G. Amarendra, B. Viswanathan, A. Bharathi, K.P. Gopinathan, Nucleatrion and growth of helium bubbles in Ni as studied by positron annihilation spectroscopy. Phys. Rev. B **45**, 10231 (1992)
26. G. Amarendra, B. Viswanathan, G. Venugopal Rao, J. Parimala, B. Purniah, Variable low energy positron beam for depth resolved defect spectroscopy in thin film structures. Curr. Sci. **73**, 409–417 (1997); R. Ramachandran, C. David, P. Magudapathy, R. Rajaraman, R. Govindaraj, G. Amarendra, Study of defect complexes and their evolution with temperature in hydrogen and helium irradiated RAFM steel using positron annihilation spectroscopy. Fusion Eng. Des. **142**, 55–62 (2019)
27. A. Jancso et al., TDPAC and β-NMR applications in chemistry and biochemistry. J. Phys. G: Nucl. Part. Phys. **44**, 064003 (2017). https://doi.org/10.1088/1361-6471/aa666b
28. G.S. Collins, S.L. Shropshire, J. Fan, Perturbed $\gamma - \gamma$ angular correlation: a spectroscopy for point defects in metals and alloys. Hyp. Int. **62**(1990), 1 (2001); R. Govindaraj, K.P. Gopinathan, B. Viswanathan. Hf impurity and defect interactions in helium-implanted NiHf, Nucl. Instrum. Methods Phys. Res. B: Beam Interact. Mater. At. **179**, 200–208. https://doi.org/10.1016/S0168-583X(01)00474-8

29. R. Govindaraj, C.S. Sundar, Linear relationship between magnetic hyperfine field and electric field gradient at Mn sites in lanthanum strontium manganite. Phys. Rev. **B70**, 220405 (R) (2004). https://doi.org/10.1103/PhysRevB.70.220405
30. R. Govindaraj, C.S. Sundar, Atomic scale study of oxidation of hafnium: formation of hafnium core and oxide shell. J. Appl. Phys. **100**, 084318 (2006). https://doi.org/10.1063/1.2360148
31. R. Govindaraj, C.S. Sundar, A.K. Arora, Atomic scale study of negative thermal expansion in zirconium tungstate. Phys. Rev. B **76**, 012104 (2007). https://doi.org/10.1103/PhysRevB.76.012104
32. R. Sielemann, R. Govindaraj, M. Muller, L. Wande, G. Weyer, Microscopic observation of vacancy and self-interstitial and formation of Frenkel pairs in InSb by Mossbauer spectroscopy. Physica B **404**, 5045–5049 (2009). https://doi.org/10.1016/j.physb.2009.09.051
33. G.F. Knoll, *Radiation Detection and Measurement* (Wiley, NY, 1979)
34. S.S. Kapoor, V.S. Ramamurthy, *Nuclear Radiation Detectors* (New Age International (P) Ltd., 2005)
35. K. Siegbhan (ed.), *Alpha-, Beta-, and Gamma Ray Spectroscopy* (North-Holland, Amsterdam, 1968)
36. P.J. Ouseph, *Introduction to Nuclear Radiation Detectors* (Plenum press, NY, 1975)
37. W.H. Tait, *Radiation Detection* (Butterworth, London, 1980)
38. H. Kanitsubo, K. Husimi, T. Doke, Nucl. Rad. Detectos, Nucl. Inst. Methods **196**, 1 (1982)
39. R.M. Keyser, T.W. Raudorf, Germanium radiation detector manufacturing: process and advances. Nucl. Instrum. Methods. **A286**, 357 (1990). https://doi.org/10.1016/0168-9002(90)90877-9
40. E.L. Hull et al., Temperature sensitivity of surface channel effects on high-purity germanium detectors. Nucl. Instrum. Methods **A364**, 488 (1995). https://doi.org/10.1016/0168-9002(95)00339-8
41. Z. Kis et al., Comparison of efficiency functions for Ge gamma—Ray detector in a wide energy range. Nucl. Instrum. Methods **A418**, 374 (1998). https://doi.org/10.1016/S0168-9002(98)00778-5
42. K.D. Irwin, An application of electrothermal feedback for high resolution cryogenic particle detection. Appl. Phys. Lett. **66**, 1998 (1995). https://doi.org/10.1063/1.113674
43. K.D. Irwin, G.C. Hilton, Transition-edge sensors, in *Cryogenic Particle Detection*, ed. by C. Enss (Springer, 2005). https://doi.org/10.1007/10933596-3
44. D.A. Bennett et al., A high resolution gamma-ray spectrometer based on superconducting microcalorimeters. Rev. Sci. Instrum. **83**, 093113 (2012)
45. A.S. Hoover et al., Microcalorimeter arrays for ultra-high energy resolution X- and gamma ray detection. J. Radioanal. Nucl. Chem. **282**, 227–232 (2009). https://doi.org/10.1007/s10967-009-0223-3
46. J.B. Birks, *The Theory and Practice of Scintillation Counting* (Pergamon press, Oxford 1964)
47. L.W. Weathers, M.B. Tsang, Fabrication of thin scintillator foils. Nucl. Istrum. Methods **A382**, 567 (1996)
48. M. Hirschberg et al., Precise measurement of Birks kB parameter in plastic scintillator. IEEE Trans. Nucl. **39**(4), 511 (1992)
49. J. Bortfeldt et al., Timing performance of micro-channel—Plate photomultiplier tube. Nucl. Instrum. Methods Phys. Res. A. **960**, 163592 (2020). https://doi.org/10.1016/j.nima.2020.163592
50. M. Jardin, M. Lambrecht, A. Rempel, Y. Nagai, E. Van Walle, A. Almazouzi, Digital positron lifetime spectrometer for measurements of radioactive materials. Nucl. Instrum. Methods. **568**, 716–722 (2006). https://doi.org/10.1016/j.nima.2006.08.087
51. D. Bosnar et al., Digitized positron lifetime spectrometer for the simultaneous recording of time and energy information, Nucl. Instrum. Methods. **581**, 91–93 (2007). https://doi.org/10.1016/j.nima.2007.07.035
52. M. Jager, K. Iwig, T. Butz, A user-friendly digital TDPAC spectrometer. Hyperfine Interact. **198**, 167–172 (2010). https://doi.org/10.1007/s10751-010-0201-8

Chapter 6
Nuclear-Powered Wireless Sensor System

Suresh Kaluvan, Tom Scott, and G. R. Mackenzie

6.1 Betalight-Powered Sensor Systems: An Overview

In the world of nuclear energy, nuclear batteries remain exotic because of their high cost, demand of nuclear source, and time-consuming power production. However, when compared to other types of chemical batteries, nuclear batteries have higher volumetric energy density, long lifetime, and stronger endurance in extremely harsh conditions [1]. The batteries that harvest energy from the nuclear decay of isotopes may produce very low levels of current, but it can last for decades without any replacement. These types of batteries, called betavoltaics, are being developed by various companies and research institutes [2, 3].

Betalight technology utilizes Gaseous Tritium Light Sources (GTLS) to power energy harvesting systems, offering a reliable and long-lasting power source for wireless sensor networks. These systems capture the electrons emitted by the radioactive decay of tritium, converting them into electrical energy to power low-consumption devices. This innovative approach of harvesting energy promises to enhance the sustainability and autonomy of wireless sensors, particularly in remote or inaccessible locations where traditional power sources are unfeasible [4–10].

In this chapter, the betalight-powered wireless sensor system is designed to measure the ambient temperature and humidity. The system comprises two units. Namely, a unit which senses and transmits, including a betalight-powered Bluetooth wireless temperature sensor, and a receiver unit which collects and displays real-time

S. Kaluvan (✉) · T. Scott · G. R. Mackenzie
University of Bristol, Bristol, UK
e-mail: sureshmooncity@gmail.com

T. Scott
e-mail: T.B.Scott@bristol.ac.uk

G. R. Mackenzie
e-mail: gm12172@bristol.ac.uk

U. Mangalanathan et al. (eds.), *Nuclear Instrumentation and Control*,
https://doi.org/10.1007/978-981-97-1283-0_6

sensor data. The sensor unit employs an ultra-low power energy harvesting module (EH300) that can receive energy generated by radiation sources and store it to operate standard 3.3 V electrical systems. It also includes a power management IC (PMIC-S6AE101A), which is used to regulate the power from the energy harvester module and bias the temperature sensor and Bluetooth module in a pulse mode.

The wireless sensor system powered by the betalight incorporates a temperature sensor node that runs on radiation and uses Bluetooth Low Energy (BLE) to send sensor data. The sensor data is displayed in real time and its key features are

- Operates using tritium-based GTLS.
- Supports BLE communication.
- A BLE Beacon sends measured data every 2.5 h and it performs well when the surrounding temperature is below 20 °C.
- The motherboard of the sensor units has an extendable terminal that can accommodate sensor interfaces such as I^2C (Inter-Integrated Circuit), UART (Universal Asynchronous Receiver/Transmitter), and SPI (Serial Peripheral Interface).

6.2 Design of Betalight-Powered Sensor Unit

The design shown in Fig. 6.1 depicts a wireless sensor unit that operates as a BLE Beacon. This self-contained device is powered by a Betalight battery, which offers a reliable and long-lasting energy source. The unit incorporates a highly compact Bluetooth Low Energy (BLE) radio transmitter that broadcasts signal within an estimated

Fig. 6.1 Betalight-powered wireless sensor unit

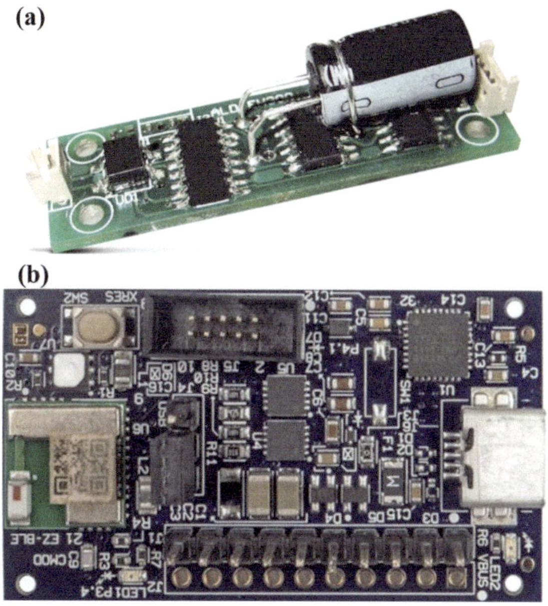

Fig. 6.2 **a** EH300 energy harvesting module and **b** S6AE101A energy harvesting and PMIC module

range of 10 m. These signals are designed to be detected and interpreted by a Linux PC equipped with Bluetooth capabilities, thereby facilitating real-time monitoring and data acquisition. The sensor unit is composed of several critical components that work in unison to achieve its intended functionality:

- **Beta cell (custom-made)**: At the core of the device is a custom-made Betalight cell. This component is the powerhouse of the unit, providing a steady and maintenance-free power supply derived from the radioactive decay of tritium. The energy output is both consistent and safe, with the cell encapsulated in a robust housing to shield its surroundings from radiation.
- **Energy Harvesting Module (EH300 module)**: The EH300/EH301 shown in Fig. 6.2a complements the Betalight cell by harnessing additional energy from environmental sources, thereby enhancing the overall efficiency and longevity of the power supply. This module is capable of receiving energy from various electrical energy sources and storing it to operate standard 3.3 and 5.0 V electrical circuits and systems. These modules can handle instantaneous input voltages from 0.0 V to $\pm$500 V AC or DC, as well as input currents from 200 nA to 400 mA. It converts ambient energy into electrical energy, ensuring that even in conditions of low power from the beta cell, the sensor unit remains operational.
- **Power Management IC (S6AE101A)**: The Power Management Integrated Circuit (PMIC) shown in Fig. 6.2b plays a pivotal role in ensuring efficient energy distribution and usage within the unit. It is responsible for managing the charge and discharging cycles, voltage regulation, and power-saving modes, which are crucial for maintaining the longevity and reliability of the sensor.

- **Bluetooth Low Energy Module (CYBLE-022001-00)**: This module is the communication heart of the sensor unit. It utilizes BLE technology to transmit data with low power consumption, making it ideal for applications where energy efficiency is paramount. Its compact design does not compromise on performance, providing robust wireless connectivity within its operational radius.
- I^2C **Temperature and Humidity Sensor (Si7020-A10)**: The inclusion of this sensor allows the unit to monitor environmental conditions with precision. The Si7020-A10 communicates via the I^2C protocol, a popular choice for embedded systems due to its simplicity and effectiveness. This sensor can measure a wide range of temperatures and humidity levels, providing valuable data that can be utilized for various analytical purposes.

6.2.1 Beta Cell Design

All the components listed above are assembled in a plastic case with dimensions of 110 × 60 × 30 mm. The sensor unit is always ON due to the nature of the power source, and there would be no advantage in including an ON/OFF switch. The betalight (GTLS) properties and dimensions are given in Table 6.1. The GTLS emits a constant glow by emitting beta particles as tritium decays, which in turn excite the phosphor. A conventional photo-voltaic cell collects the optical photons from the betalight and produces an electrical current at the voltage of the photovoltaic (PV) cell.

To create a proposed betalight-powered wireless temperature sensor system, three beta cells are required. Each beta cell comprises 15 GTLS vials and 2 PV cells. Based on the seller's specifications, each GTLS vial has an activity of around 13.6 gigabecquerels. The design procedure is: 15 GTLS vials are sandwiched between two photovoltaic (PV) panels as shown in Fig. 6.3 and wrapped with aluminum foils and encapsulated on a heat-shrunk rubber tubing (to protect from accidental drop) as shown in Fig. 6.4a, b. Each beta cell produces open-circuit voltage of approximately 5 V and short-circuit current of 0.7 μA output.
The beta cell is connected to the energy harvester and Bluetooth module and the schematic of the circuit is shown in Fig. 6.5, and it contains EH300 energy harvester and S6AE101A BLE module onboard.

Table 6.1 Properties and dimension of GTLS and PV cell

Part no.	Description
T 5678-1/I	Trigalight GTLS, 3.5 × 25.0 mm, Green
AM-1801CA	Panasonic photovoltaic cell

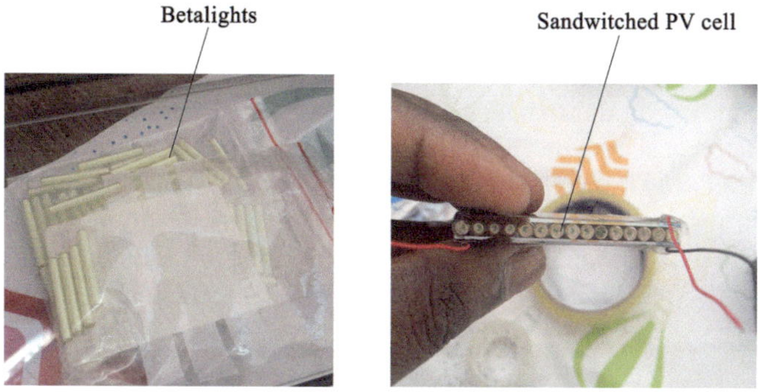

Fig. 6.3 GTLS vials and PV cell assembly

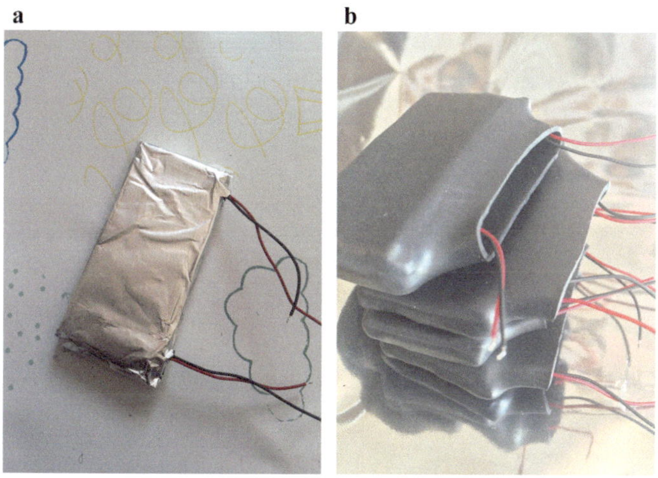

Fig. 6.4 **a** Aluminum foil-wrapped betalight cell and **b** Beta cell with heat-shrink cushion

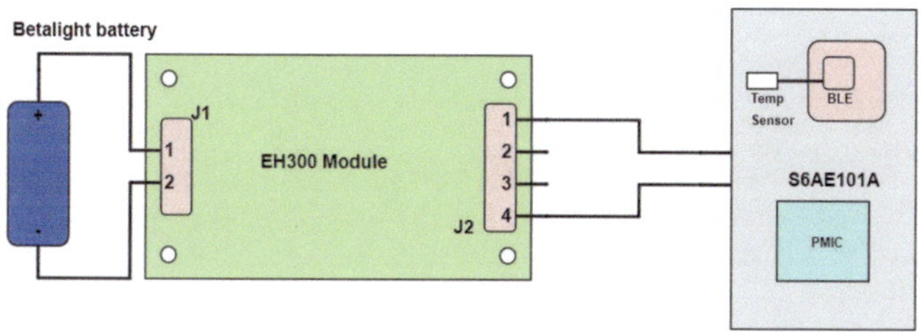

Fig. 6.5 Schematic of beta cell energy harvester

6.2.2 Custom Energy Harvester Circuit for Beta Cell

The schematic circuit shown in Fig. 6.6 details an energy harvesting system powered by beta cell. It consists of an energy harvester module with a photovoltaic cell interfacing with a UB20M voltage detector IC as power gating switch. The UB20M chip was chosen for this energy harvesting application due to its exceptional ability to function with ultra-low power sources, such as the GTLS. Its low start-up voltage of approximately 0.65 V allows it to activate even with the minimal voltages expected from GTLS-powered photovoltaic cells. Moreover, the chip is capable of handling

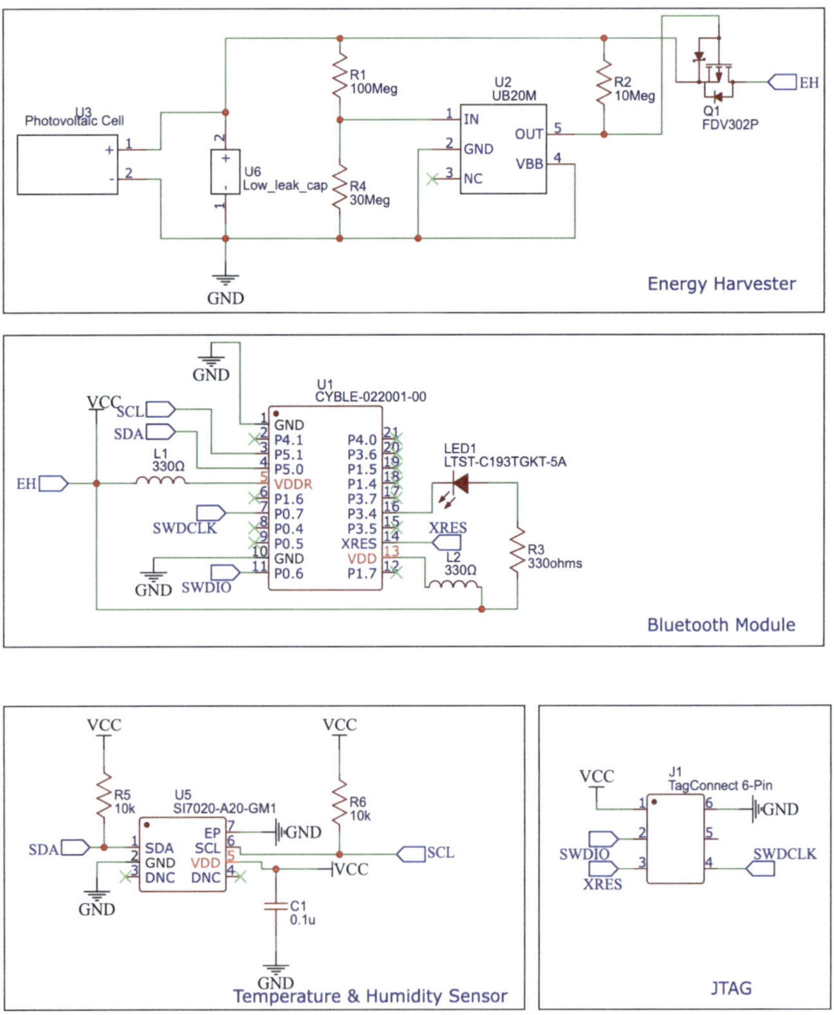

Fig. 6.6 Schematic of low power, power gating energy harvester

input voltages up to 20 V, providing flexibility in the design. The UB20M's high-impedance input and low quiescent current are crucial for maximizing the efficiency of energy harvesting from the high-impedance output of GTLS, making it an ideal choice for powering a low-energy Bluetooth module and sensors.

The Bluetooth module uses a CYBLE-022001-00 for wireless communication and includes an LED indicator. The temperature and humidity sensor (Si7020-A20-GM1) communicates via I^2C, with 10 kΩ resistors for pull-up. Finally, a JTAG (Joint Test Action Group) interface is present for programming, using a standard 6-pin connector. The custom circuit has a 1.5 μW minimum cold start input power, which is sufficient to power the temperature and humidity sensor.

6.3 Receiver Unit Design

The receiver unit is a simple display unit based on the Raspberry Pi, which is configured to search for the transmission from its paired sensor unit, indicating that the display unit consistently looks for the pre-configured Bluetooth MAC address of the sensor unit. The receiver unit contains the following components:

- Raspberry Pi 3B with Debian 10.1 (Buster) Raspbian operating system.
- 32GB micro-SD card.
- 7 Raspberry Pi display unit.
- A custom-made firmware to scan and display sensor data.

Once the receiver unit is powered using the given power supply adapter, it will automatically establish the connection to its sensor unit pair and collects the temperature/humidity data and stores them. The live streaming temperature and humidity data can be visualized on the Raspberry Pi display.

6.3.1 Algorithm on Receiver Unit

The receiver runs a Python script which uses two threads, one for scanning the specific BLE device's MAC address using hcitool and hcidump and the other for monitoring BLE data intervals. When the receiver is on, it searches for the transmitter and displays "waiting for sensor data" until it finds it. The transmitter sends 7 data packets every 1.5 s, but only the last packet of each batch is shown on the display graph; only the long-term trend is being tracked, not short-term fluctuations. The transmitter wakes up approximately every 2.5 h. The timer thread waits for the transmitter to finish sending its data packets. If no new data comes in for 2 min, it updates the display graph with the last packet from that wake-up. All the measured temperature sensor data are stored locally in a CSV file.

6.4 The Diamond Gammavoltaic Device—An Overview

Gammavoltaics are devices which generate electricity when irradiated with gamma rays, or high-energy photons from other sources. As with betavoltaics, they may be split into two types: direct conversion and indirect conversion. As with the betalight voltaic discussed in Sects. 6.1–6.3, indirect-conversion gammavoltaics use scintillating materials to convert the incident ionizing radiation to visible light, and this light is then collected by a photovoltaic cell optimized for visible light. An impressive recent demonstration of an indirect-conversion gammavoltaic was published by [11] in 2021 , wherein such a device was driven with laboratory X-rays and proved capable of generating sufficient power to power a sensor pod similar to that outlined in Sects. 6.2 and 6.3. Indirect-conversion gammavoltaics often produce more power than direct-conversion gammavoltaics. At present, direct-conversion gammavoltaics seem of most utility where the gamma ray energy is high enough, or targeted deployment duration long enough, that scintillator materials will degrade.

Unlike betavoltaics, gammavoltaics can be considered under two different paradigms: that of the battery and that of the collector. The battery paradigm is the same as that under which alpha- and betavoltaics operate; a radioisotope source is packaged alongside the device for converting the radiation into electricity. This package is then, at least to some extent, portable. The collector paradigm is more akin to that of a solar cell, reflecting the fact that unlike other alpha- and betavoltaics, gammavoltaics are photovoltaics also. Under this paradigm, the gammavoltaic is not fabricated containing any radioisotope, but rather is brought to an existing source of gamma rays. There are benefits and drawbacks to each paradigm. In particular, battery-like gammavoltaics have much more freedom in the choice of radioisotope, meaning they are freer to use an indirect-conversion design. On the other hand, fabrication using radioisotopes is hazardous, and the eventual product must usually comply with a much more stringent set of safety regulations (country-dependent).

Our work focuses on direct-conversion gammavoltaics. This is because our eventual targeting application is long-term, unpowered gamma sensing in hard-to-access locations; it may also one day be feasible to power sensor pods directly from gamma fields in nuclear waste stores using future radiation-hard electronics or MEMS devices. Gamma dose rates in such stores can reach 1,200 Sv/h near waste canisters [12],[1] with ambient dose rates about one order of magnitude lower. The modal photon energy is 662 keV, emitted indirectly by Cs-137 [13]. The high energy of these photons means that they are liable to pass through gammavoltaic devices without interacting, making energy conversion more difficult, and also that when they do interact, they are able to cause damage. We used diamond for our device, which as a material is good for radiation hardness [14–16], but among the poorer choices for radiation capture. This is because gamma photons at 662 keV interact primarily via

[1] A similar value was privately communicated to us as present at facilities run by Sellafield Ltd. in the UK.

Compton scattering [17],[2] which differs in cross section between materials depending on their electron density. As diamond is an allotrope of carbon, which has only six electrons per atom, it is a poorer absorber of gamma rays than many candidate gammavoltaic materials. We were willing to make this compromise because longevity is more important than power generating efficiency for our target applications. That said, the situation is alleviated somewhat by the fact that diamond has the highest atomic density of any known material [19]. It thus, for example, has over one and a half times the electron density of silicon, despite having less than half the number of electrons per atom.[3]

The principle of our diamond gammavoltaic (DGV) design is to use highly pure (electronic grade) single crystal diamond to capture gamma rays. Conduction of the resulting charge carriers, however, is affected around the surface rather than through the bulk, to mitigate the effect of parasitic resistance brought on by the high resistivity of pure diamond otherwise. This is achieved by making use of the surface transfer-doping effect, whereby a diamond surface terminated with hydrogen and surrounded by atmospheric water vapour in ambient conditions becomes conductive [20]. As we do not wish the surface to be fully conductive, as this may short-circuit the device, we apply a 30% hydrogen coverage, with the rest being oxygen. Contacts are placed on opposing surfaces of the diamond and have only oxygen underneath them, to aid adhesion. The metals are aluminum on the one side and nichrome on the other, chosen due to the mismatch of electrostatic barriers they induce. This mismatch allows a voltage to develop. The contacts are further coated with gold to prevent oxidation changing the electrical properties. A schematic of the cell design is shown in Fig. 6.7. Our fabrication method is described in detail in Refs. [21, 22].

Each diamond has dimensions of 4.5 mm × 4.5 mm × 0.5 mm, with the metal contacts being around 100 nm thick. As such the volume of each cell is approximately 10.1 mm^3. This is small compared to the volumes available for deployment in waste stores, though scaling up by increasing the number of cells would, at present, be costly: each diamond costs over £2k (US $2.7k at the current exchange rate).

Circuit design for direct-conversion gammavoltaics is a particular challenge because as a technology, direct-conversion gammavoltaics are nanopower devices. The purpose of the rest of this section is to give details of a test combining the energy harvesting circuitry described above with a direct-conversion DGV. Access to Cs-137 is rare and can be dangerous, with the well-known ALARP safety principle (As Low As Reasonably Practicable) suggesting that X-ray generators and lower energy

---

[2] Strictly speaking, Compton scattering is the dominant interaction mechanism for 662 keV photons incident on any element up to thorium (Z = 90). For heavier elements, photoelectric absorption remains dominant to higher energies, as it is at lower energies. This information can be found in, for example, the XCOM database provided by the US National Institute of Standards and Technology (NIST) [18].

[3] The mass densities ρ, atomic numbers Z, and molar masses M of diamond and silicon are ρ_D = 3.52 g/cm^3, Z_D = 6, M_D = 12.0 g/mol and ρ_{Si} = 2.33 g/cm^3, Z_{Si} = 14, M_{Si} = 28.1 g/mol. The electron densities, found according to $\delta = \rho N_A Z/M$ where N_A is Avogadro's constant, are thus $\delta_D = 1.1 \times 10^{24}$/cm^3 and $\delta_{Si} = 7.0 \times 10^{23}$/cm^3.

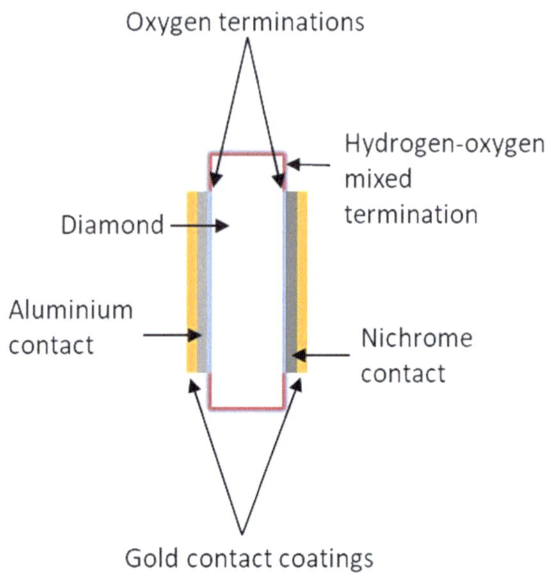

Fig. 6.7 A schematic of the design used for gammavoltaics cells in this work. Not to scale

photons should be used where possible to test gammavoltaics. Access to X-ray generators is also easier and cheaper, and allows a quicker development cycle. As such, we use laboratory X-rays here. However, it should be noted that Cs-137 is a required final testing standard before a device such as ours could be considered mature. Most significantly, the onset energy for damage to the diamond lattice from photons is 324 keV, as this is the energy required to create the Compton electrons above a threshold of 180 keV [17] known to be necessary to cause damage [23]. As such, due to our use of laboratory X-rays, we do not attempt to investigate radiation tolerance here.

6.4.1 Electrical Testing of the Diamond Gammavoltaic Devices

DGVs were made up using first two cells, and then three cells (with the latter containing the same two cells as the former). They were installed into a Zeiss Xradia Versa 520 X-ray Tomography Microscope (XRT) to conduct electrical testing under irradiation. The XRT contains a sealed, transmission-type tungsten X-ray tube ($K\alpha_1$ = 59.3188 keV and $K\alpha_2$ = 57.9819 keV [24, 25]), X-ray detectors for imaging, and positioning apparatus capable of micrometer precision in the three linear ranges of motion and of sub-degree rotation precision in the horizontal rotational direction. These high precisions do, however, belie the accuracy, as the main source of error in the position of the DGVs were how they were clamped, which was a manual process and thus subject to uncertainties on the order of millimeters. Although the imaging

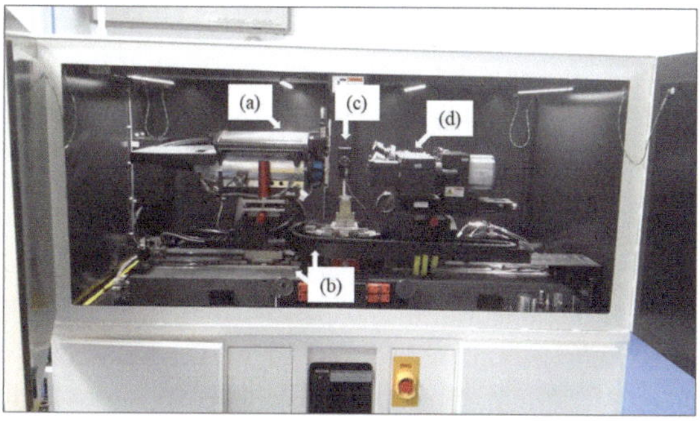

Fig. 6.8 A photograph of the X-ray tomography microscope used for electrical testing of the diamond gammavoltaic cell under irradiation. Labels highlight **a** the X-ray tube enclosure, **b** the linear and rotational movement apparatus for the X-ray source and sample stage, **c** the empty sample stage with visible light camera for rough positioning, and **d** the X-ray detectors and battery of focal lenses, used for fine positioning

apparatus was not relevant for the irradiation test itself, it did provide a convenient manner of centrally locating the DGV in the X-ray beam with good accuracy. Figure 6.8 shows a labeled photograph of the XRT.

In each case, the device was positioned as close to the X-ray source as possible using the motorized sample stage, and the tube voltage set to 160 kV, operating at 10 W via a tube current of 62.5 μA. The dose rate was not known and so had to be estimated. The precise construction of the XRT is proprietary and the dose rate curves are not provided. This is in any case a difficult thing to specify, as dose rates from X-ray tube sources can be very tube dependent and also depend on the history (i.e., uptime) of the tube. To estimate the dose rates used in these tests it was therefore necessary to make some assumptions. Firstly, it was assumed that the source position as given by the machine was the position of the transmission window of the tube. Secondly, it was assumed that the dose rates would accord to the relevant British standard [26], the relevant section of which is based on data provided by the US National Bureau of Standards. Finally, it was assumed that of the dose rate curves given in that standard, the one most closely matching the tube in the XRT would be that relating to a tube with transmission window made of 1-mm-thick beryllium in addition to the thin tungsten layer. This is the most transmissive in the standard, so the dose rate estimate is more likely to be an overestimate than an underestimate. The software RadProCalculator [27], which uses the data from the standard, was used to calculate the estimate. This gave a value of approximately 8 kSv/h, for a source—DGV distance of approximately 12 mm. However, it should be noted that this close to the source, differences in position of only 1 mm can cause the dose rate estimate to change by up to 1 kSv/h. Given this is the same order of magnitude as the uncertainty in position caused by the manual process of clamping the DGV in place, it must be assumed that the uncertainty on the dose rate estimate is large.

Electrical characterization took the form of current–voltage *(I–V)* curves, as is also common practice for solar photovoltaics. These were taken using a Keithley 6517A multimeter, communicating via an RS232-USB cable to a computer running measurement acquisition code written in Python and based on the PyVISA [28] library. The voltage was stepped between –5 and 5 V in steps of 0.05 V, with current measurements allowed to auto-range. Ten measurements were taken at each voltage step and their mean average used as the value, with the standard deviation used as the uncertainty. In practise, the uncertainties were too small to be visible on plots and so are omitted. The sign convention used for the *I–V* curves was for positive voltages to bias the high-barrier aluminum contact over the low-barrier nichrome contact, in a manner analogous to diode *I–V* curves. The opposite convention is sometimes seen in photovoltaic literature.

Power–voltage (*P–V*) curves were calculated from the *I–V* curves. Because of the sign convention used in the latter, $P = IV$ is positive when power is being dissipated and negative when power is being generated. As such, the *P–V* curves, showing only where power in generated, were calculated according to $P = -IV$.

6.4.1.1 The Two-Celled Device

It is common to provide dark *I–V* curves for voltaics. In the case of our DGVs, they behave very differently when not irradiated, acting more or less like resistors. This brings into question the utility of showing the behavior. Nevertheless, Fig. 6.9 shows a dark *I–V* curve of the two-celled DGV for reference. Although the points pass through a power-generating quadrant of the graph rather than through the origin, there was no source of power incident on the device and so this is most likely due to systematic error in the multimeter, given the very small currents.

When the X-ray generator was turned on, an *I–V* curve showed the $I_{SC} = -0.7\ \mu A$ and the $V_{OC} = 1.36$ V. The resulting *P–V* curve showed that at the maximum power-point, the power $P_{MPP} = 0.31\ \mu W$, from a current $I_{MPP} = -0.42\ \mu A$, and a voltage $V_{MPP} = 0.74$ V. Taken together these values give a fill factor, a common metric of photovoltaic quality, of $FF = P_{MPP}/(-I_{SC} V_{OC}) = 0.31$, where the minus sign again accounts for the sign convention. This is fairly poor compared to mature voltaic technologies: the range for the parameter is $0.25 \leq \text{FF} < 1$. The curves are shown in Fig. 6.10.

6.4.1.2 The Three-Celled Device

For the three-celled device, an *I–V* curve showed the $I_{SC} = -0.51\ \mu A$ and the $V_{OC} = 2.07$ V. The resulting *P–V* curve showed that at the maximum power-point, the power $P_{MPP} = 0.30 \mu W$, from a current $I_{MPP} = -0.30 \mu A$, and a voltage $V_{MPP} = 1.01$ V. Taken together, these values gave a fill factor FF = 0.28, near the theoretical minimum, showing that although the device successfully produced power, there is likely a lot

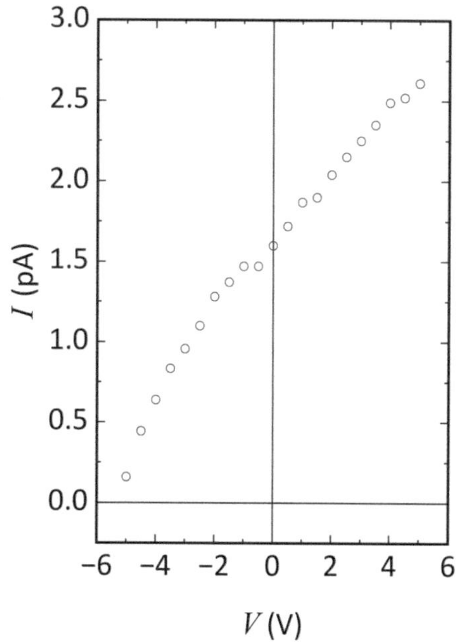

Fig. 6.9 Electrical test data from the two-celled diamond gammavoltaic device with no irradiation. The fact the curve does not pass through the origin is likely due to systematic error in the multimeter

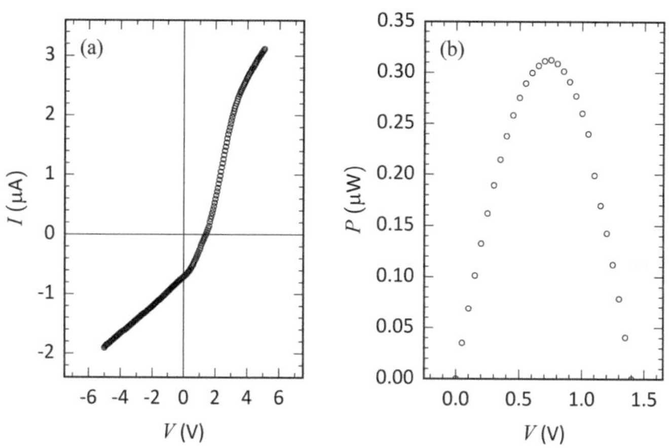

Fig. 6.10 Electrical test data from the two-celled diamond gammavoltaic under laboratory X-ray irradiation. Panes show **a** *I*–*V* data and **b** *P*–*V* data, where $P = -IV$ is the power generated and the voltage range has been constrained to the region where this quantity is positive

of optimization that could be done to improve performance. The curves are shown in Fig. 6.11.

While including the third cell in series did increase the output voltage as expected, it also reduced the output current. We believe there are three factors which combine to cause this effect: firstly, simple variation in the fabrication process used to create

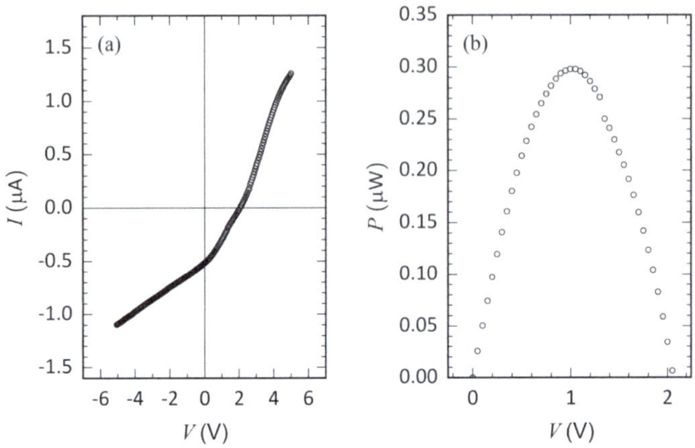

Fig. 6.11 Electrical test data from the three-celled diamond gammavoltaic under laboratory X-ray irradiation. Panes show **a** I–V data and **b** P–V data, where $P = -IV$ is the power generated and the voltage range has been constrained to the region where this quantity is positive

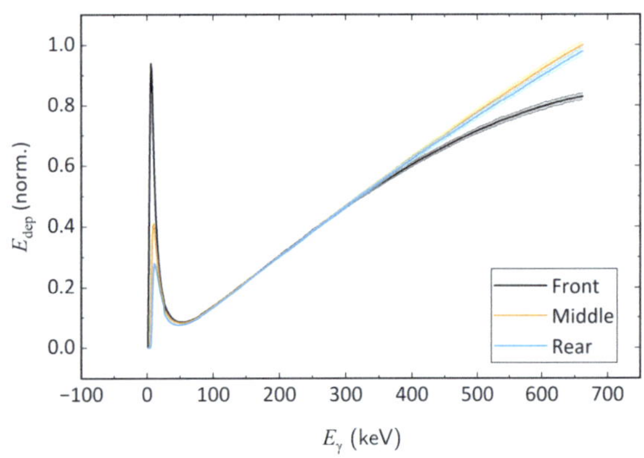

Fig. 6.12 A GEANT4 simulation of energy deposition into each of the cells of the three-celled diamond gammavoltaic for different photon energies between 1 and 700 keV. Uncertainties are given as bands. The front cell is the cell closest to the photon source

the cells means that series connections of them will, all other things being equal, be constrained by the current output of the worst cell. This is the cell which was added to create the three-cell device. The second factor is that unlike solar photovoltaics, the DGV cells have high series resistance, which increases as cells are stacked in series. Finally, the third factor is that energy deposition into the cells will vary due to attenuation for different photon energies. Figure 6.12 shows a GEANT4 [29] simulation, adapted from code provided with Ref. [30], of normalized energy deposition from monochromatic photons of varying energies, from 1 to 700 keV, thus including both the range of energies from the XRT and the emission energy of Cs-137. Ten million primarily photons were simulated at each energy to reduce uncertainty.

This simulation quantifies the knowledge that lower energy photons from the XRT energy spectrum will be absorbed primarily in the front cell, closest to the source. The energy deposition becomes mostly equal across the cells above about 25 keV, but the disparity in the lower energies will mean that photon for photon, the rear cell will be producing the least current and thus will be constraining the current output of the device. This has reversed by the time the photon energy reaches the Cs-137 emission energy at 662 keV.

Overall, the choice between the two-cell and three-cell device was based on the voltage and current requirements of the circuit to be powered. A higher current would charge the capacitor in the EH300 energy harvesting module quicker, but would bring it to a lower maximum voltage, potentially causing it to fail to activate the rest of the circuit.

6.4.2 *Energy Harvesting from Diamond Gammavoltaics*

As previously mentioned, the tests we undertook were to assess the suitability of the circuit design to the gammavoltaic, and the gammavoltaic to the circuit design, equally. Direct-conversion gammavoltaics are producing increasing power, and low-power energy harvesting circuitry is advancing to take advantage of less power, in such a manner that it was worth investigating whether the two could meet in the middle. Indeed at the time of investigation in January and February 2020, the results of [11] were not yet published, so we believed that no-one had powered a circuit with any type of gammavoltaic, direct or indirect.

A variant of the transmitter circuit shown in Fig. 6.5 was used to account for the lower voltage output of the DGV compared to the solar photovoltaics in the betalight voltaic apparatus. The three-celled DGV was used to maximize the available output voltage. An EH4295 voltage-booster module was used to increase this further, along with manual switching to the Bluetooth transmitter. This use of manual switching marks the gammavoltaic and its circuit as less developed than the betalight voltaic setup, but very low power switches do exist that could be envisioned in use. The result was as shown schematically in Fig. 6.13.

The DGV was loaded into the XRT using the same parameters and positioning as in Sect. 6.4.1. The energy harvesting circuit was located outside the XRT enclosure,

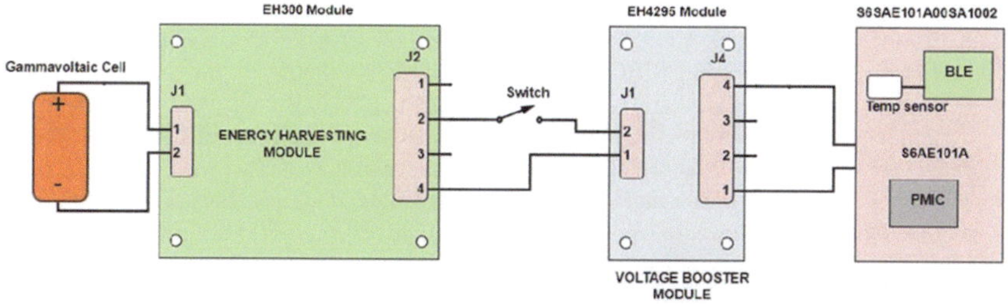

Fig. 6.13 The circuit used to harvest energy from the three-celled diamond gammavoltaic device to send a temperature reading. The switching was performed manually

both so that the switch could be thrown and so that the XRT enclosure did not act like a Faraday cage. The circuit was allowed to charge from the irradiated DGV for 22.5 h. It was found that the sensor circuit was loading the DGV in such a way that the voltage output was 1.25 V, placing it reasonably close to the maximum power-point voltage of 1.01 V despite the lack of power-point tracking circuitry. The EH300 capacitor bank had reached 0.92 V. Upon throwing the switch, a temperature reading was successfully made and transferred to the receiver unit via Bluetooth. It is not yet clear how little time could be used for charging and still sending a reading.

In summary, the work with the energy harvesting circuit and DGV was a partial success. It provided a demonstration of the first instance of a direct-conversion gammavoltaic powering a circuit that we are aware of, and the second published instance for gammavoltaics powering circuits in general. Being able to send a reading after less than 1 day is respectable; we are aware of one waste facility where similar measurements, made using manned equipment, are made roughly annually due to the cost and safety considerations of doing so. Increasing the resolution of ambient sensing information in such locations would increase the safety of operating the stores. However, the DGV had a poor fill factor and could not be tested for longevity due to the energies of the X-rays involved being below the damage threshold for diamond, where the gamma rays from Cs-137 are above it. Furthermore, the switching on the circuit needed to be done manually—the power generated by the DGV and the power required by the circuit did not quite match up when automatic switching was included. We hope that further developments of the DGV will solve this issue, but, at present, more work is needed. The dose rate used was about 6.7 times higher than would be found in the most active areas of waste stores we are aware of, although increasing the number of cells in a DGV would solve this. Doing so would similarly solve the issue of power requirement for automatic switching, but it must be determined whether this would be worth the cost.

References

1. M.L. Terranova, Nuclear batteries: current context and near-term expectations. Int. J. Energy Res. **46**(14), 19368–19393 (2022). https://doi.org/10.1002/er.8539
2. CityLabs. https://citylabs.net/
3. K. Bower, Y.A. Barbanel, Y.G. Shreter, G.W. Bohnert, *Polymers, Phosphors and Voltaics for Radioisotope Microbatteries* (CRC Press, 2002). 978-0-367-39605-3
4. R. Walton, C. Anthony, M. Ward, N. Metje, D.N. Chapman, radioisotopic battery and capacitor system for powering wireless sensor networks. Sens. Actuators A: Phys. **203**, 405–412 (2013). ISSN 0924-4247. https://doi.org/10.1016/j.sna.2013.09.010
5. T. Jiang, Z. Xu, X. Tang, Z. Yuan, H. Wang, M. Bian, Comparison and study of the preparation methods for phosphor layer in nuclear battery. Int. J. Energy Res. **45**(8), 11712–11720 (2021). https://doi.org/10.1002/er.5526
6. Y. Haim, Y. Marciano, G. DeBotton, Tunable direct beta-radiation harvester at the nanowatt scale. Sens. Actuators A: Phys. **283**, 228–234 (2018)
7. Y. Haim, G. Debotton, Experimental study of the direct-charge beta radiation method for energy harvesting. IEEE Trans. Electron Dev. **67**(11), 5076–5081 (2020) (IEEE)

8. W. Chen, X. Tang, Y. Liu, X. Zhiheng, Y. Zicheng, Z. Zhang, K. Liu, Radioluminescent nuclear battery containing cspbbr3 quantum dots: application of a novel wave-shifting agent. Int. J. Energy Res. **43**(9), 4520–4533 (2019)
9. C. Zhao, J. Ren, L. lei, F. Liao, K. Liu, Y. Zhao, X-ray radioluminescent battery with near milliwatt output power using CsI: Tl single crystal scintillator. Appl. Phys. Lett. **121**, 123906 (2022)
10. Z. Xu, Y. Liu, Z. Zhang, W. Chen, Z. Yuan, K. Liu, X. Tang, Enhanced radioluminescent nuclear battery by optimizing structural design of the phosphor layer. Int. J. Energy Res. **42**(4), 1729–1737 (2018)
11. C. Zhao, J. Ren, L. Lei, F. Liao, X. Shi, D. Zhou, K. Liu, Y. Zhao, Tenfold efficiency improvement of x-ray radioluminescent batteries basing on GAGG: Ce single crystal scintillators. Appl. Phys. Lett. **119**(22) (2021)
12. J.R. White, D. Kinsman, T.M. Regan, L.M. Bobek, Novel nuclear powered photocatalytic energy conversion. Technical report, UMass-Lowell Radiation Laboratory (University of Massachusetts Lowell, 2005)
13. M.M. Be, V. Chiste, C. Dulieu, E. Browne, C. Baglin, V. Chechev, N. Kuzmenco, R. Helmer, F. Kondev, D. MacMahon, K.B. Lee, Cs-137, in *A = 3 to 244*. volume 3 of Monographie BIPM-5 - Table of Radionuclides. (Bureau International des Poids et Mesures, Sevres, 2006), pp. 91–98
14. T. Behnke, M. Doucet, N. Ghodbane, A. Imhof, C. Martinez, W. Zeuner, Electromagnetic radiation hardness of diamond detectors. Nucl. Instrum. Methods Phys. Res. Sect. A: Accel. Spectrom. Detect. Assoc. Equip. **489**(1–3), 230–240 (2002)
15. A diamond based neutron spectrometer for diagnostics of deuterium-tritium fusion plasmas. Rev. Sci. Instrum. **85**(11) (2014)
16. N.R. Erasmus, Simulation of silicon and diamond detector systems by GEANT4 simulation techniques. Ph.D. thesis (University of Western Cape, 2014)
17. A.H. Compton, A quantum theory of the scattering of X-rays by light elements. Phys. Rev. **21**(5), 483–502 (1923)
18. M.J. Berger, J.H. Hubbell, S.M. Seltzer, J. Chang, J.S. Coursey, R. Sukumar, D.S. Zucker, K. Olsen, NIST XCOM: photon cross sections database
19. J.C. Angus, Structure and thermochemistry of diamond, in *The Physics of Diamond*, vol. 135, ed. by A. Paoletti, A. Tucciarone (IOS Press, Oxford, 1997), pp. 9–30
20. K.G. Crawford, I. Maini, D.A. Macdonald, D.A.J. Moran, Surface transfer doping of diamond: a review. Progr. Surf. Sci. **96**(1), 100613 (2021)
21. G.R. Mackenzie, S. Kaluvan, P.G. Martin, C. Hutson, T. Connolley, M. Cattelan, H. Dominguez-Andrade, T.L. Martin, N.A. Fox, T.B. Scott, A diamond gammavoltaic cell utilizing surface conductivity and its response to different photon interaction mechanisms. Mater. Today Energy **21**, 100688 (2021)
22. G.R. Mackenzie, *A Diamond Gammavoltaic Cell* (University of Bristol, Ph.D. 23)
23. J. Koike, D.M. Parkin, T.E. Mitchell, Displacement threshold energy for type IIa diamond. **60**(12), 1450–1452 (1998)
24. R.D. Deslattes, E.G. Kessler, Experimental evaluation of inner-vacancy level energies for comparison with theory (Springer US, Boston, MA, 1985), pp. 181–235
25. R.D. Deslattes, E.G. Kessler Jr., P. Indelicato, L. de Billy, E. Lindroth, J. Anton, J.S. Coursey, D.J. Schwab, J. Chang, R. Sukumar, K. Olsen, R.A. Dragoset, X-ray transition energies database (2005)
26. BS 4094-21971 Recommendation for data on shielding from ionizing radiation. Shielding from X-radiation. Technical report (British Standards Institute, 1971)
27. R. McGinnis, RadProCalculator (2009)
28. PyVISA: control your instruments with Python
29. S. Agostinelli, J. Allison, K. Amako, J. Apostolakis, H. Araujo, P. Arce, MAsai, D. Axen, S. Banerjee, G. Barrand, F. Behner, L. Bellagamba, J. Boudreau, L. Broglia, A. Brunengo, H. Burkhardt, S. Chauvie, J. Chuma, R. Chytracek, G. Cooperman, G. Cosmo, P. Degtyarenko, A. Dell'acqua, G. Depaola, D. Dietrich, R. Enami, A. Feliciello, C. Ferguson, H. Fesefeldt, G. Folger, F. Foppiano, A. Forti, S. Garelli, S. Giani, R. Giannitrapani, D. Gibin, J.J. Ç´omez

Cadenas, I. Gonźalez, G. Gracia Abril, G. Greeniaus, W. Greiner, V. Grichine, A. Grossheim, S. Guatelli, P. Gumplinger, R. Hamatsu, K. Hashimoto, H. Hasui, A. Heikkinen, A. Howard, V. Ivanchenko, A. Johnson, F.W. Jones, J. Kallenbach Aa, N. Kanaya, M. Kawabata, Y. Kawabata, M. Kawaguti, S. Kelner, P. Kent, A. Kimura, T. Kodama, R. Kokoulin, M. Kossov, H. Kurashige, E. Lamanna, T. Lampen, V. Lara, V. Lefebure, F. Lei, M. Liendl, W. Lockman, F. Longo, S. Magni, M. Maire, E. Medernach, K. Minamimoto, P. Mora De Freitas, Y. Morita, K. Murakami, M. Nagamatu, R. Nartallo, P. Nieminen, T. Nishimura, K. Ohtsubo, M. Okamura, S. O'neale, Y. Oohata, K. Paech, J. Perl, A. Pfeiffer, M.G. Pia, F. Ranjard, A. Rybin, S. Sadilov, E. Di Salvo, G. Santin, T. Sasaki, N. Savvas, Y. Sawada, S. Scherer, S. Sei, V. Sirotenko, D. Smith, N. Starkov, H. Stoecker, J. Sulkimo, M. Takahata, S. Tanaka, E. Tcherniaev, E. Safai Tehrani, M. Tropeano, P. Truscott, H. Uno, L. Urban, P. Urban, M. Verderi, A. Walkden, W. Wander, H. Weber, J. P. Wellisch, T. Wenaus, D.C. Williams, D. Wright, T. Yamada, H. Yoshida, D. Zschiesche, Geant4-a simulation toolkit. Nucl. Instrum. Methods Phys. Res. A **506**, 250–303 (2003)

30. J.A. Davis, K. Ganesan, A.D.C. Alves, S. Guatelli, M. Petasecca, A.A. Jeremy Davis, J. Livingstone, M.L.F. Lerch, D.A. Prokopovich, M.I. Reinhard, R.M. Siegele, S. Prawer, D. Jamieson, Z. Kuncic, V.L Pisacane, J.F. Dicello, J. Ziegler, M. Zaider, A.B. Rosenfeld, Characterization of a novel diamond-based microdosimeter prototype for radioprotection applications in space environments. IEEE Trans. Nucl. Sci. **59**(6), 3110–3116 (2012)

Chapter 7
Measurement Techniques Using Inorganic Scintillation Detectors

G. Anil Kumar, V. Ranga, and S. Panwar

7.1 Introduction

Since the invention of spinthariscope around the year 1900, the technological advancements in detection technology have made it possible to use scintillation detectors in various fields such as studying the nuclear structure, nuclear power plants, particle physics, environmental radioactivity, astronomy, earth sciences, mining, medical imaging, security, agriculture, food products, etc. The detailed chronological events pertinent to the dawn of scintillation counting can be found in [1].

Scintillation detectors work on the principle of production of visible light due to the interaction of radiation within the active volume of the detector followed by the detection of the light by a photosensor (e.g., photomultiplier tube, silicon photomultiplier, photodiode, avalanche photodiode). Inorganic scintillation detectors, in particular, are well known to detect gamma rays with better efficiency and reasonably good resolution compared to High Purity Germanium (HPGe) semiconductor detectors. To name a few, NaI:Tl, $PbWO_4$, $LaCl_3$:Ce, $CaWO_4$, $CdWO_4$, Lu_2SiO_5, $LaBr_3$:Ce, CsI:Tl, $LuPO_4$:Ce, $CeBr_3$, YAP, SrI_2:Eu, BaF_2, BGO, Garnet-based crystals have proved to be beneficial in the applications as mentioned above. Some of the essential properties of inorganic scintillation detectors include energy resolution, decay time, transparency to the visible light, ability to discriminate different types of nuclear radiation based on their pulse shape or pulse height, energy linearity, radiation hardness, efficiency (both total detection and full energy absorption), homogeneity, ease of growing them, temperature stability, internal radioactivity, mechanical strength, hygroscopic nature, and, of course, the final cost. Most of the applications of a particular inorganic scintillation detector depend on how efficiently it can (i) resolve

G. Anil Kumar (✉) · V. Ranga · S. Panwar
Radiation Detectors and Spectroscopy Laboratory, Department of Physics, Indian Institute of Technology Roorkee, Roorkee, Uttrakhand, India
e-mail: anil.gourishetty@ph.iitr.ac.in

U. Mangalanathan et al. (eds.), *Nuclear Instrumentation and Control*,
https://doi.org/10.1007/978-981-97-1283-0_7

nuclear radiation energies closely separated from each other, (ii) detect and discriminate different types of nuclear radiation, (iii) work in wider temperature regions, and (iv) resolve two particles whose interaction times are close to each other (fast action). Considering the requirement of continuous monitoring of nuclear radiation inside and near nuclear power plants (NPPs), it is quite natural to prefer cheaper and reliable inorganic scintillator-based spectrometers over the HPGe detectors [2]. The present chapter discusses a few measurement techniques to study the energy resolution and detection efficiency of inorganic scintillation detectors for gamma rays, keeping in mind their applications in nuclear power plants (NPPs). In addition, a measurement method to study the *nuclear reaction cross sections* relevant for the boron-based next-generation fusion reactor is presented.

7.2 Energy Resolution

In the NPPs, the radioactive isotopes emit gamma rays in the range of 0.05 MeV (corresponding to the detection of Americium-241 in spent fuel) to 7.5 MeV (corresponding to the detection of ^{16}N during leakage in the core circuit). Table 7.1 summarizes the radioactive sources and the energies of gamma rays they emit in NPPs. Monitoring gamma rays in this energy range demands the usage of gamma detectors with an energy resolution of 3%. Therefore, the study of factors affecting energy resolution and measurement techniques to improve energy resolution are key factors in designing and installing radiation monitoring devices.

The overall energy resolution R_{total} of a scintillation detector can be written as a quadrature sum of different parameters associated with the scintillation material, photosensor, and coupling between both of them, i.e.:

$$R_{\text{total}}^2 = R_{\text{intrinsic}}^2 + R_{\text{transfer}}^2 + R_{\text{statistical}}^2 + R_{\text{noise}}^2$$

where $R_{\text{intrinsic}}$ denotes the resolution associated with the intrinsic properties of the scintillation material; R_{transfer} is the resolution associated with the transfer of scintillation photons emitted by the scintillation material to the photocathode; $R_{\text{statistical}}$ is the resolution associated with the three processes, namely, the production of photoelectrons by the photocathode, collection of photoelectrons at the first dynode of the PMT and multiplication of photoelectrons; and R_{noise} is the resolution associated with the dark noise arising due to the detector current and due to the electronic noise. Extensive theoretical and experimental results suggest that the contribution

Table 7.1 Radioisotopes (gamma emitters) whose contents are to be determined in NPPs

Source	^{144}Ce	^{141}Ce	^{51}Cr	^{103}Ru	^{134}Cs	^{196}Ru	^{137}Cs	^{54}Mn	^{58}Co	^{59}Fe	^{95}Zr	^{65}Zn	^{60}Co
Gamma energies (keV)	134	145	320	497	605, 796	622	662	810, 835	811	1099, 1291	1100	1150	1173, 1332

of $R_{\text{intrinsic}}$ to the R_{total} is a major one [3]. One of the measurement techniques to measure the $R_{\text{intrinsic}}$ is presented here.

7.2.1 Compton Coincidence Technique

Compton scattering is a phenomenon in which a photon (X- or γ-ray) gets scattered by a charged particle, usually an electron, resulting in the partial transfer of energy and momentum from the photon to the electron. The partial energy transfer to electrons gives rise to the Compton continuum in the measured γ-ray energy spectrum. Compton scattering is one of the three primary interactions through which a γ-ray deposits energy in a detector. The other two interactions are the photoelectric effect and the pair production [4]. Compton scattering finds practical applications in many areas, such as the determination of the depth of a radioactive source in a material [5, 6], tomography [7], random coincidence rejection [8], Compton continuum suppression [9, 10], measurement of the energy resolution of low-Z scintillators [11], energy calibration of plastic scintillators [12], and measurement of electron response of low-Z scintillators [13]. A review of theoretical approaches to investigate Compton scattering is given in [14–16]. As stated earlier, the most widely used detectors for γ-rays are scintillation detectors, and hence the study of the response of scintillators to γ-rays is not possible without a thorough knowledge of the Compton scattering phenomenon. A tremendous amount of research has been done to understand the mechanism of the scintillation process in inorganic scintillators. However, the scintillation mechanism is not fully understood.

Understanding the intrinsic resolution of inorganic scintillation crystals [17] is still in progress. The Compton Coincidence Technique (CCT), proposed in [18], can be used to study the linearity of electron response and the intrinsic energy resolution of Compton electrons in scintillation crystals [18, 19]. The CCT requires a test detector and a high-energy resolution reference detector (usually HPGe) arranged so that the γ-ray originating from the radiation source gets Compton scattered from the test detector to the reference detector, as shown in Fig. 7.1 [20]. Interaction of γ-rays within the scintillator of the test detector generates optical photons. The photocathode of a PMT collects these photons and generates an electrical pulse. The acquisition system records the pulses from the test and reference detectors if they arrive within a pre-specified time interval, called the coincidence time window. A typical coincidence spectrum of the test detector is shown in Fig. 7.2. The Compton scattered γ-rays detected by the reference detector are scattered over a range of angles because of the finite surface area of the reference detector [17]. Events in the reference detector that deposited energy in a particular energy window are filtered out during offline analysis, and the corresponding correlated events in the test detector are identified. Consequently, a spectrum containing the response of the test detector for the particular energy of electrons is obtained, as shown in Fig. 7.3. The energy resolution of the reference detector constraints the preciseness of the energy window. High Purity Germanium (HPGe) detector is used for precise energy gating

Fig. 7.1 Experimental setup of CCT

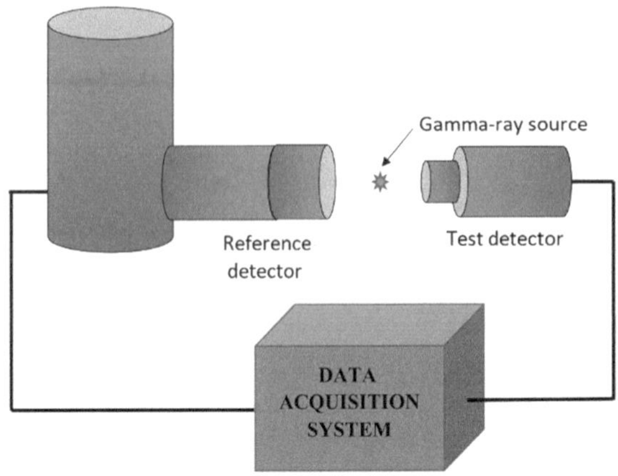

on events. The interaction event will not be recorded if the γ-ray photon does not interact with both the test and the HPGe detector. As a result, the events occurring via the photoelectric effect in the test detector will not be recorded unless the accidental coincidence events result in a photopeak, as shown in Fig. 7.2. Due to this, the CCT enables the rejection of the response of electrons generated by the photoelectric effect and measures the Compton electron's response [18]. Therefore, the CCT can be employed to study the scintillators where the photoelectric effect is not prominent. For example, a liquid scintillation detector having a very low effective atomic number cannot give a photopeak in the measured gamma spectrum. The probability of the photoelectric effect is negligible for such detectors, and the photopeak does not occur in the spectrum. Only a limited number of mono-energetic γ-ray sources are usually available in the laboratory, limiting the measurement of energy resolution at only those γ-ray energies. The CCT can be modified to record the spectrum at various angles rather than only at 180°. This variation in angle gives rise to variation in energy of recoil electron according to Compton scattering formula. Hence, the energy resolution measurement for arbitrary electron energy can be done. The whole energy region can be scanned without extrapolation to low energies. More details of measurement of intrinsic resolution employing CCT with the help of analog and digital techniques can be found in [20, 21].

The electron response is considered to be more fundamental in studying scintillation processes than the photon response due to the influence of multiple Compton scattering in the full absorption peak of the energy spectrum. Electron beams can also be used to study the electron response at various energies. However, these experiments are prone to surface effects [22, 23]. Normally, two techniques are used to determine the electron response of a scintillator. First, the analytical deconvolution of the electron response from the measured γ-ray response. Second, using external electron sources (e.g., accelerators). These experiments are affected by surface effects, particularly at energies below 20 keV. The CCT is a powerful technique that

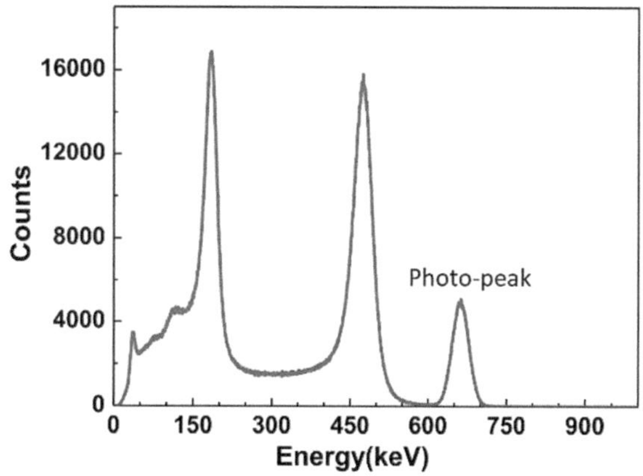

Fig. 7.2 Coincidence spectrum from test detector

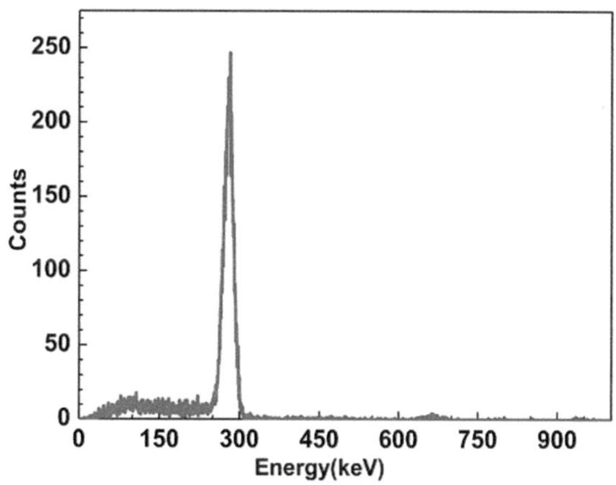

Fig. 7.3 Gated spectrum from test detector

directly gives us the electron response for a scintillator without the need for complex deconvolution techniques and electron accelerators.

7.3 Detection Efficiency

For the practical application of any scintillation detector for gamma rays in NPPs, detection efficiency is of paramount importance. The plot of detection efficiency versus gamma energy, i.e., efficiency calibration curve of a detector plays a crucial role in the determination of contents of radionuclides listed in Table 7.1. The detection efficiency can be understood in terms of both absolute total detection efficiency (ratio of the total number of gamma rays that have deposited non-zero energy within the detector active volume and the total number of gamma rays emitted by the source in all directions) and absolute full energy absorption efficiency (ratio of the total number of gamma rays that have deposited their full energy within the detector active volume and the total number of gamma rays emitted by the source in all directions). In the literature, it is found that the full energy absorption efficiency is also referred to as the photopeak efficiency which may denote the process involved in the full deposition of gamma energy via the photoelectric effect alone. However, multiple Compton scattering also contributes to the full deposition of gamma energy in a considerable amount. This should not be neglected unless the size of the detector's active volume is very less. The method to measure the full energy absorption efficiency of a detector for mono-energetic rays is straightforward because of the fact that the number of full energy absorption peaks in the mono-energetic gamma-induced spectrum is only one irrespective of the type of detector material. Unfortunately, the number of mono-energetic gamma sources used in the laboratories is quite less which makes the generation of the efficiency calibration curve for any detector quite difficult. This demands the extraction of absolute detection efficiencies (both total detection and full energy absorption) using available gamma sources that emit more than one gamma ray. For example, ^{60}Co is a widely used gamma source, which emits 1173 and 1332 keV gamma rays in cascade, for the energy calibration of a scintillation detector. The full energy deposition of these two gamma rays within the active volume of a scintillation material gives rise to two peaks in the measured spectrum. In addition, a third peak (called sum peak) also arises due to the true coincidence summing of these two gamma rays. This coincidence effect becomes prominent if the solid angle submitted by the detector at the source is very large or the energies of individual gamma rays are relatively less. As the sum peak occurs at the cost of loss of individual counts in the peaks corresponding to 1173 and 1332 keV, the determination of absolute detection efficiencies (both total detection and full energy absorption) of the detector is not possible for these two energies. To circumvent this problem, T. Vidmar has proposed a simple and powerful method to correct the apparent efficiencies for true coincidence summing in order to calculate the absolute efficiencies (both total detection and full energy absorption) in [24]. The advantage of the method lies in the fact that the correction of apparent efficiencies for true coincidence summing and subsequent estimation of true efficiencies (both total detection and full energy absorption) can be done simply by using the counts under the two individual peaks and the sum peak. The method was successfully tested with an HPGe detector considering several double gamma emitters [24]. The method was also checked, using experimental measurements and GEANT4 simulations, to extract the absolute efficiencies of $LaBr_3$:Ce detectors of different sizes and of large arrays of NaI(Tl) in soccer-ball and castle geometries for double gamma emitters and positron emitter [25–27]. This method can generate an efficiency calibration curve of any gamma detector up to around 3 MeV provided the detector resolution is less than or equal to 6%. Naturally, there is a demand to extend the work to generate the efficiency calibration curve beyond 3 MeV gamma energy.

7.4 Cross-sectional Measurement in the Fusion-Based Reactor

For the past few decades, intensive work has been going on to generate electricity using fusion-based nuclear reactors. Some of the ideal characteristics of fuel material in such types of reactors are

- Reaction with fuel should not emit neutrons or radioactive isotopes as products.
- Energy output in the form of the kinetic energy of charged particles for efficient direct conversion.
- Low cost and high abundance of fuel.

Boron-11 is considered to be a major candidate as a fuel in fusion-based reactors because of its role in the reaction $^{11}B(p, \alpha)\alpha\alpha$ which is an aneutronic source of energy production and charged particles as products of the reaction. However, natural boron contains 19.78% of ^{10}B and the rest is ^{11}B. If natural boron is used as a fuel, then the two radioactive products, ^{11}C and ^{7}Li, will be produced due to the reaction of protons with ^{10}B. Though enriched ^{11}B is the simple solution for this problem, for a comprehensive assessment of this case, the detailed nuclear reaction cross sections are needed for $^{10}B(p, \alpha)^{7}Be(\beta^{+})^{7}Li$ and $^{10}B(p, \gamma)^{11}C(\beta^{+})^{11}B$ reactions. Both ^{7}Be and ^{11}C are positron emitters with half-lives of 53.3 days and 20.4 min, respectively. How serious can be the radioactive contamination in using natural boron as a fuel in reactor devices is described by the cross sections, and in turn by the reaction rates, of these two reactions. The measurement of total capture cross section demands detection of all the γ-rays using a full array of detectors and detailed analysis including branching ratios, intensities, etc. We present here an elegant technique to measure the total capture cross section of $^{10}B(p, \gamma)^{11}C$ reaction employing two large volume scintillation detectors. More details can be found in [28] in which the measurements were reported using the proton beams of 2–6 MeV.

As stated earlier, the ground state of ^{11}C decays to ^{11}B by the emission of a β^{+} particle with a half-life of 20.4 min. The positron so produced will interact with the nearest electron and undergoes annihilation to produce two 511 keV γ-rays. The measurement of the total number of the two coincident 511 keV γ-rays provides the total capture cross section without any requirement of detecting the γ-rays decaying from excited states of ^{11}C. Two large volume cylindrical $LaBr_3$:Ce scintillation detectors of dimensions 3.5″ ×6 ″ were placed at an angle of 90° with respect to the beam direction to measure 511 keV γ-rays emitted in opposite direction. From the events with full energy deposition in both detectors, a one-dimensional TAC spectrum with full energy (photopeak) gates on both the detectors serves the purpose to record the decay of ^{11}C nuclei. The counts under the peak corrected for geometric and photopeak efficiencies will correspond to the total number of positrons emitted or one can say the total number of ^{11}C nuclei formed. The time span of irradiation of the target was defined by the typical half-life of the reaction product. After the irradiation of about three or four half-lives, the number of isotopes produced gets saturated. After each irradiation is stopped, the data acquisition can be started after

2 min to avoid the background contribution from other positron emitters with very short half-lives, if any. For example, a source of background could be a short-lived ^{17}F isotope with a half-life of 64.5 s which is less than 2 min time gap. This technique has the additional advantage of eliminating the beam-induced background. During the time of irradiation, the formation and decay of ^{11}C occur simultaneously. While after irradiation is stopped, only decay will occur. Therefore, a special differential equation, as given below, governs the production of ^{11}C:

$$\frac{dN(t)}{dt} = \sigma \frac{N_B}{A} \frac{dQ(t)}{dt} \odot (t_i - t) - \lambda N(t)$$

where N denotes the number of ^{11}C nuclei formed at any time t, σ denotes the production cross section of ^{11}C nuclei, $\frac{N_B}{A}$ denotes the areal density of the target, $\odot(t_i - t)$ denotes the Heaviside function, t_i denotes the time of irradiation, and $\frac{dQ(t)}{dt}$ denotes the incident proton current with $Q(t)$ the integrated charge and λ the decay constant for the positron emitter ^{11}C. By integrating the above equation, the total capture cross section can be determined.

References

1. Z.I. Kolar, W. Den Hollander, Appl. Radiat. Isot. **61**, 261–266 (2004)
2. M.P. Belousov et al., Instrum. Exp. Tech. **60**, 1–19 (2017)
3. M. Moszyński et al., Nucl. Instrum. Methods A **805**, 35–35 (2016)
4. G.F. Knoll, *Radiation Detection and Measurement*, 3rd ed. (Wiley, New York, 2000)
5. R.B. Oberer et al., Nucl. Instrum. Methods A **722**, 65–70 (2013)
6. H. Liu et al., Nucl. Instrum. Methods A **949**, 162917 (2020)
7. S.J. Norton, J. Appl. Phys. **76** (1994)
8. M. Kolstein, M. Chmeissani, J. Instrum. **11**, C12017 (2016)
9. C. Chung et al., Nucl. Instrum. Methods A **243**, 102 (1986)
10. I. Bikit et al., Phys. Procedia **31**, 84 (2012)
11. K. Roemer et al., in *2009 IEEE Nuclear Science Symposium Conference Record (NSS/MIC)* (2009), pp. 6–11
12. N. Kudomi, Nucl. Instrum. Methods A **430**, 96 (1999)
13. L. Swiderski et al., J. Instrum. **7**, P06011 (2012)
14. C.-K. Qiao et al., Crystals **11**, 525 (2021)
15. R.H. Pratt et al., Radiat. Phys. Chem. **79**, 124 (2010)
16. P.M. Bergstrom, R.H. Pratt, Radiat. Phys. Chem. **50**, 3 (1997)
17. M. Moszynski et al., Nucl. Instrum. Methods A **805**, 25 (2016)
18. J.D. Valentine, B.D. Rooney, Nucl. Instrum. Methods A **353**, 37 (1994)
19. B.D. Rooney, J.D. Valentine, in *IEEE Nuclear Science Symposium and Medical Imaging Conference Record*, vol. 1 (1995), pp. 404–408
20. V. Ranga et al., IEEE Trans. Nucl. Sci. **65**, 616 (2018)
21. L. Swiderski et al., IEEE Trans. Nucl. Sci. **57**, 1697 (2010)
22. C.D. Zerby et al., Nucl. Instrum. Methods **12**, 115 (1961)
23. W.C. Kaiser et al., IRE Trans. Nucl. Sci. **9**, 22 (1962)
24. T. Vidmar et al., Nucl. Instrum. Methods A **508**, 404 (2003)
25. G. Anil Kumar et al., Nucl. Instrum. Methods A **609**, 183–186 (2009)
26. G. Anil Kumar et al., Nucl. Instrum. Methods A **611**, 76–83 (2009)
27. G. Anil Kumar et al., Appl. Rad. Isot. **118**, 32–37 (2016)
28. Kafkarkou et al., Phys. Rev. C **89**, 014601 (2014)

Chapter 8
Point Kinetics Model of Nuclear Reactors

A. P. Tiwari

Mathematical modeling is an important step in study of physical systems. Mathematical model is a set of equations characterizing the transient and steady-state behavior of the system. This chapter begins with a brief review of current nuclear energy scenario. Then, the well-known "Point Kinetics Model" will be introduced. Subsequently, the equations will be cast in "State-space form," which is a standard step in representation of physical dynamical systems by control experts. The chapter will be concluded with discussions on limitations of the point kinetics model.

8.1 Current Nuclear Energy Scenario

On December 2, 1942, a group of scientists led by Enrico Fermi successfully demonstrated the first self-sustaining fission chain reaction in a nuclear reactor at University of Chicago, Illinois. Among the peaceful use of nuclear energy is the generation of electricity, production of radioisotopes for medical applications, and production of artificial fissile isotopes, such as Uranium-233 $\left({}^{233}_{92}\mathrm{U}\right)$ and Plutonium-239 $\left({}^{239}_{94}\mathrm{Pu}\right)$. Nuclear energy production from fission is carried out in a nuclear reactor. A system used to transform nuclear energy into electrical energy is called a nuclear power plant (NPP), which consists of associated systems for heat transfer, and turbo generator. The first NPP to produce electricity was the small Experimental Breeder Reactor-1 (EBR-1) designed and operated by Argonne National Laboratory, Idaho, USA.

As on December 2021, there are around 450 nuclear power plants in 30 countries with total capacity of about 380,000 MWe, contributing about 11% of electricity generation in the world. About 60 more reactors are under construction. More than

A. P. Tiwari (✉)
School of Computing and Electrical Engineering, Indian Institute of Technology Mandi (IIT Mandi), Kamand, Mandi 175005, HP, India
e-mail: akhilanand@iitmandi.ac.in; aptiwari.barc@gmail.com

U. Mangalanathan et al. (eds.), *Nuclear Instrumentation and Control*,
https://doi.org/10.1007/978-981-97-1283-0_8

50 countries operate about 240 reactors and further about 180 nuclear reactors power ships and submarines. In India, 22 nuclear power reactors are under operation producing 6,780 MWe. One 700 MWe unit at Kakrapar, Gujarat was recently commissioned and is being synchronized to the grid, three 700 MWe units are under construction at Kakrapar and Kota in Rajasthan. Two 1000 MWe units are under construction at Kudankulum in Tamil Nadu.

8.2 Main Components of a Nuclear Reactor

Nuclear reactor is a device in which controlled fission chain reaction takes place. Fuel, coolant, moderator, reflector, control rods, and neutron flux sensors are the main components of a nuclear reactor, as shown in Fig. 8.1. Fuel is the essential component of a nuclear reactor. Often Uranium and Plutonium, in metallic, oxide, and carbide forms, are used as fuel material. Fuel in the form of pellets is contained in clad which ensures that fission products are not released in coolant.

Nuclear fission reaction releases energy in the form of kinetic energy of fission fragments. This kinetic energy gets converted into heat and can be transferred through conduction across clad, to coolant, which converts into two phase mixture of steam and water. Steam is separated from the mixture and it can drive a turbine. In other types of designs, coolant transfers its heat to water into a heat exchanger, also called steam generator, and the steam thus generated is used to drive turbine. A neutron

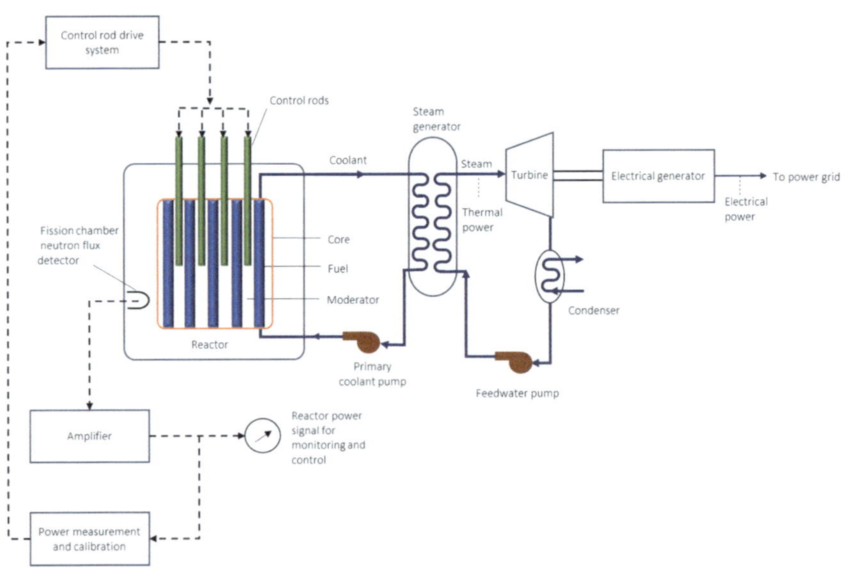

Fig. 8.1 Schematic representation of a nuclear power plant

flux detector measures the operating flux level of the reactor and generates a signal proportional to it. The signal is further amplified and displayed in the control room. The control rods are maneuvered on the basis of the signal sensed by the flux detector to adjust the operating flux through manual or automatic control actions.

Nuclear fission reaction gives rise to production of two to three neutrons per fission event. These neutrons have high energy (upto 10 MeV). Thus they are called fast neutrons. The fast neutrons need to be slowed down as the slow neutrons are capable of carrying more fissions than the fast neutrons. The material which can slow down the fast neutrons is called moderator. Typical moderator materials are water, heavy water, graphite, and berellium.

Fission reactions taking place into the core are regulated and stopped, respectively, by means of control devices and shutdown devices, often in the form of rods or plates. The usually used materials for control and shutdown are cadmium and boron.

Reflector is used to send back the neutrons, which would otherwise have escaped, into the core. Typical moderator materials also serve as reflectors.

Neutron flux sensors are used to derive information regarding operating conditions and rate of fission reactions.

8.3 Types of Nuclear Reactors

Depending upon the arrangement and selection of fuel, coolant, and moderator, there can be different types of reactors. Different types of designs have their own advantages and drawbacks. For example, light water is a good coolant and it can also be used as moderator but due to its large thermal neutron absorption cross section, uranium needs to be enriched if light water is to be used as coolant and moderator. Also, light water has low boiling point. Thus, to use it as coolant, the operating pressure needs to be increased.

If heavy water is used as coolant and moderator, natural uranium can be used as fuel. This is so because heavy water has very low thermal neutron absorption cross section and it is a very good moderator.

Reactors, which are used for power production, are often classified as follows:

1. Light water reactors, i.e., the reactors which use light water as coolant and moderator along with enriched uranium fuel. These are again of two types: Boiling Water Reactor (BWR) and Pressurized Water Reactor (PWR).
2. Heavy water reactors, i.e., the reactors which use heavy water as moderator and coolant along with natural uranium fuel. Some heavy water moderated reactors use light water as coolant along with enriched fuel.
3. Gas-cooled reactors, which use enriched fuel and carbon dioxide or helium as coolant and graphite as moderator.
4. Boiling water graphite-moderated reactors or Reaktor Bolshoy Moshchnosty Kanalny (RBMK) reactors, which use graphite moderator and boiling light water coolant along with enriched fuel.

5. Fast breeder reactors, in which fission chain reaction is sustained by fast neutrons. They require fuel, generally plutonium, with very high enrichment and sodium as coolant.

In addition to the above basic concepts, there are many advanced designs, e.g., APR-1000 developed by Korea Hydro and Nuclear Power, and EPR (European Pressurized Reactor, also known as Evolutionary Power Reactor) reactors, AHWR developed by Bhabha Atomic Research Centre, India, MSBR (Molten Salt Breeder Reactor), and HTGR (High Temperature Gas—cooled Reactor).

8.4 Preliminary Concepts

In this section, we briefly discuss some preliminary concepts from nuclear reactor physics which are easily found in standard text books.

8.4.1 Nuclear Interaction Cross Sections

Neutrons can interact with target material in various ways, e.g., scattering, absorption, and fission. Nuclear cross sections are used to characterize the probability of various types of neutron nuclear reactions to occur.

8.4.1.1 Microscopic Cross Sections

Consider a beam of neutrons all traveling with the same velocity (i.e., the same speed and direction). Suppose this neutron beam is incident normally upon and uniformly across the face of a target material, which is assumed to be very thin, say of the order of one atomic layer. In this case, the neutron nuclear reactions in the target will take place at the rate $\sigma I N_A$, where I and N_A are, respectively, the incident neutron beam intensity and the number of target nuclei per unit area, and σ is called the microscopic cross section. It has the units of area, i.e., cm^2. However, the microscopic cross sections are usually very small. So, they are specified in a much smaller unit, called barn (1 barn = 10^{-24} cm^2).

Depending upon the type of neutron–nuclear interaction, such as scattering, absorption, or fission, the cross section is denoted by adding a suffix with σ. Thus, σ_f denotes the fission cross section, σ_a the absorption cross section, σ_s denotes the scattering cross section, and σ_t denotes the total cross section. The microscopic cross sections are usually different for neutrons of different speeds.

8.4.1.2 Macroscopic Cross Sections

The neutron nuclear reaction cross sections in thicker targets are characterized by macroscopic cross sections, Σ, given as

$$\Sigma = N\sigma,$$

where N denotes the atomic number density (i.e., the number of nuclei of target material per unit volume). Σ is measured in cm^{-1} and is usually accompanied with a suffix to specify the type of nuclear reaction. Thus, Σ_f, Σ_a, Σ_s, and Σ_t, respectively, denote the macroscopic fission, absorption, scattering, and total cross sections. Larger cross section will obviously mean a larger rate of neutron–nuclear interaction.

8.4.2 Fission

Fission is a nuclear reaction in which the excited compound nucleus of a fissile isotope formed by absorption of a neutron breaks up into two lighter nuclei, called "fission fragments." The importance of fission, from the standpoint of the utilization of nuclear energy, lies in two facts. First, a large amount of energy per unit mass of the nuclear fuel is released in the fission reaction, and second, the fission reaction, which is initiated by a neutron, is accompanied by the liberation of 2 or 3 neutrons. Fission reaction is schematically represented in Fig. 8.2.

The average number of neutrons liberated per fission is called as neutron yield and denoted usually by ν. It depends on the mass number of the nucleus undergoing fission and energy of the incident neutron absorbed to form the compound nucleus. The average number of neutrons liberated per neutron absorbed (in fission as well as in nonfission processes) called as utilization factor is given by

$$\eta = \frac{\nu \Sigma_f}{\Sigma_a}. \tag{8.1}$$

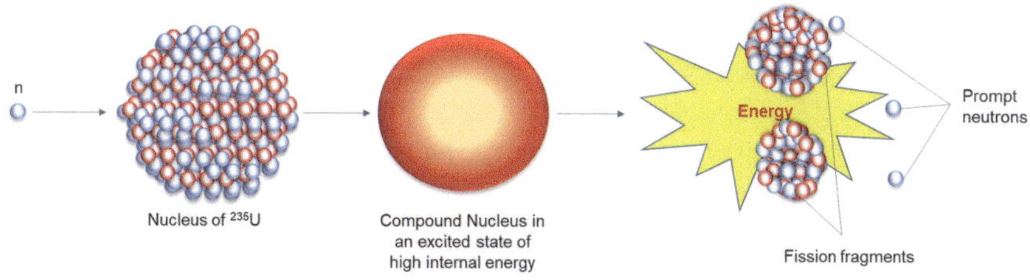

Fig. 8.2 Schematic representation of nuclear fission

Resonance escape probability, usually denoted as p, is fraction of neutrons slowing down past the resonance threshold of fertile isotope, ^{238}U. Thermal utilization factor, denoted usually by f, is expressed as

$$f = \frac{\nu_F \Sigma_F \phi_F}{\nu_F \Sigma_F \phi_F + \nu_M \Sigma_M \phi_M + \ldots}. \tag{8.2}$$

8.4.3 Prompt and Delayed Neutrons

The neutrons released along with the splitting of the compound nucleus or within a short time (of the order of 10^{-14} s) after the formation of the compound nucleus are called prompt neutrons. On the contrary, delayed neutrons are released long time after splitting of the compound nucleus into fragments. They are produced by decay of delayed neutron precursors which are some of the fission fragments, usually characterized by a delay of the order of a few milliseconds to up to a few seconds.

8.4.4 Neutron Generation Time and Neutron Lifetime

The prompt neutron lifetime is the average time elapsed between formation of a neutron by slowing down of fast neutrons and its capture. It is defined as

$$\ell_\infty = \frac{1}{v \Sigma_a} \tag{8.3}$$

where v denotes the average speed and Σ_a denotes the macroscopic absorption cross section of thermal neutrons. In a finite medium, neutrons also leak out of the system. Hence, the term prompt neutron generation time, which is more useful in study of reactor kinetics, is introduced. It is defined as

$$\ell = \frac{1}{v \Sigma_a} P_{\mathrm{NL}} \tag{8.4}$$

where P_{NL} denotes the neutron nonleakage probability. It is dependent upon the neutron diffusion length, L and the buckling, B, which in turn depends upon the material in the reactor core and its geometry. Mathematically, it is given as

$$P_{\mathrm{NL}} = \frac{1}{1 + L^2 B^2}. \tag{8.5}$$

8.4.5 *Fission Rate and Power*

The rate of nuclear fissions taking place per unit volume of a nuclear reactor is given as

$$\text{Fission rate} = \Sigma_f \phi \tag{8.6}$$

where Σ_f denotes the macroscopic fission cross section and ϕ denotes the neutron flux. The macroscopic fission cross section itself is the product of number density of fissile nuclei, N, and the microscopic fission cross section, σ_f, i.e.:

$$\Sigma_f = N\sigma_f. \tag{8.7}$$

On average, each fission event results into release of 190 MeV or equivalently 3×10^{-11} W of energy. Let us denote this factor by E_{eff}. Then, the thermal power produced from nuclear fission can be expressed as

$$P_{\text{th}} = E_{\text{eff}} \Sigma_f \phi V \tag{8.8}$$

where V denotes the core volume.

8.4.6 *Infinite and Effective Multiplication Factors*

In an infinite medium the neutrons produced from fission are lost only because of absorption in fuel and in other material, such as coolant, moderator, and structure. The infinite multiplication factor denoted by k_∞ is defined as the ratio of number of neutrons produced in a generation to the number of neutrons absorbed in the preceding generation. It is mathematically expressed using the so-called four factor formula as

$$k_\infty = \eta \epsilon p f \tag{8.9}$$

where ϵ denotes the fast fission factor, which is the factor accounting also for neutrons produced through fast neutron fission of ^{238}U. Other factors, namely, η, p, and f, have already been introduced. k_∞ also characterizes a finite-sized core if the leakage of neutrons out of the core is ignored.

In a finite-sized core, the neutrons are also lost due to leakage out of core. Hence, a more meaningful quantity called effective multiplication factor, k_{eff}, is used to characterize the neutron multiplication in a finite medium. Mathematically,

$$k_{\text{eff}} = k_\infty P_{NL}. \tag{8.10}$$

For a critical core configuration in which the number of neutrons is constant, the effective multiplication factor is unity, i.e., $k_{\mathrm{eff}} = 1$; for a subcritical configuration in which the neutron population decreases with time, it is less than unity, i.e., $k_{\mathrm{eff}} < 1$; and for a supercritical configuration in which the neutron population grows with time, it is larger than unity, i.e., $k_{\mathrm{eff}} > 1$.

8.5 Point Kinetics Model Equations

In a nuclear reactor, the complex phenomena of production of neutrons, their absorption, and leakage are dependent on neutron energy or speed, space coordinates, and time. For example, the neutrons liberated from fission have energies ranging from a few keV to a MeV. A fast reactor is designed to operate with fast neutrons, i.e., all interactions such as fission, absorption, and leakage of neutrons happen at high energies. In contrast to this, in a thermal reactor, a very small fraction of the fast neutrons produced may get absorbed or escape out of the core and another very small fraction causes fast fission of ^{238}U but the majority undergo slowing down and attain thermal energies. Subsequently, fission, absorption, and leakage happen at thermal energies. Also, a practical reactor is large enough compared to the neutron slowing down length and their mean free path giving rise to spatial variations in the fission, absorption, and slowing down phenomena.

The multigroup neutron diffusion equations are used to represent the complex phenomena taking place in a nuclear reactor accurately. However, a simpler mathematical model known as the "point kinetics model" is often used in control studies. It consists of a set of equations characterizing the time variation of neutron density and delayed neutron precursors' concentration, given as

$$\frac{dn(t)}{dt} = \frac{\rho(t) - \beta}{\ell^*} n(t) + \sum_{i=1}^{m_d} \lambda_i c_i(t), \tag{8.11}$$

and
$$\frac{dc_i(t)}{dt} = \frac{\beta_i}{\ell^*} n(t) - \lambda_i c_i(t), \; i = 1, 2, \ldots, m_d, \tag{8.12}$$

where $n(t)$ denotes the neutron density, $c_i(t)$ denotes the concentration of ith group of delayed neutron precursor,

$$\rho(t) = \frac{k_{\mathrm{eff}} - 1}{k_{\mathrm{eff}}}, \tag{8.13}$$

$$\ell^* = \frac{\ell}{k_{\mathrm{eff}}} \tag{8.14}$$

and ℓ and k_{eff} are already defined, respectively, by (8.4) and (8.10). The quantity ℓ^*, defined by (8.14), is called the prompt neutron generation time and it is approximately equal to prompt neutron lifetime close to criticality.

A few comments about the point kinetics equations are in order.

1. Generally the neutron flux in a nuclear reactor is both space and time dependent. However, the spatial variation is ignored and also it is assumed that the time variation of neutron flux density at all different locations in the reactor core is identical, i.e., the one given by $n(t)$.
2. Thus, it would be sufficient to study any one point in the core based on Eqs. (8.11) and (8.12) as all other points are characterized by the same variation in time. This signifies the word *point* in the name *point kinetics equations.*
3. Both the equations are first-order differential equations. However, they are coupled as the rate of variation of neutron density is dependent on the neutron density and delayed neutron precursors' concentration both. Similarly, the rate of variation of delayed neutron precursors' concentration depends on both these quantities.
4. The equation giving rate of variation of neutron density with time is nonlinear as one of the terms in right-hand side of it involves product of two variables, namely, $\rho(t)$ and $n(t)$.
5. The nuclear reactor is described by the above $m_d + 1$ equations. Generally $m_d =$ 6, i.e., delayed neutrons fall in 6 groups. Hence, in the control theory perspective the order of nuclear reactor is $m_d + 1$ or 7.
6. An alternate form of point kinetics equations, in which the normalized or relative values of neutron density and delayed neutron precursors' concentrations are used, is also very popular. Let n_{FP} denote the rated neutron density and $c_{i,\text{FP}}$ the corresponding steady-state value of delayed neutron precursors' concentrations, i.e.:

$$c_{i,\text{FP}} = \frac{\beta_i n_{\text{FP}}}{\lambda_i \ell^*}, i = 1, 2, \ldots, m_d. \tag{8.15}$$

Now dividing (8.11) by n_{FP}, and (8.12) by $c_{i,\text{FP}}$ we have

$$\frac{d}{dt}\left(\frac{n(t)}{n_{\text{FP}}}\right) = \frac{\rho(t) - \beta}{\ell^*}\frac{n(t)}{n_{\text{FP}}} + \sum_{i=1}^{m_d} \lambda_i \frac{c_i(t)}{c_{i,\text{FP}}}\frac{c_{i,\text{FP}}}{n_{\text{FP}}}, \tag{8.16}$$

$$\frac{d}{dt}\left(\frac{c_i(t)}{c_{i,\text{FP}}}\right) = \frac{\beta_i}{\ell^*}\frac{n(t)}{c_{i,\text{FP}}} - \lambda_i \frac{c_i(t)}{c_{i,\text{FP}}}. \tag{8.17}$$

Note that the second term in the right-hand side of the first equation in the above set is multiplied and divided by $c_{i,\text{FP}}$ to make further mathematical manipulations simpler. Now, the values of $c_{i,\text{FP}}$ in terms of n_{FP} as given by (8.15) are substituted and simplifications carried out to obtain

$$\frac{dn_r(t)}{dt} = \frac{\rho(t) - \beta}{\ell^*} n_r(t) + \sum_{i=1}^{m_d} \frac{\beta_i}{\ell} c_{i,r}(t) \tag{8.18}$$

$$\text{and } \frac{dc_{i,r}(t)}{dt} = \lambda_i n_r(t) - \lambda_i c_{i,r}(t) \tag{8.19}$$

where $n_r(t) = \frac{n(t)}{n_{\text{FP}}}$ and $c_{i,r}(t) = \frac{c_i(t)}{c_{i,\text{FP}}}$ denote the relative values of neutron density and delayed neutron precursors' concentrations.

7. Usually, one effective group of delayed neutrons is used whereby the set of Eqs. (8.11) and (8.12) simplifies to

$$\frac{dn(t)}{dt} = \frac{\rho(t) - \beta}{\ell^*} n(t) + \lambda_{\text{eq}} c_{\text{eq}}(t), \tag{8.20}$$

$$\frac{dc_{\text{eq}}(t)}{dt} = \frac{\beta}{\ell^*} n(t) - \lambda_{\text{eq}} c_{\text{eq}}(t), \tag{8.21}$$

and the system order reduces to 2. In terms of relative values of neutron density and effective one group delayed neutron precursor concentration, the equations governing the reactor are

$$\frac{dn_r(t)}{dt} = \frac{\rho(t) - \beta}{\ell^*} n_r(t) + \beta c_{\text{eq},r}(t), \tag{8.22}$$

$$\text{and } \frac{dc_{\text{eq},r}(t)}{dt} = \lambda_{\text{eq}} n(t) - \lambda_{\text{eq}} c_{\text{eq},r}(t). \tag{8.23}$$

8.6 Reactor at Steady-State Equilibrium

At equilibrium, the neutron density and delayed neutron precursors' concentrations would be constant, i.e., the rate of change of neutron density and delayed neutron precursors' concentrations with time will be zero. Hence, from the set of Eqs. (8.11) and (8.12), we have

$$\frac{\rho_{\text{ss}} - \beta}{\ell^*} n_{\text{ss}} + \sum_{i=1}^{m_d} \lambda_i c_{i,\text{ss}} = 0, \tag{8.24}$$

$$\frac{\beta_i}{\ell^*} n_{\text{ss}} - \lambda_i c_{i,\text{ss}} = 0, \tag{8.25}$$

where the subscript 'ss' associated with the variable indicates its value at the steady state. From (8.25), we have

$$c_{i,\text{ss}} = \frac{\beta_i}{\lambda_i \ell^*} n_{\text{ss}}, \quad (i = 1, 2, \ldots, m_d). \tag{8.26}$$

Substituting the above values of $c_{i,\text{ss}}$ in (8.24) and simplifying, we have

$$\rho_{\text{ss}} n_{\text{ss}} = 0. \tag{8.27}$$

From the above, it is clearly evident that at steady state, either $n_{ss} = 0$ or $\rho_{ss} = 0$. From these, the former is only of mathematical significance as in a nuclear reactor system, number of neutrons cannot be zero though it can be small. The second one, i.e., $\rho_{ss} = 0$ is an important relation from which it can be interpreted that at steady-state equilibrium, the reactivity must be zero while neutron density can take any nonzero value. Alternatively speaking, in a theoretical sense, the steady-state operating neutron density in a nuclear reactor is not limited from neutronic considerations. Practically, of course, there will be a limit based on heat removal consideration.

Sometimes, a neutron source is placed in the reactor core to raise the neutron density in the system in the shutdown condition, to sufficiently high value so as to be able to bring the measuring instruments on scale. In such a situation, the concentration of delayed neutron precursors will be given by (8.26) but the neutron density at the steady state will satisfy the relation

$$\frac{\rho_{ss} - \beta}{\ell^*} n_{ss} + \sum_{i=1}^{m_d} \lambda_i C_{i,ss} + q_e = 0, \tag{8.28}$$

which can be simplified using (8.25) to obtain

$$n_{ss} = -\frac{q_e \ell^*}{\rho_{ss}}. \tag{8.29}$$

From the above, it is clear that if the reactor is subcritical, the neutron density will reach a steady-state value. If degree of subcriticality is small, the neutron density corresponding to the subcritical steady state will be correspondingly large.

8.7 Point Kinetics Equations in Standard State-Space Form

8.7.1 State-Space Representation of General Dynamical Systems

A standard way of representing a dynamical system mathematically is the state-space representation, which has the form:

$$\dot{\mathbf{x}}(t) = \frac{d\mathbf{x}(t)}{dt} = \mathbf{f}(\mathbf{x}(t), \mathbf{u}(t), t) \tag{8.30}$$

$$\mathbf{y}(t) = \mathbf{h}(\mathbf{x}(t), \mathbf{u}(t), t), \tag{8.31}$$

where $\mathbf{x}$ called state vector or simply state is a set of physical variables the knowledge of which at $t = t_0$ along with the knowledge of inputs, denoted by the vector $\mathbf{u}$, is sufficient to determine the evolution of the system for $t \geq t_0$; $\mathbf{y}$ is the output vector or the set of output variables of the system. The functions **f** and **h** are vector-valued functions of the state vector, input vector, and time. Written explicitly

$$\mathbf{x} = \begin{bmatrix} x_1 \\ x_2 \\ \vdots \\ x_n \end{bmatrix}, \ \mathbf{u} = \begin{bmatrix} u_1 \\ u_2 \\ \vdots \\ u_m \end{bmatrix}, \ \mathbf{y} = \begin{bmatrix} y_1 \\ y_2 \\ \vdots \\ y_p \end{bmatrix} \tag{8.32}$$

and

$$\dot{x}_1(t) = f_1(x_1, x_2, \ldots, x_n, u_1, u_2, \ldots, u_m, t), \tag{8.33}$$
$$\dot{x}_2(t) = f_2(x_1, x_2, \ldots, x_n, u_1, u_2, \ldots, u_m, t,) \tag{8.34}$$
$$\vdots \quad \vdots \tag{8.35}$$
$$\dot{x}_n(t) = f_n(x_1, x_2, \ldots, x_n, u_1, u_2, \ldots, u_m, t), \tag{8.36}$$
$$y_1(t) = h_1(x_1, x_2, \ldots, x_n, u_1, u_2, \ldots, u_m, t), \tag{8.37}$$
$$y_2(t) = h_2(x_1, x_2, \ldots, x_n, u_1, u_2, \ldots, u_m, t), \tag{8.38}$$
$$\vdots \quad \vdots \tag{8.39}$$
$$y_p(t) = h_p(x_1, x_2, \ldots, x_n, u_1, u_2, \ldots, u_m, t). \tag{8.40}$$

Note that the above equations express the time derivative of each of the state variables as general functions of all the state variables, inputs and time. The vector-valued functions **f** and **h** have $n + m + 1$ arguments. When time t does not appear explicitly in none of the functions f_i and h_i, the system is said to be *time invariant*. Otherwise, it is a time-varying system.

An exact or fairly accurate model of a process is usually nonlinear but fortunately many processes are adequately approximated by linear models over a significant range of operation. In the state-space representation of a linear process, the general differential Eq. (8.30) and the output Eq. (8.31) take the form:

$$\dot{\mathbf{x}}(t) = \mathbf{F}(t)\mathbf{x}(t) + \mathbf{G}(t)\mathbf{u}(t) \tag{8.41}$$
$$\mathbf{y}(t) = \mathbf{H}(t)\mathbf{x}(t) \tag{8.42}$$

where $\mathbf{F}(t)$, $\mathbf{G}(t)$ and $\mathbf{H}(t)$ are $n \times n, n \times m$, and $p \times n$ matrices given, respectively, by

$$\mathbf{F}(t) = \begin{bmatrix} f_{11} & f_{12} & \cdots & f_{1n} \\ f_{21} & f_{22} & \cdots & f_{2n} \\ \vdots & \vdots & \ddots & \vdots \\ f_{n1} & f_{n2} & \cdots & f_{nn} \end{bmatrix}, \tag{8.43}$$

$$\mathbf{G}(t) = \begin{bmatrix} g_{11} & g_{12} & \cdots & g_{1m} \\ g_{21} & g_{22} & \cdots & g_{2m} \\ \vdots & \vdots & \ddots & \vdots \\ g_{n1} & g_{n2} & \cdots & g_{nm} \end{bmatrix}, \tag{8.44}$$

$$\text{and } \mathbf{H}(t) = \begin{bmatrix} h_{11} & h_{12} & \dots & h_{1n} \\ h_{21} & h_{22} & \dots & h_{2n} \\ \vdots & \vdots & \ddots & \vdots \\ h_{p1} & h_{p2} & \dots & h_{pn} \end{bmatrix}. \tag{8.45}$$

When the system is time invariant, none of the elements of the matrices **F**, **G**, and **H** depend upon time. In such a situation, the system representation changes to

$$\dot{\mathbf{x}}(t) = \mathbf{F}\mathbf{x}(t) + \mathbf{G}\mathbf{u}(t) \tag{8.46}$$

$$\mathbf{y}(t) = \mathbf{H}\mathbf{x}(t) \tag{8.47}$$

where **F**, **G**, and **H** are constant matrices.

8.7.2 State-Space Representation of Nuclear Reactor Described by Point Kinetics Model

In case of nuclear reactor, the choice of neutron density and delayed neutron precursor concentrations as state variables is quite obvious, and the reactivity is considered as input and the neutron density is considered as output, i.e.:

$$\mathbf{x}(t) = \begin{bmatrix} n(t) \\ c_1(t) \\ c_2(t) \\ \vdots \\ c_{m_d}(t) \end{bmatrix}; \quad \mathbf{u}(t) = \rho(t); \quad \mathbf{y}(t) = n(t). \tag{8.48}$$

Then, the system is represented in standard state-space form (8.41) and (8.42), where

$$\mathbf{F}(t) = \begin{bmatrix} -\frac{\beta}{\ell*} & \lambda_1 & \lambda_2 & \dots & \lambda_{m_d} \\ \frac{\beta_1}{\ell*} & -\lambda_1 & 0 & \dots & 0 \\ \frac{\beta_2}{\ell*} & 0 & -\lambda_2 & \dots & 0 \\ \vdots & \vdots & \vdots & \ddots & \vdots \\ \frac{\beta_{m_d}}{\ell*} & 0 & 0 & \dots & -\lambda_{m_d} \end{bmatrix}; \tag{8.49}$$

$$\mathbf{G}(t) = \begin{bmatrix} \frac{n(t)}{\ell^*} \\ 0 \\ 0 \\ \vdots \\ 0 \end{bmatrix}; \quad \text{and} \tag{8.50}$$

$$\mathbf{H}(t) = \begin{bmatrix} 1 \; 0 \; 0 \ldots 0 \end{bmatrix}. \tag{8.51}$$

If relative neutron density and relative delayed neutron precursors' concentrations are chosen as state variables, while retaining reactivity as input and the neutron density as the output, i.e., if

$$\mathbf{x}(t) \equiv \mathbf{x_r}(t) = \begin{bmatrix} n_r(t) \\ c_{1,r}(t) \\ c_{2,r}(t) \\ \vdots \\ c_{m_d,r}(t) \end{bmatrix}; \; \mathbf{u}(t) = \rho(t); \; \mathbf{y}(t) = n(t) \tag{8.52}$$

then the system, input, and output matrices are given, respectively, as

$$\mathbf{F}(t) \equiv \mathbf{F_r}(t) = \begin{bmatrix} -\frac{\beta}{\ell^*} & \frac{\beta_1}{\ell^*} & \frac{\beta_2}{\ell^*} & \ldots & \frac{\beta_{m_d}}{\ell^*} \\ \lambda_1 & -\lambda_1 & 0 & \ldots & 0 \\ \lambda_2 & 0 & -\lambda_2 & \ldots & 0 \\ \vdots & \vdots & \vdots & \ddots & \vdots \\ \lambda_{m_d} & 0 & 0 & \ldots & -\lambda_{m_d} \end{bmatrix}; \tag{8.53}$$

$$\mathbf{G}(t) \equiv \mathbf{G_r}(t) = \begin{bmatrix} \frac{n_r}{\ell^*} \\ 0 \\ 0 \\ \vdots \\ 0 \end{bmatrix}; \quad \text{and} \tag{8.54}$$

$$\mathbf{H}(t) \equiv \mathbf{H_r}(t) = \begin{bmatrix} n_{FP}(t) \; 0 \; 0 \ldots 0 \end{bmatrix}. \tag{8.55}$$

8.8 Linearization of Point Kinetics Model

As already pointed out in the preceding, the model of the nuclear reactor is nonlinear, which is difficult to analyze except for some simpler situations. One resorts to linearization and uses the well-developed and popular literature in the area of linear control theory to analyze and design the reactor control systems from the perspective of control engineering.

We consider a small perturbation in reactivity, neutron density, and delayed neutron precursors' concentration around their respective steady-state values, i.e., let $n(t) = n_{\text{ss}} + \delta n(t)$, $c_i(t) = c_{i,\text{ss}} + \delta c_i(t)$, $(i = 1, 2, \ldots, m_d)$, and $\rho(t) = 0 + \delta\rho(t)$, where steady-state values of neutron density and delayed neutron precursors' concentrations are nonzero but the steady-state value of reactivity is zero, which characterizes a critical reactor in the absence of external neutron source. Then, from (8.11) we have

$$\begin{aligned}\frac{d\left(n_{\text{ss}} + \delta n(t)\right)}{dt} &= \frac{0 + \delta\rho(t) - \beta}{\ell^*}\left(n_{\text{ss}} + \delta n(t)\right) \\ &+ \sum_{i=1}^{m_d} \lambda_i \left(c_{i,\text{ss}} + \delta c_i(t)\right). \end{aligned} \tag{8.56}$$

If the perturbation $\delta\rho(t)$ is small and the change in neutron density $\delta n(t)$ is also small, then the product $\delta\rho(t)\delta n(t)$ will be much small and can be ignored in comparison to the other terms. With this approximation and using the steady-state relationship (8.24), the above equation can be simplified to obtain

$$\frac{d\delta n(t)}{dt} = \frac{\delta\rho(t) - \beta}{\ell^*}\delta n(t) + \sum_{i=1}^{m_d} \lambda_i \delta c_i(t). \tag{8.57}$$

Similarly from (8.12), we have

$$\frac{d\left(c_{i,\text{ss}} + \delta c_i(t)\right)}{dt} = \frac{\beta_i}{\ell^*}\left(n_{\text{ss}} + \delta n(t)\right) - \lambda_i\left(c_{i,\text{ss}} + \delta c_i(t)\right) \tag{8.58}$$

which can be simplified using (8.24) to obtain

$$\frac{d\delta c_i(t)}{dt} = \frac{\beta_i}{\ell^*}\delta n(t) - \lambda_i \delta c_i(t), \; i = 1, 2, \ldots, m_d. \tag{8.59}$$

Equations (8.57) and (8.59) are linearized set of equations corresponding to the nonlinear point kinetics, Eqs. (8.11) and (8.12). Identifying the incremental changes in neutron density and delayed neutron precursors' concentration as state variables; change in reactivity as input and change in neutron density as output, i.e.:

$$\begin{aligned} \mathbf{x}(t) &= \left[\delta n(t)\; \delta c_1\; \delta c_2 \ldots \delta c_{m_d}\right]^T, \\ \mathbf{u}(t) &= \delta\rho(t) \;=\; \rho(t), \\ \text{and } \mathbf{y}(t) &= \delta n(t), \end{aligned}$$

the model is expressed in standard state-space form as

$$\dot{\mathbf{x}}(t) = \mathbf{F}_{\text{lin}}\mathbf{x}(t) + \mathbf{G}_{\text{lin}}\mathbf{u}(t) \tag{8.60}$$

$$\mathbf{y}(t) = \mathbf{H_{lin}}\mathbf{x}(t) \tag{8.61}$$

where

$$\mathbf{F_{lin}}(t) = \begin{bmatrix} -\frac{\beta}{\ell^*} & \lambda_1 & \lambda_2 & \dots & \lambda_{m_d} \\ \frac{\beta_1}{\ell^*} & -\lambda_1 & 0 & \dots & 0 \\ \frac{\beta_2}{\ell^*} & 0 & -\lambda_2 & \dots & 0 \\ \vdots & \vdots & \vdots & \ddots & \vdots \\ \frac{\beta_{m_d}}{\ell^*} & 0 & 0 & \dots & -\lambda_{m_d} \end{bmatrix}; \tag{8.62}$$

$$\mathbf{G_{lin}}(t) = \begin{bmatrix} \frac{n_{ss}}{\ell^*} \\ 0 \\ 0 \\ \vdots \\ 0 \end{bmatrix}; \quad \text{and} \tag{8.63}$$

$$\mathbf{H_{lin}}(t) = \begin{bmatrix} 1\ 0\ 0 \dots 0 \end{bmatrix}. \tag{8.64}$$

If instead of multiple delayed neutron groups, only one effective delayed neutron group is modeled, then the change in neutron density and the change in the one effective group delayed neutron precursors concentration will be the two state variables for the reactor, i.e.: $\mathbf{x}(t) = \begin{bmatrix} \delta n(t)\ \delta c_{eq} \end{bmatrix}^T$ and the linear model will be obtained in the standard state-space form (8.60), (8.61), where

$$\mathbf{F_{lin}}(t) = \begin{bmatrix} -\frac{\beta}{\ell^*} & \lambda_{eq} \\ \frac{\beta}{\ell^*} & -\lambda_{eq} \end{bmatrix}; \tag{8.65}$$

$$\mathbf{G_{lin}}(t) = \begin{bmatrix} \frac{n_{ss}}{\ell^*} \\ 0 \end{bmatrix}; \quad \text{and} \tag{8.66}$$

$$\mathbf{H_{lin}}(t) = \begin{bmatrix} 1\ 0 \end{bmatrix}. \tag{8.67}$$

If changes in relative neutron density and relative delayed neutron precursors' concentrations are chosen as state variables, while retaining change in reactivity as input and change in the neutron density as the output, i.e., if

$$\mathbf{x}(t) = \begin{bmatrix} \delta n_r(t) \\ \delta c_{1,r}(t) \\ \delta c_{2,r}(t) \\ \vdots \\ \delta c_{m_d,r}(t) \end{bmatrix}; \quad \mathbf{u}(t) = \delta\rho(t); \quad \mathbf{y}(t) = \delta n(t) \tag{8.68}$$

then for the linearized model of the reactor, the system, input, and output matrices are given, respectively, as

$$\mathbf{F_{lin}}(t) = \begin{bmatrix} -\frac{\beta}{\ell^*} & \frac{\beta_1}{\ell^*} & \frac{\beta_2}{\ell^*} & \cdots & \frac{\beta_{m_d}}{\ell^*} \\ \lambda_1 & -\lambda_1 & 0 & \cdots & 0 \\ \lambda_2 & 0 & -\lambda_2 & \cdots & 0 \\ \vdots & \vdots & \vdots & \ddots & \vdots \\ \lambda_{m_d} & 0 & 0 & \cdots & -\lambda_{m_d} \end{bmatrix}; \tag{8.69}$$

$$\mathbf{G}(t) = \begin{bmatrix} \frac{n_{ss}}{n_{FP}} \frac{1}{\ell^*} \\ 0 \\ 0 \\ \vdots \\ 0 \end{bmatrix}; \quad \text{and} \tag{8.70}$$

$$\mathbf{H}(t) = \begin{bmatrix} n_{\text{FP}}(t) \; 0 \; 0 \ldots 0 \end{bmatrix}. \tag{8.71}$$

8.9 Concluding Remarks

The essence of the chapter is the point kinetics model which was introduced and then its representation in the standard state-space form, which is the most suitable form for control analysis and design studies, has been given. However, as we conclude we make mention of a few limitations of the point kinetics model.

The most important simplifying assumption made in the derivation of point kinetics equation is that the neutron flux shape does not vary in time. In a practical nuclear reactor and more so in large reactors, however, the neutron flux shape may vary with time. Hence, the application of point kinetics model is generally limited to small nuclear reactors in which the flux shape does not undergo appreciable variation. It is also applicable to large nuclear reactors around the typical operating condition in which flux shape is maintained close to the "design flux shape."

Also, the neutron interaction cross sections undergo variations owing to variations in neutron energy with the operating power level. During the course of operation of the nuclear reactor, various types of nuclei are produced. The build-up of such nuclei also causes changes in nuclear interaction cross sections. Such aspects, which are often referred to as internal feedback effects, have been ignored. Because of these idealizations, the point kinetics model introduced here is also called as model of zero power, cold and clean reactor. The terms "zero power" and "cold" signify that changes in neutron interaction cross sections due to changes in temperature and densities of fuel, coolant, moderator, and other structural material have not been accounted for. The term "clean" accounts for the fact that formation and accumulation of various fission product nuclei has been ignored.

This chapter provides for the foundation to explore literature on reactor dynamics and space-time reactor kinetics. In reactor dynamics studies, the effect of variations of fuel temperature, coolant temperature and density, moderator temperature and density, and fission product poisons is studied. Space-time kinetics study is intended to understand the behavior or large-sized nuclear reactors.

Chapter 9
Nuclear Reactor Characteristics

A. P. Tiwari

In the preceding chapter, the mathematical model of the nuclear reactor was derived. This chapter is devoted to understanding some basic characteristics, such as the behavior of the reactor to change in reactivity, and its stability, controllability, and observability.

The first part of the chapter will be devoted to obtaining approximate analytical solution of the nonlinear point kinetics model. Expressions for variation in neutron density caused due to step change in reactivity will be obtained. The concepts of stable reactor period, prompt jump, and in-hour equation will be introduced. Further, the set of linearized point kinetics equations will be solved for step change in reactivity and comparison between responses of linear and nonlinear systems will be made to draw some important conclusions.

Later, the prompt-jump approximation, which is an important mathematical tool for simplifying the point kinetics model equations, will be introduced. Comparison between variation of neutron density obtained with prompt-jump approximation and that obtained with nonlinear point kinetics model will be made.

With the help of characteristic equation of the reactor, its stability will be explained. Finally, controllability and observability characteristics of the nuclear reactor on the basis of the point kinetics model will be investigated.

Generally, as we will rely on hand calculations and simpler interpretations, the second-order mathematical model of the nuclear reactor which is introduced in Chap. 8, i.e., the point kinetics model with one effective group of delayed neutrons will be used.

The chapter will be concluded with recall of some important results and remark about directions for further studies.

A. P. Tiwari (✉)
School of Computing and Electrical Engineering, Indian Institute of Technology Mandi (IIT Mandi), Kamand, Mandi (HP) 175005, India
e-mail: akhilanand@iitmandi.ac.in; aptiwari.barc@gmail.com

U. Mangalanathan et al. (eds.), *Nuclear Instrumentation and Control*,
https://doi.org/10.1007/978-981-97-1283-0_9

9.1 Solution of Nonlinear Point Kinetics Equations

Mathematical model of a physical system is very useful in understanding the behavior of the system. In this chapter, we solve the reactor equations and understand the effect of changes in reactivity by looking at the resulting change in the neutron density and delayed neutron precursors' concentrations.

9.1.1 Solution of General State Equation

We start with the state-space representation of a general linear time-invariant dynamical system, given in the form of Eqs. (8.46) and (8.47) in Chap. 8. Recall that **F**, **G**, and **H** are constant matrices. The general solution of this is given as

$$\mathbf{x}(t) = \Phi(t - t_0)\mathbf{x}(t_0) + \int_{t_0}^{t} \Phi(t - \tau)\mathbf{G}(\tau)\mathbf{u}(\tau)\mathrm{d}\tau, \quad t \geq t_0. \tag{9.1}$$

The above equation, called the state transition equation, expresses the value of the state at any arbitrary time $t \geq t_0$ as two components. The first component, called as the zero input response, is the one resulting solely from the value of state at the initial time t_0. The second component called as the zero state response results from the effect of the input. Both of the components contain the term, Φ, which is called as the state transition matrix. It is given as

$$\Phi(\tau) = \mathcal{L}^{-1}\left[\Phi(s)\right] = \mathcal{L}^{-1}\left[(s\mathbf{I} - \mathbf{F})^{-1}\right] \tag{9.2}$$

where s is the Laplace operator, $\mathcal{L}^{-1}(.)$ denotes the inverse Laplace transform of (.), and **I** denotes an identity matrix. The output can be determined using

$$\mathbf{y}(t) = \mathbf{H}\Phi(t - t_0)\mathbf{x}(t_0) + \int_{t_0}^{t} \mathbf{H}\Phi(t - \tau)\mathbf{G}(\tau)\mathbf{u}(\tau)\mathrm{d}\tau, \quad t \geq t_0. \tag{9.3}$$

Generally, the initial time $t_0 = 0$. In such a case, Eqs. (9.1) and (9.3) simplify as

$$\mathbf{x}(t) = \Phi(t)\mathbf{x}(0) + \int_{0}^{t} \Phi(t - \tau)\mathbf{G}(\tau)\mathbf{u}(\tau)\mathrm{d}\tau, \quad t \geq 0 \tag{9.4}$$

$$\text{and } \mathbf{y}(t) = \mathbf{H}\Phi(t)\mathbf{x}(0) + \int_{0}^{t} \mathbf{H}\Phi(t - \tau)\mathbf{G}(\tau)\mathbf{u}(\tau)\mathrm{d}\tau, \quad t \geq 0. \tag{9.5}$$

9.1.2 Solution of Point Kinetics Model with One Effective Delayed Neutron Group for Constant Reactivity Input

The most common exploratory study of the reactor behavior is to find its response for a step change in reactivity with the help of point kinetics model with one effective delayed neutron group, given by set of Eqs. (8.18) and (8.19) in Chap. 8. The reactor is assumed to be operating at steady-state critical condition for long time. Let the neutron density be n_{ss}. The effective one group delayed neutron precursor concentration would have achieved the equilibrium value given by

$$c_{eq,ss} = \frac{\beta}{\lambda_{eq}\ell^*} n_{ss}. \tag{9.6}$$

Now at $t = 0^+$, reactivity is changed to $\bar{\rho}$. In such a situation, the equations describing the reactor take the form

$$\frac{dn(t)}{dt} = \frac{\bar{\rho}-\beta}{\ell^*} n(t) + \lambda_{eq} c_{eq}(t), \quad n(0) = n_{ss} \tag{9.7}$$

$$\frac{dc_{eq}(t)}{dt} = \frac{\beta}{\ell^*} n(t) - \lambda_{eq} c_{eq}(t), \quad c_{eq}(0) = c_{eq,ss}. \tag{9.8}$$

In the standard state space form, they may be written as

$$\dot{\mathbf{x}}(t) = \mathbf{F}\mathbf{x}(t), \quad \mathbf{x}(0) = \begin{bmatrix} n_{ss} \\ c_{eq,ss} \end{bmatrix}, \tag{9.9}$$

where $\mathbf{F} = \begin{bmatrix} \frac{\bar{\rho}-\beta}{\ell^*} & \lambda_{eq} \\ \frac{\beta}{\ell^*} & -\lambda_{eq} \end{bmatrix}$ and $\mathbf{x}(t) = \begin{bmatrix} n(t) & c_{eq}(t) \end{bmatrix}^T$. Because the reactivity is constant, the terms have been rearranged as in an autonomous equation. As discussed in Sect. 3.1, the solution of the above equation will be

$$\mathbf{x}(t) = \Phi(t)\mathbf{x}(0), \quad \Phi(t) = \mathcal{L}^{-1}[\Phi(s)] \tag{9.10}$$

where

$$\Phi(s) = (s\mathbf{I} - \mathbf{F})^{-1} = \left(\begin{bmatrix} s & 0 \\ 0 & s \end{bmatrix} - \begin{bmatrix} \frac{\bar{\rho}-\beta}{\ell^*} & \lambda_{eq} \\ \frac{\beta}{\ell^*} & -\lambda_{eq} \end{bmatrix} \right)^{-1} \tag{9.11}$$

$$= \begin{bmatrix} s - \frac{\bar{\rho}-\beta}{\ell^*} & -\lambda_{eq} \\ -\frac{\beta}{\ell^*} & s + \lambda_{eq} \end{bmatrix}^{-1} \tag{9.12}$$

$$= \frac{1}{d(s)} \begin{bmatrix} s + \lambda_{eq} & \lambda_{eq} \\ \frac{\beta}{\ell^*} & s - \frac{\bar{\rho}-\beta}{\ell^*} \end{bmatrix}. \tag{9.13}$$

In the above expression, $d(s)$, that denotes the characteristic polynomial of the system matrix $\mathbf{F}$, is given by

$$d(s) = s^2 - \left(\frac{\bar{\rho} - \beta}{\ell^*} - \lambda_{\text{eq}}\right) s - \frac{\lambda_{\text{eq}}\bar{\rho}}{\ell^*} = (s + s_1)(s + s_2), \tag{9.14}$$

where s_1 and s_2 are the roots of the characteristics equation

$$s^2 - \left(\frac{\bar{\rho} - \beta}{\ell^*} - \lambda_{\text{eq}}\right) s - \frac{\lambda_{\text{eq}}\bar{\rho}}{\ell^*} = 0. \tag{9.15}$$

Thus we have

$$s_1,\ s_2 = \frac{1}{2}\left(\frac{\bar{\rho} - \beta}{\ell^*} - \lambda_{\text{eq}}\right) \pm \sqrt{\frac{1}{4}\left(\frac{\bar{\rho} - \beta}{\ell^*} - \lambda_{\text{eq}}\right)^2 + \frac{\lambda_{\text{eq}}\bar{\rho}}{\ell^*}}. \tag{9.16}$$

The parameters β, λ_{eq}, and ℓ^* are always positive. Hence, the quantity $\frac{1}{4}\left(\frac{\bar{\rho}-\beta}{\ell^*} - \lambda_{\text{eq}}\right)^2 + \frac{\lambda_{\text{eq}}\bar{\rho}}{\ell^*}$ will be positive for all values of $\bar{\rho}$. Therefore, both the roots s_1 and s_2 are always real. Also, the roots are distinct, i.e., $s_1 \neq s_2$. Thus, we can write

$$\Phi(s) = \begin{bmatrix} \frac{s+\lambda_{\text{eq}}}{(s+s_1)(s+s_2)} & \frac{\lambda_{\text{eq}}}{(s+s_1)(s+s_2)} \\ \frac{\frac{\beta}{\ell^*}}{(s+s_1)(s+s_2)} & \frac{s-\frac{\bar{\rho}-\beta}{\ell^*}}{(s+s_1)(s+s_2)} \end{bmatrix}. \tag{9.17}$$

Expressing each of the terms of $\Phi(s)$ into partial fractions and taking inverse Laplace transform, we get

$$\Phi(\tau) = \frac{1}{s_2 - s_1}\begin{bmatrix} (\lambda_{\text{eq}} - s_1)e^{-s_1\tau} - (\lambda_{\text{eq}} - s_2)e^{-s_2\tau} & \lambda_{\text{eq}}\left(e^{-s_1\tau} - e^{-s_2\tau}\right) \\ \frac{\beta}{\ell^*}\left(e^{-s_1\tau} - e^{-s_2\tau}\right) & \left(-\frac{\bar{\rho}-\beta}{\ell^*} - s_1\right)e^{-s_1\tau} - \left(-\frac{\bar{\rho}-\beta}{\ell^*} - s_2\right)e^{-s_2\tau} \end{bmatrix}. \tag{9.18}$$

Thus the solutions to neutron density and delayed neutron precursor concentration are obtained as

$$\begin{bmatrix} n(t) \\ c_{\text{eq}}(t) \end{bmatrix} = \Phi(t)\begin{bmatrix} n_{\text{ss}} \\ c_{\text{eq,ss}} \end{bmatrix} \tag{9.19}$$

from which we have

$$n(t) = \frac{n_{ss}}{s_2 - s_1}\left[(\lambda_{eq} - s_1)e^{-s_1 t} - (\lambda_{eq} - s_2)e^{-s_2 t}\right] + \frac{\lambda_{eq} c_{eq,ss}}{s_2 - s_1}\left(e^{-s_1 t} - e^{-s_2 t}\right)$$

in which we substitute the values of $c_{eq,ss}$ in terms of n_{ss} from (9.6) to obtain

$$n(t) = \frac{n_{ss}}{s_2 - s_1}\left[(\lambda_{eq} - s_1)e^{-s_1 t} - (\lambda_{eq} - s_2)e^{-s_2 t} + \frac{\beta}{\ell^*}\left(e^{-s_1 t} - e^{-s_2 t}\right)\right].$$

Simplifying further, we have

$$n(t) = \frac{n_{ss}}{s_2 - s_1}\left[\left(\frac{\beta}{\ell^*} + \lambda_{eq} - s_1\right)e^{-s_1 t} - \left(\frac{\beta}{\ell^*} + \lambda_{eq} - s_2\right)e^{-s_2 t}\right]. \tag{9.20}$$

9.1.3 Response for Small Positive Reactivity Inputs

For very small values of positive reactivity such that $\bar{\rho} << \beta$ and with certain other approximations, the analytical solutions for s_1 and s_2 in (9.16) are obtained as follows:

$$s_1 \approx \frac{\bar{\rho} - \beta}{\ell^*}, \quad s_2 \approx \frac{\lambda_{eq}\bar{\rho}}{\bar{\rho} - \beta}. \tag{9.21}$$

Substituting the values of s_1 and s_2 in (9.20), we obtain the approximate analytical solution of neutron density as

$$n_{approx}(t) = n_{ss}\left[-\frac{\bar{\rho}}{\beta - \bar{\rho}}e^{\frac{\bar{\rho}-\beta}{\ell^*}t} + \frac{\beta}{\beta - \bar{\rho}}e^{\frac{\lambda_{eq}\bar{\rho}}{\beta-\bar{\rho}}t}\right]. \tag{9.22}$$

For ^{235}U fueled light water reactors, typical values of β, λ_{eq}, and ℓ^* are 0.0065, 0.08 s^{-1}, and 10^{-6} s, respectively. For heavy water reactors, these parameters are typically 0.0075, 0.08 s^{-1}, and 10^{-3} s, respectively. Hence for very small values of reactivity, $\frac{\bar{\rho}-\beta}{\ell^*}$ is very large negative and $\frac{\lambda_{eq}\bar{\rho}}{\beta-\bar{\rho}}$ is very small in magnitude but positive. Hence, the first term in the expression for $n_{approx}(t)$ given above by (9.22) will quickly decay to zero. The second term, however, slowly rises. Hence, a few seconds later, the variation of neutron density will be characterized dominantly by the second exponent, i.e.:

$$n_{\text{approx}}(t) \approx n_{\text{dom}}(t) = n_{\text{ss}} \frac{\beta}{\beta - \bar{\rho}} e^{\frac{\lambda_{\text{eq}}\bar{\rho}}{\beta - \bar{\rho}} t}. \tag{9.23}$$

Or by introducing

$$T_p = \frac{\beta - \bar{\rho}}{\lambda_{\text{eq}}\bar{\rho}}, \tag{9.24}$$

we have

$$n_{\text{dom}}(t) = n_{\text{ss}} \frac{\beta}{\beta - \bar{\rho}} e^{\frac{t}{T_p}}. \tag{9.25}$$

T_p is called as stable reactor period. At $t = T_p, 2T_p, \ldots, kT_p, (k+1)T_p, \ldots$, the values of neutron density will be $e, e^2, \ldots, e^k, e^{k+1}, \ldots$ times, respectively, of its value at $t = 0$. Thus, the stable reactor period is the duration of time in which the neutron density will rise to e-fold of its initial value. Consequently, the stable reactor period is also called as e-folding time.

For values of reactivity much smaller than β, the expression given by (9.24) may be further simplified to obtain

$$T_p \approx T_{p,\text{dom}} = \frac{\beta}{\lambda_{\text{eq}}\bar{\rho}}, \tag{9.26}$$

from which it is evident that the stable reactor period is approximately inversely proportional to the magnitude of positive reactivity inserted into the reactor.

Example
For $\bar{\rho} = 0.001$, $\beta = 0.0075$, $\lambda_{\text{eq}} = 0.08\ \text{s}^{-1}$, and $\ell^* = 10^{-3}$, the variation of neutron density occurs as shown in Fig. 9.1. In this, plots of $n(t)$ defined by (9.20), $n_{\text{approx}}(t)$ defined by (9.22), and $n_{\text{dom}}(t)$ defined by (9.25) versus time are shown. There is a good agreement between $n(t)$ and $n_{\text{approx}}(t)$ in the initial short portion of the transient during which the neutron density has risen rapidly from its initial value of 1 to approximately 1.15. Subsequently also, the variation of $n_{\text{approx}}(t)$ is similar to $n(t)$, though the former is slightly larger than the later. The variation of $n_{\text{dom}}(t)$ matches with $n(t)$ only after initial phase of approximately 1 s, as expected.

9.1.4 Response for Small Negative Reactivity Inputs

The expressions for $n(t)$, $n_{\text{approx}}(t)$ and $n_{\text{dom}}(t)$ given, respectively, by (9.20), (9.22), and (9.25) are applicable for negative reactivities also. For interpreting the reactor characteristics for negative reactivities, let us consider the expression for $n_{\text{approx}}(t)$, by substituting $\bar{\rho} = -\bar{\mu}$, $\bar{\mu} > 0$ in it. Thus we have

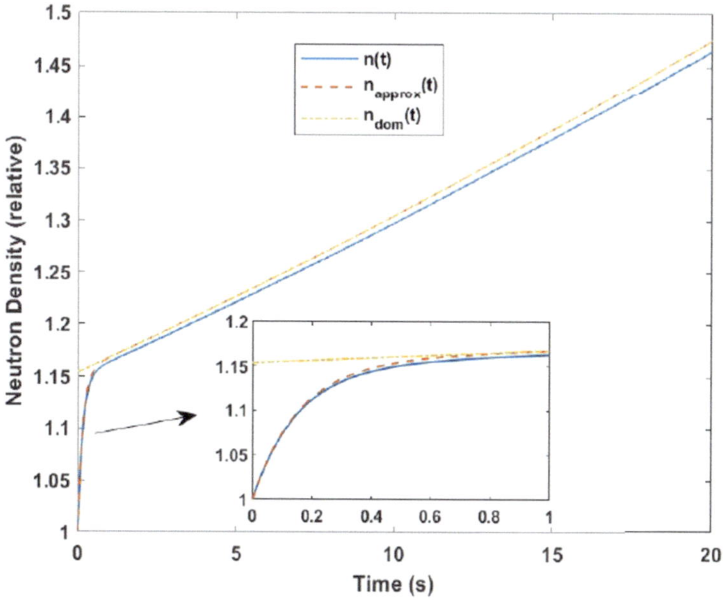

Fig. 9.1 Variation of neutron density with time subsequent to insertion of small positive reactivity $\bar{\rho} = 0.001$

$$n_{\text{approx}}(t) = n_{\text{ss}} \left[\frac{\bar{\mu}}{\beta + \bar{\mu}} e^{-\frac{\beta+\bar{\mu}}{\ell^*} t} + \frac{\beta}{\beta + \bar{\mu}} e^{-\frac{\lambda_{\text{eq}}\bar{\mu}}{\beta+\bar{\mu}} t} \right]. \tag{9.27}$$

Note that the first term will decay much faster compared to that in case of positive reactivity. After a sufficiently long time since the negative reactivity change was introduced, so that the first term would have decayed almost completely, the neutron density variation will take place as given by

$$n_{\text{dom}}(t) = n_{\text{ss}} \frac{\beta}{\beta + \bar{\mu}} e^{-\frac{\lambda_{\text{eq}}\bar{\mu}}{\beta+\bar{\mu}} t}. \tag{9.28}$$

Now, for small values of reactivity, i.e, if $\bar{\mu} << \beta$, then

$$n_{\text{dom}}(t) \approx n_{\text{ss}} e^{-\frac{\lambda_{\text{eq}}\bar{\mu}}{\beta} t} = n_{\text{ss}} e^{-\frac{t}{T_p}}, \tag{9.29}$$

where $T_p = \frac{\beta}{\lambda_{\text{eq}}\bar{\mu}} = |\frac{\beta}{\lambda_{\text{eq}}\bar{\rho}}|$. Again, the stable reactor period is approximately inversely proportional to the magnitude of reactivity inserted into the reactor.

However, for very large negative reactivities, i.e., if $\bar{\mu} >> \beta$ such that $\beta + \bar{\mu} \approx \bar{\mu}$, then

$$n_{\text{dom}}(t) \approx n_{\text{ss}} \frac{\beta}{\beta + \bar{\mu}} e^{-\lambda_{\text{eq}} t} = n_{\text{ss}} e^{-\frac{t}{T_{pn}}}, \tag{9.30}$$

where $T_{pn} = \frac{1}{\lambda_{eq}}$. The above expression signifies that the rate of decay of neutron density is independent of the magnitude of reactivity in case of very large negative reactivity changes. In addition, the rate of decay is characterized only by the decay constant for the delayed neutron precursor. This result has an important bearing in case of reactor shutdown, in which a large negative reactivity is inserted in the core by means of shut-off rods leading to appreciable quick drop in neutron density immediately but as evident from the above result, the neutron density will not become zero instantly, rather it will decay only slowly.

Example

For $\bar{\rho} = -0.001$, $\beta = 0.0075$, $\lambda_{eq} = 0.08\ s^{-1}$, and $\ell^* = 10^{-3}$, the variation of neutron density occurs as shown in Fig. 9.2. In this case too, the plots of $n(t)$ defined by (9.20), $n_{approx}(t)$ defined by (9.27), and $n_{dom}(t)$ defined by (9.29) versus time are shown. There is a good agreement between $n(t)$ and $n_{approx}(t)$ throughout the transient with excellent match during the initial phase of the transient characterized by rapid drop of neutron density from the initial value of 1 to approximately 0.88. However, as expected, $n_{dom}(t)$ matches with $n(t)$ after $t = 0.6$ s.

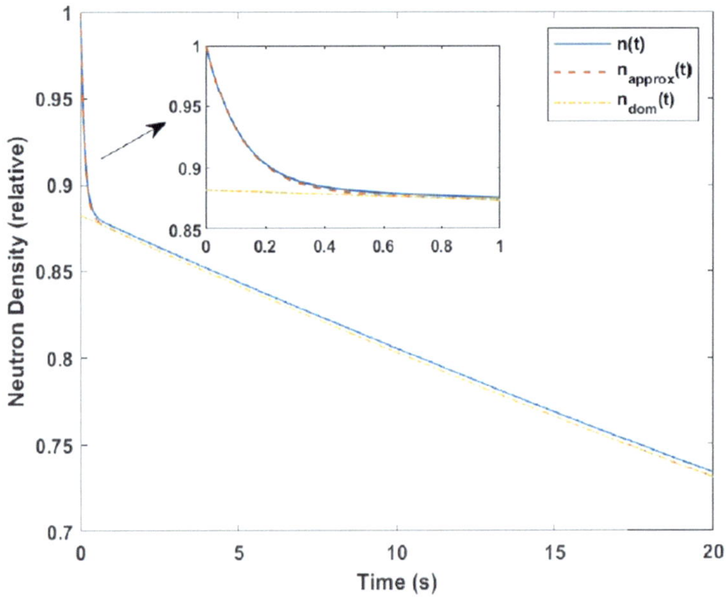

Fig. 9.2 Variation of neutron density with time subsequent to insertion of small negative reactivity $\bar{\rho} = -0.001$

9.2 In-Hour Equation

The various terms of the characteristic Eq. (9.15) can be rearranged as

$$\bar{\rho} = s\ell^* + \frac{s\beta}{s + \lambda_{\mathrm{eq}}}. \tag{9.31}$$

Long time after insertion of reactivity, as was seen in the preceding, the reactor behavior is governed by the dominant root s_2 in (9.21). Further, it is observed that the stable reactor period $T_p = \frac{1}{s_2}$. Hence, the above equation can also be written as

$$\bar{\rho} = \frac{\ell^*}{T_p} + \frac{\beta}{1 + \lambda_{\mathrm{eq}} T_p}. \tag{9.32}$$

This equation relates the stable reactor period with constant reactivity input and is popularly called as the in-hour equation.

9.3 Prompt-Jump Phenomenon and Prompt-Jump Approximation

As evident from (9.22), within a very short time of the order of a few tens of prompt neutron lifetime, subsequent to the reactivity change, the neutron density attains the value $\frac{\beta n_{\mathrm{ss}}}{\beta - \bar{\rho}} = \frac{n_{\mathrm{ss}}}{1 - \rho/\beta}$. Thereafter, it varies slowly as governed by the stable reactor period on the time scale of inverse decay constant of one effective delayed neutron group. This phenomenon in which subsequent to a reactivity change, the neutron density changes initially very sharply on the time scale of prompt neutron lifetime, is called as the prompt jump. The fast variation occurring during the prompt jump causes numerical ill-conditioning and to avoid it, the numerical solution to set of Eqs. (9.7) and (9.8) should be attempted with small time steps. However, if one is not interested in the details of variation during the prompt jump, the set of Eqs. (9.7) and (9.8) can be approximated by only one differential equation using the so-called "prompt-jump approximation" whereby it is assumed that

$$\ell^* \frac{\mathrm{d}n}{\mathrm{d}t} = 0. \tag{9.33}$$

This assumption is justified as

1. ℓ^* is very small, and
2. for small values of reactivity, the rate of change of neutron density with time is expected to remain small too.

Now, substituting the above in (9.7), we obtain

$$(\bar{\rho} - \beta)\, n_{pj}(t) + \lambda_{\text{eq}} \ell^* c_{\text{eq}}(t) = 0,$$

in which n_{pj} denotes the neutron density obtained based on the prompt-jump approximation. Simplifying, we have

$$n_{pj}(t) = -\frac{\lambda_{\text{eq}} \ell^*}{\bar{\rho} - \beta} c_{\text{eq}}(t). \tag{9.34}$$

Substituting n_{pj} from (9.34) in place of n in (9.8) and simplifying, we have

$$\frac{\mathrm{d}c_{\text{eq}}(t)}{\mathrm{d}t} = -\frac{\lambda_{\text{eq}}}{1 - \beta/\bar{\rho}} c_{\text{eq}}(t). \tag{9.35}$$

The above equation is a first-order equation which can be solved easily for $c_{\text{eq}}(t)$ and the solution so obtained can be substituted in (9.34) to obtain the time variation of neutron density based on the prompt-jump approximation. Thus, we have

$$n_{pj}(t) = \frac{n_{\text{ss}}}{1 - \bar{\rho}/\beta} e^{\frac{\lambda_{\text{eq}}\bar{\rho}}{\beta - \bar{\rho}} t}. \tag{9.36}$$

The plot of $n_{pj}(t)$ for the values of $\bar{\rho}$, β, etc. as mentioned earlier is shown in Fig. 9.3. For facilitating quick comparison, plot of $n(t)$ is also shown in it. Except for the initial short period of approximately 1 s which is of the order of $1000\ell^*$, $n_{pj}(t)$ is seen to be very close to $n(t)$.

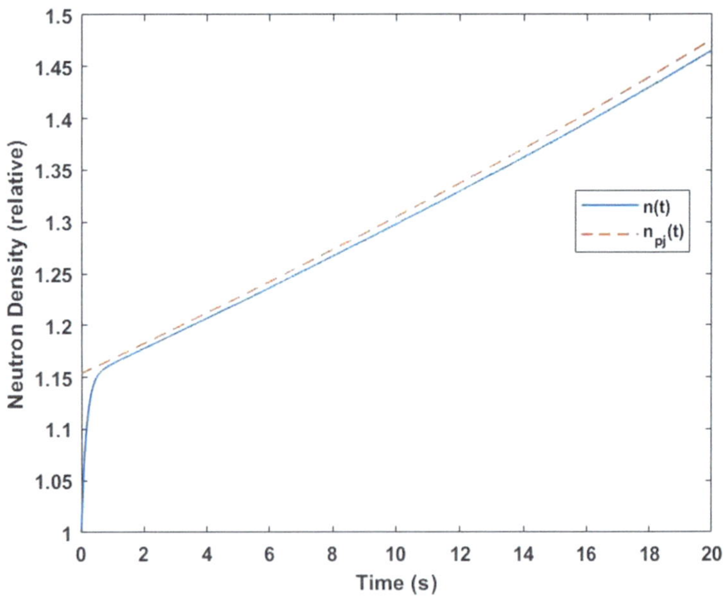

Fig. 9.3 Variation of neutron density (based on prompt-jump approximation) with time subsequent to insertion of small positive reactivity

9.4 Solution of Linearized Point Kinetics Model

Now, we turn our attention to the set of linearized point kinetics equations in standard state-space form introduced in Chap. 8. Recall that the system has been linearized around a steady-state condition in which the neutron density is assumed to be n_{ss} and the effective one group delayed neutron concentration given by

$$c_{\text{eq,ss}} = \frac{\beta}{\lambda_{\text{eq}}\ell^*} n_{\text{ss}}. \tag{9.37}$$

The reactivity input $\delta\rho(t) = \bar{\rho}(t)$ as the steady-state reactivity is 0. The system, input, and output matrices are given by Eqs. (8.65), (8.66), and (8.67) in Chap. 8. The solution will be given by (9.1) in which the quantities $\Phi(\tau)$ and $\Gamma(\tau)$ are to be evaluated for $\mathbf{F}_{\text{lin}}$ and $\mathbf{G}_{\text{lin}}$. Thus, we have

$$\begin{aligned}\Phi(t) &= \mathcal{L}^{-1}\left(sI - \mathbf{F}_{\text{lin}}\right)^{-1} \\ &= \mathcal{L}^{-1}\left(\begin{bmatrix} s + \frac{\beta}{\ell^*} & -\lambda \\ -\frac{\beta}{\ell^*} & s + \lambda_{\text{eq}} \end{bmatrix}\right)^{-1}\end{aligned}$$

which after some mathematical manipulations takes the form

$$\Phi(t) = \frac{1}{\frac{\beta}{\ell^*} + \lambda_{\text{eq}}} \begin{bmatrix} \lambda_{\text{eq}} + \frac{\beta}{\ell^*} e^{-\left(\frac{\beta}{\ell^*} + \lambda_{\text{eq}}\right)t} & \lambda_{\text{eq}}\left(1 - e^{-\left(\frac{\beta}{\ell^*} + \lambda_{\text{eq}}\right)t}\right) \\ \frac{\beta}{\ell^*}\left(1 - e^{-\left(\frac{\beta}{\ell^*} + \lambda_{\text{eq}}\right)t}\right) & \frac{\beta}{\ell^*} + \lambda_{\text{eq}} e^{-\left(\frac{\beta}{\ell^*} + \lambda_{\text{eq}}\right)t} \end{bmatrix}. \tag{9.38}$$

Now, for constant reactivity input, the second term is evaluated as

$$\begin{aligned}&\int_0^t \Phi(t-\tau)\mathbf{G}_{\text{lin}}\delta\rho d\tau \\ &\quad = \int_0^t \frac{n_{\text{ss}}}{\ell^*} \begin{bmatrix} \lambda_{\text{eq}} + \frac{\beta}{\ell^*} e^{-\left(\frac{\beta}{\ell^*} + \lambda_{\text{eq}}\right)(t-\tau)} \\ \frac{\beta}{\ell^*}\left(1 - e^{-\left(\frac{\beta}{\ell^*} + \lambda_{\text{eq}}\right)(t-\tau)}\right) \end{bmatrix} \delta\rho d\tau.\end{aligned} \tag{9.39}$$

After mathematical simplifications, we obtain

$$\begin{aligned}&\int_0^t \Phi(t-\tau)\mathbf{G}_{\text{lin}}\delta\rho d\tau \\ &\quad = \int_0^t \frac{n_{\text{ss}}\delta\rho}{\ell^*} \begin{bmatrix} \lambda_{\text{eq}} t + \frac{\beta/\ell^*}{\frac{\beta}{\ell^*} + \lambda_{\text{eq}}}\left[1 - e^{-\left(\frac{\beta}{\ell^*} + \lambda_{\text{eq}}\right)t}\right] \\ \frac{\beta}{\ell^*} t - \frac{\beta/\ell^*}{\frac{\beta}{\ell^*} + \lambda_{\text{eq}}}\left[1 - e^{-\left(\frac{\beta}{\ell^*} + \lambda_{\text{eq}}\right)t}\right] \end{bmatrix},\end{aligned} \tag{9.40}$$

from which the solution to change in neutron density is obtained as

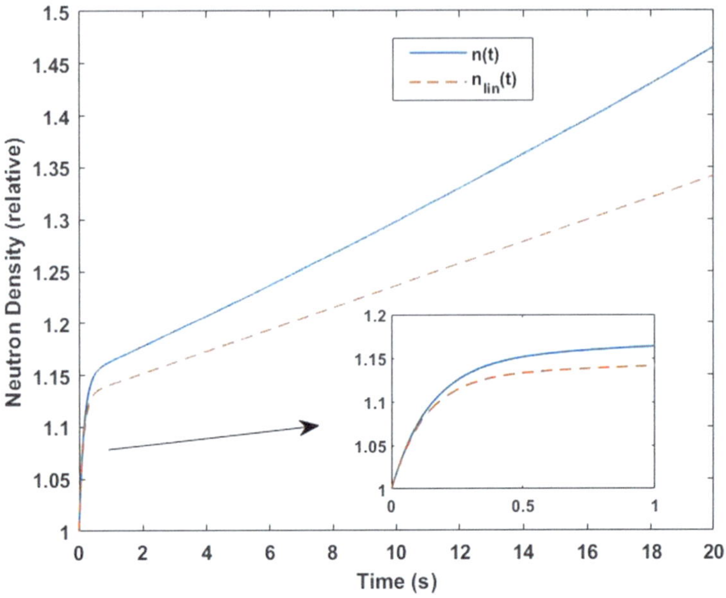

Fig. 9.4 Time variation of neutron density based on Linearized point kinetics model and the actual neutron density subsequent to insertion of small positive reactivity

$$\delta n(t) = \frac{n_{ss}\delta\rho}{\ell^*}\left[\lambda_{eq}t + \frac{\beta/\ell^*}{\frac{\beta}{\ell^*} + \lambda_{eq}}\left(1 - e^{-(\frac{\beta}{\ell^*}+\lambda_{eq})t}\right)\right]. \tag{9.41}$$

Long time after change in reactivity, i.e., for very large values of t, the term $e^{-(\frac{\beta}{\ell^*}+\lambda_{eq})t}$ becomes negligibly small. Hence, the change is neutron density after long time subsequent to change in reactivity can be approximated as

$$\delta n(t) \approx \delta n_{approx}(t) = \frac{n_{ss}\delta\rho}{\ell^*}\left[\lambda_{eq}t + \frac{\beta/\ell^*}{\frac{\beta}{\ell^*} + \lambda_{eq}}\right]. \tag{9.42}$$

Example

We consider the data as in the example in Sect. 9.2 earlier, i.e., $\delta\rho = 0.001$, $\beta = 0.0075$, $\lambda_{eq} = 0.08\ \text{s}^{-1}$, and $\ell^* = 10^{-3}$. The variation of neutron density occurs as shown in Fig. 9.4. In this, time variation of $n_{lin}(t) = n_{ss} + \delta n(t)$ where $\delta n(t)$ is defined by (9.41) is shown. For comparison, the plot of neutron density predicted based on (9.20) versus time is also shown there. The neutron density predicted based on the linearized point kinetics model is seen to be in agreement with the actual neutron density during a short time duration of the order of fraction of a second. Subsequently, the deviation of $n_{lin}(t)$ from $n(t)$ becomes appreciable. From this, it is obvious that linearized model will have limited applicability.

9.5 Reactor Stability

Information about reactor stability can be obtained by looking at the roots of its characteristic Eq. (9.15) for different values of reactivity. Some observations about the nature of roots are already stated in the preceding sections. However, in Table 9.1, the values of both the roots for different values of $\bar{\rho}$ ranging from about -100β to 0.5β are given. It must be reckoned that during normal operation of the reactor, the reactivity values on positive side are limited to about $\frac{1}{3}$ β. From the values, the following observations are evident:

1. For all values of reactivity, the roots are real.
2. For large negative values of reactivity several times of β, which usually is the case during the reactor shutdown, both the roots are negative and large in magnitude.

Table 9.1 Variation of roots of characteristic equation with reactivity

S. no.	Reactivity, β	First root	Second root
1	−100.000	−0.079208	−656.5
2	−50.000	−0.078431	−331.5
3	−25.000	−0.076922	−169.0
4	−10.000	−0.072720	−71.507
5	−5.000	−0.066644	−39.013
6	−2.500	−0.057085	−22.773
7	−1.000	−0.039877	−13.04
8	−0.500	−0.026521	−9.8035
9	−0.250	−0.015875	−8.1891
10	−0.100	−0.007199	−7.2228
11	−0.050	−0.003767	−6.9012
12	−0.020	−0.001550	−6.7084
13	−0.005	−0.0003932	−6.6121
14	−0.002	−0.0001577	−6.5928
15	−0.001	-7.90×10^{-5}	−6.5864
16	0	0	−6.58
17	0.001	7.91×10^{-5}	−6.5736
18	0.002	0.0001584	−6.5672
19	0.005	0.0003971	−6.5479
20	0.020	0.0016120	−6.4516
21	0.050	0.0041539	−6.2592
22	0.100	0.008756	−5.9388
23	0.250	0.026099	−4.9811
24	0.500	0.076329	−3.4063

3. For small negative values of reactivity, in the range $-\beta$ to < 0, the roots continue to remain negative though they tend to become smaller and smaller in magnitude.
4. For $\bar{\rho} = 0$, one of the roots is 0 while the other negative and still quite large in magnitude.
5. For positive values of reactivity less than β, one of the roots is positive while the other negative. And, if the magnitude of reactivity increases, the magnitude of the positive root increases too. But, the negative root tends to saturate at λ_{eq}.

Based on the above observations, it can be stated that the nuclear reactor described by point kinetics model is stable for negative reactivities. It is stable for zero reactivity too but for positive reactivities the reactor is unstable.

9.6 Controllability and Observability of Reactor

Controllability and observability properties are distinct to modern control. If a system is controllable it implies that a controller can be designed to stabilize and obtain any desired response from the system. Similarly, observability implies that the system state variables may be determined based on output measurements taken over a finite time interval. Standard text books in the area of modern control deal with these topics in detail.

There are numerous methods for test of controllability and observability but here we apply the algebraic controllability and observability criteria to derive results about reactor controllability and observability properties. As per the algebraic criterion, the linear system described by Eqs. (8.65), (8.66), and (8.67) in Chap. 8 will be controllable if

$$\text{rank}\ [\mathbf{G}_{\text{lin}}\quad \mathbf{F}_{\text{lin}}\mathbf{G}_{\text{lin}}] = 2 \tag{9.43}$$

and observable if

$$\text{rank}\ \left[\mathbf{H}_{\text{lin}}^{\text{T}}\quad \mathbf{F}_{\text{lin}}^{\text{T}}\mathbf{H}_{\text{lin}}^{\text{T}}\right] = 2. \tag{9.44}$$

Substituting the values of the different matrices, we have the controllability test matrix as

$$[\mathbf{G}_{\text{lin}}\quad \mathbf{F}_{\text{lin}}\mathbf{G}_{\text{lin}}] = \frac{n_{\text{ss}}}{\ell^*}\begin{bmatrix} 1 & -\frac{\beta}{\ell^*} \\ 0 & \frac{\beta}{\ell^*} \end{bmatrix} \tag{9.45}$$

which is an upper triangular matrix. Its rank will be 2 if n_{ss} is nonzero. Hence, the nuclear reactor will be controllable if the neutron density in it will be nonzero. It is an important result which mathematically establishes the well-known fact that there must be some initial neutrons in the system to be able to adjust the speed of the chain reaction and thereby the neutron population in the reactor.

Next, consider the observability matrix, which is obtained as

$$\begin{bmatrix} \mathbf{H}_{\text{lin}}^{\text{T}} & \mathbf{F}_{\text{lin}}^{\text{T}} \mathbf{H}_{\text{lin}}^{\text{T}} \end{bmatrix} = \begin{bmatrix} 1 & -\frac{\beta}{\ell^*} \\ 0 & \frac{\beta}{\ell^*} \end{bmatrix}. \tag{9.46}$$

Rank of this matrix is 2 and hence the nuclear reactor is an observable system.

9.7 Concluding Remarks

In this chapter, the reactor behavior has been studied for step change in reactivity with the help of nonlinear as well as the linearized model. Of course, use of the simple second-order model based on one effective group of delayed neutrons has been made in the analysis. The methodology can be extended to mathematical model of higher order with the six delayed neutron groups and for arbitrary reactivity variations. Numerical schemes for solution of the equations may be deployed under the complex analysis situations.

It has been established that for any nonzero reactivity input, the neutron density in the reactor would undergo continuous unbounded variation, i.e., it will either continue increasing or decreasing with time, indicating the instability of the system in bounded-input bounded-output (BIBO) sense. It has also been established that the reactor is a controllable and observable physical system. Thus, it would be possible to design suitable controllers and observers for the reactor to stabilize it and also to obtain the desired performance.

As we conclude the chapter, we make a mention that the situation in a practical reactor system is far more complex. Although the topics presented here help one to understand the basic characteristics, their application to real-life situations should be made with care. In a practical system operating on any meaningful power level, the fuel, coolant, and moderator, besides being arranged heterogeneously, operate at different temperatures. As a result, there are temperature-dependent internal feedback effects. Changes in density also lead to variation in parameters and internal feedback effects. During the operation of the reactor, accumulation and decay of different types of fission products takes place. Some of these fission products can result into changes in the neutronic interaction cross sections, bringing internal reactivity feedback effects. Fortunately, a rich pool of knowledge is available in the existing literature to satisfy the quest of inquisitive readers.

References

1. J.J. Duderstadt, L.J. Hamilton, *Nuclear Reactor Analysis* (Wiley, New York, 1976)
2. S. Glasstone, A. Sesonske, *Nuclear Reactor Engineering* (CBS Publishers & Distributers, Delhi, 1986)

3. W.M. Stacey, Jr., *Space Time Nuclear Reactor Kinetics* (Academic Press, New York, 1969)
4. W.M. Stacey, *Nuclear Reactor Physics* (WILEY-VCH Verlag GmbH & Co. KGaA, Weinheim, 2007)
5. J.R. Lamarsh, *Introduction to Nuclear Reactor Theory* (Addison-Wesley, Reading, MA, 1983)
6. M. Ash, *Nuclear Reactor Kinetics* (McGraw-Hill, New York, 1965)
7. C.T. Chen, *Linear System Theory and Design* (Oxford University Press, New York, 1999)
8. https://www.npcil.nic.in
9. https://www.world-nuclear.org

Chapter 10
Interval Methods in Nuclear Reactor Control

Amitava Gupta, Debayan Bose, and Shohan Banerjee

10.1 Introduction

Power in a nuclear reactor is controlled by altering the reactivity input through the use of various reactivity control devices like control rods, liquid poisons light water compartments, etc. A positive reactivity increases the effective neutron multiplication rate, thus increasing the reactor power, while a negative reactivity reduces the reactor power through decrease in effective neutron multiplication rate. The control system in a nuclear reactor generates the inputs which modulate the reactivity control devices to alter the reactivity input to the reactor in order to control the power.

A nuclear reactor is a nonlinear plant with time-variant (e.g. variations associated with a fresh and an equilibrium core), power dependent characteristics along with several parameters which cannot be modelled with sufficient accuracy e.g. the estimation of various parameters associated with the thermal-hydraulics of the plant [1]. Thus, robustness to parametric variations, in addition to stability robustness, is a necessary condition to be fulfilled by a reactor's power control system for a stable operation of the plant. The general schematic of power control is presented in Fig. 10.1.

Secondly, a Nuclear Power Plant (NPP) is mostly constrained to act as base-load station [2] due to the limits on the rate and quantum of reactivity addition which prevent the plant to follow a rapidly changing power demand curve. In the present

A. Gupta (✉)
Department of Power Engineering and School of Nuclear Studies and Applications, Jadavpur University, Kolkata, India
e-mail: amitava.gupta@jadavpuruniversity.in

D. Bose
Department of Computer Science Engineering, IIT Kharagpur, Kharagpur, India
e-mail: debayanbose@iitkgp.ac.in

S. Banerjee
School of Built Environment, Engineering, and Computing, Leeds Beckett University, Leeds, UK

U. Mangalanathan et al. (eds.), *Nuclear Instrumentation and Control*,
https://doi.org/10.1007/978-981-97-1283-0_10

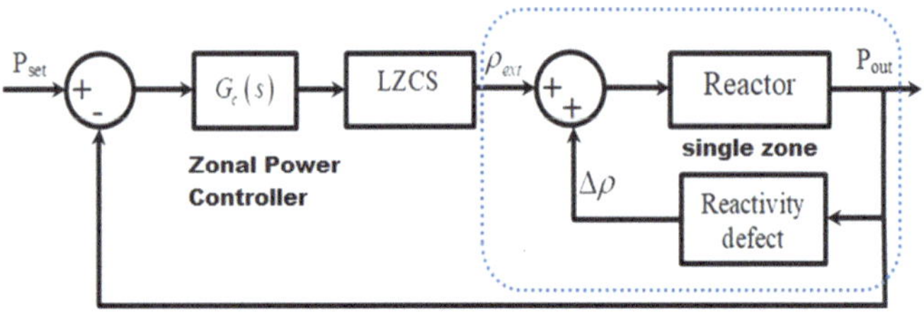

Fig. 10.1 General power control loop of a nuclear reactor

context of market-driven conditions, for example, Availability Based Tariff (ABT) [3] regime, NPPs acting as bulk power suppliers to the grids may also need to function as load-followers with a limited number of bulk load variations per day. Also according to the European Utilities Requirement (EUR), an NPP must be capable of daily load cycling operation between 50 and 100% of its Full Power (FP), with the ramping rate of 3–5% FP per minute [2].

Some studies e.g. in [3] have explored the applicability of a load-following NPP as a balancing component in a grid with fluctuating energy sources such as wind energy. An inherent requirement for a load-following NPP is a plant with self-regulating characteristics and thus, under-moderated reactors like a Pressurized Water Reactor (PWR), for example, which exhibit self-regulating features under varying load conditions, appear as a natural choice for load-following NPPs. However, contemporary research shows that a Pressurized Heavy Water Reactor (PHWR) is also capable of being used in a limited load-following mode with a robust controller.

Contemporary researchers have applied several classical robust control approaches like H_∞ based controller design methodology and μ synthesis as reported in [4–6] to realize load-following controllers for NPPs. These methodologies usually result in higher order controllers which are difficult to realize in practice and often require order reduction, with substantial performance deviation [6]. Another approach e.g. in [7] proposes a State Feedback Assisted Control (SFAC) which is based on identification of a plant's dominant eigenvalues and hence sensitivity of these eigenvalues to variations in plant parameters. The method has been applied to control a PWR. The work in [8] considers the thermal-hydraulic effects and parametric uncertainties caused by a changing reactor power for a 540 MWe Indian PHWR and proposes a de-centralized controller design methodology based on Periodic Output Feedback which produces an acceptable and robust response. However, this methodology requires a fast sampler, as the input needs to be changed multiple number of times within an output sample interval, which may be difficult to realize. Fuzzy Logic Controllers (FLCs) are known for their uncertainty handling capabilities and some researchers [9] have also proposed use of FLCs for control of PHWRs. But use of such controllers for control of nuclear reactors needs rigorous validation from stability point of view and these methods are largely restricted to simulation studies due to associated regulatory concerns. It is a well-known fact that for most physical systems, a reasonable parametric robustness can be achieved by flattening the phase curve around the Gain Crossover Frequency (GCF) and this principle has been used in [1] using a Fractional Order (FO) phase shaper in conjunction with a PID for robust control of a PHWR. In [10] have extended this using Fractional Order PID (FOPID) controllers, but these controllers mainly to take care of variation is system gain with varying operating power of the reactor and cannot take care of physical uncertainties. Moreover, these methodologies require implementation of a FO element which is not easy to realize and analyse for stability and performance requirements.

Since 90% of industrial controllers are Proportional-Integral-Derivative (PID) type and these controllers are known for their inherent good tracking and robustness, control of power in a 540 MWe PHWR with PI controllers has been explored by some researchers e.g. in [9]. The use of discrete PID controllers integrated over a data network as a Networked Control System (NCS) for control of a 540 MWe PHWR has also been proposed in [11]. Sliding Mode Control (SMC) is another technique with established capability of producing controllers for robust control of both linear and nonlinear plants [12], in presence of system uncertainties. However, these SMC based controllers exhibit the chattering in transient state which may affects the stability of the systems. A recent trend in power level control of nuclear reactors involves the use of Nonlinear Dynamic Inversion (NDI) as an approach for nonlinear controller design [13, 14] for the control of a PWR in the load following mode. This is essentially a feedback linearization approach.

It follows as a natural corollary, therefore, that a control system which is able to produce satisfactory response when the reactor's parameters vary within an interval, can be explored for realization of a load-following NPP. The principle behind the control system is simple: the designed controller should be such that if the system parameters vary within a bounded interval, for example, due to power variation, then the system performance parameters also remain within a specified interval with known bounds. An example of this approach is the work in [15] which uses the Kharitonov theorem to design a robust controller for a PWR starting with an inexact transfer function and hence with parametric uncertainties. In another study, a robust controller for a Boiling Water Reactor (BWR) to achieve integrated control of turbine power, water level and throttle pressure has been proposed in [16] using an interval approach.

In this chapter, the use of interval methods in realization of robust controllers for NPPS, reported by the authors in recent publications are reported in a consolidated manner. While the techniques proposed in [17, 18] rely on conventional PID controllers, the work done in [13] rely on nonlinear feedback linearization schemes like the NDI which is known to be very effective in the domain of aircraft control, with proven robustness both in terms of stability and parametric uncertainties.

In the sections that follow in this chapter, the basics of Interval mathematics is first presented, which is followed by a general thermal hydraulic model of a nuclear reactor extended to include the appropriate reactivity feedbacks relevant for operation of a PHWR and a PWR based plant to illustrate the full scope of a robust controller design. These models are then used to explain the control mechanisms and the responses they achieve and to further identify the scope for further research in this domain.

10.2 Basic of Interval Mathematics

An interval system with bounded parametric uncertainties can be described using the following general transfer function representation defined as [15]

$$\left.\begin{aligned} G(s,\mathbf{p}) &= \frac{f_0+f_1S^m+\ldots+f_mS^m}{g_0+g_1s+\cdots+g_ns^n} \\ \mathbf{p} &= [f_0, f_1, \ldots, f_m; g_0, g_1, \ldots, g_n]; \end{aligned}\right\} \tag{10.1}$$

The set $\mathbf{p}$ is the set of all parameters with admissible uncertainties and is represented by the set of all interval variables in $G(s, \mathbf{p})$. A bounded interval variable Z is defined as $Z \in \mathbb{R} | \underline{Z} \le Z \le \overline{Z}$ where $\underline{Z}$ represents *infimum* of Z and $\overline{Z}$ its supremum and $[\underline{Z}, \overline{Z}]$ represents the interval of Z. Further, if $N \in [\underline{N}, \overline{N}]$ be another interval variable, then the following hold [19]:

$$\left.\begin{aligned} [\underline{Z}, \overline{Z}] + [\underline{N}, \overline{N}] &= [(\underline{Z}, +\underline{N}), (\overline{Z} + \overline{N})] \\ [\underline{Z}, \overline{Z}] - [\underline{N}, \overline{N}] &= [(\underline{Z} - \overline{N}), (\overline{Z} - \underline{N})] \\ [\underline{Z}, \overline{Z}] \times [\underline{N}, \overline{N}] &= \begin{bmatrix} \min(\underline{Z}\,\underline{N}, \underline{Z}\,\overline{N}, \overline{Z}\underline{N}, \overline{Z}\,\overline{N}), \\ \max(\underline{Z}\,\underline{N}, \underline{Z}\,\overline{N}, \overline{Z}\,\underline{N}, \overline{Z}\,\overline{N})] \end{bmatrix} \\ \frac{[\underline{Z}, \overline{Z}]}{[\underline{N}, \overline{N}]} &= [\underline{Z}, \overline{Z}] \times \left[\frac{1}{\overline{N}}, \frac{1}{\underline{N}}\right] \end{aligned}\right\} \tag{10.2}$$

The definition of interval holds for a vector or a matrix as well. For matrices, the interval operations are different from the ones defined in Eq. 10.2 [20]. It is termed as an interval matrix then $\mathbf{N}$ can be defined as

$$\mathbf{N} = \begin{bmatrix} n_{11} & n_{12} & \cdots & n_{1n} \\ n_{21} & n_{22} & \cdots & n_{2n} \\ \vdots & \vdots & \ddots & \vdots \\ n_{n1} & n_{n2} & \cdots & n_{nn} \end{bmatrix} \tag{10.3}$$

where any element $n_{ij} \in \mathbf{N}$ is an *interval variable* defined as $n_{ij} = [\underline{n}_{ij}, \overline{n}_{ij}]$. Correspondingly, the matrix $\underline{\mathbf{N}}$ denoted as the *infimum* of $\mathbf{N}$ and the matrix $\overline{\mathbf{N}}$ denoted as the supremum of $\mathbf{N}$ may be defined as

$$\underline{\mathbf{N}} = \begin{bmatrix} \underline{n}_{11} & \underline{n}_{12} & \cdots & \underline{n}_{1n} \\ \underline{n}_{21} & \underline{n}_{22} & \cdots & \underline{n}_{2n} \\ \vdots & \vdots & \ddots & \vdots \\ \underline{n}_{n1} & \underline{n}_{n2} & \cdots & \underline{n}_{nn} \end{bmatrix} \tag{10.4}$$

and

$$\overline{\mathbf{N}} = \begin{bmatrix} \overline{n}_{11} & \overline{n}_{12} & \cdots & \overline{n}_{1n} \\ \overline{n}_{21} & \overline{n}_{22} & \cdots & \overline{n}_{2n} \\ \vdots & \vdots & \ddots & \vdots \\ \overline{n}_{n1} & \overline{n}_{n2} & \cdots & \overline{n}_{nn} \end{bmatrix} \tag{10.5}$$

respectively. Again, the matrix **N** can be alternatively expressed as

$$\mathbf{N} = \mathbf{N}_c + \mathbf{N}_r \tag{10.6}$$

where 'c' as a suffix represents nominal matrix and 'r' as a suffix represents the variation of the *infimum* and/or *supremum* from the nominal matrix. Therefore,

$$\mathbf{N}_c = \frac{1}{2}(\underline{\mathbf{N}} + \overline{\mathbf{N}}) \tag{10.7}$$

$$\mathbf{N}_r = \mathbf{N}_c - \underline{\mathbf{N}} \tag{10.8}$$

Now if $\mathbf{Z} \in [\underline{\mathbf{Z}}, \overline{\mathbf{Z}}]$ be another interval matrix, then using the results presented in [20], for the product $\mathbf{W} = \mathbf{NZ}$, the following hold:

$$\mathbf{W}_r = (|\mathbf{N}_c|\mathbf{Z}_r) + \mathbf{N}_r(|\mathbf{Z}_c| + \mathbf{Z}_r) \tag{10.9}$$

$$\mathbf{W}_c = \mathbf{N}_c\mathbf{Z}_c \tag{10.10}$$

$$\overline{\mathbf{W}} = \mathbf{N}_c.\mathbf{Z}_c + \mathbf{W}_r \tag{10.11}$$

$$\underline{\mathbf{W}} = \mathbf{N}_c.\mathbf{Z}_c - \mathbf{W}_r \tag{10.12}$$

If **N** is a *point matrix* as opposed to an interval matrix, then $\mathbf{N}_c = \mathbf{N}$ and $\mathbf{N}_r = 0$. For such cases, the following hold for the product $\mathbf{W} = \mathbf{NZ}$

$$\mathbf{W}_r = (|\mathbf{N}|.\mathbf{Z}_r) \tag{10.13}$$

$$\overline{\mathbf{W}} = (\mathbf{N}.\mathbf{Z}_c + \mathbf{W}_r) \tag{10.14}$$

$$\underline{\mathbf{W}} = (\mathbf{N}.\mathbf{Z}_c - \mathbf{W}_r) \tag{10.15}$$

Again for a set of interval variables $\mathbf{w} = [w_1, w_2, \ldots w_N]$ a convex box of **w** means the Cartesian product of the intervals $[\underline{w_1}, \overline{w_1}] \times [\underline{w_2}, \overline{w_2}] \times \cdots \times [\underline{w_N}, \overline{w_N}]$.

In Eq. (10.1) the transfer function $G(s, \mathbf{p})$ represents a ratio of two polynomials in s, the coefficients of each being interval variables represented by real numbers which lie within specific bounds $[\underline{g}_i, \overline{g}_i] \forall i \in [0, m]$ and $[\underline{f}_j, \overline{f}_j] \forall j \in [0, n]$. If the denominator of the interval plant is expressed as

$$q(s, \mathbf{p}) = g_0 + g_1 s + g_2 s^2 + \cdots + g_n s^n \tag{10.16}$$

then the interval plant $G(s, \mathbf{p})$ will be Hurwitz stable if the set of following four polynomials known as Kharitonov Polynomials [21] are Hurwitz stable:

$$\left\{ \begin{array}{l} q_1(s) = \underline{g}_0 + \underline{g}_1 s + \overline{g}_2 s^2 + \overline{g}_3 s^3 + \underline{g}_4 s^4 + \underline{g}_5 s^5 \ldots \\ q_2(s) = \overline{g}_0 + \overline{g}_1 s + \underline{g}_2 s^2 + \underline{g}_3 s^3 + \overline{g}_4 s^4 + \overline{g}_5 s^5 \ldots \\ q_3(s) = \underline{g}_0 + \overline{g}_1 s + \overline{g}_2 s^2 + \underline{g}_3 s^3 + \underline{g}_4 s^4 + \overline{g}_5 s^5 \ldots \\ q_4(s) = \overline{g}_0 + \underline{g}_1 s + \underline{g}_2 s^2 + \overline{g}_3 s^3 + \overline{g}_4 s^4 + \overline{g}_5 s^5 \ldots \end{array} \right\}. \tag{10.17}$$

Hurwitz stability of Kharitonov Polynomials serve as the necessary and sufficient condition for robust stability of the interval plant $G(s, \mathbf{p})$.

10.3 General Lumped-Parameter Modelling of a Nuclear Reactor and Its Extension to Include Reactor Thermal Hydraulics

In this section, the modelling aspects associated with two distinct types of reactors viz. a PHWR and PWR are discussed. Of these, the PHWRs considered are typically 540 MWe Heavy Water (i.e. D_2O) moderated and cooled reactors fuelled by natural Uranium, while the PWRs are Light Water (i.e. H_2O) moderated and cooled, enriched Uranium fuelled reactors.

10.3.1 Mathematical Modelling of PHWR

The mathematical modelling of a typical 540 MWe PHWR is presented in this section.

Figure 10.2 shows the schematic of a standard 540 MWe PHWR plant where the reactor vessel is cylindrical containing a number of pressurized horizontal coolant channels. Each channel is a concentric fuel assembly, with the coolant (D_2O) running through these channels. The pressurized channels prevent in-core boiling and the heat from the nuclear fuel in the channel is passed on to the sub-cooled flowing coolant forced into the channels by Primary Circulating Pumps (PCPs) and this sub-system is known as the Primary Heat Transport (PHT) system. The thermal energy of the radioactive coolant in the PHT is then passed on to the secondary side through a Steam Generators (SGs) which drive a turbo-alternator through a steam header. The steam driving the turbine is Light Water (H_2O) and does not come into contact with the radioactive PHT coolant in a shell and tube type SG.

As mentioned in the previous section, a 540 MWe PHWR consists of 14 reactor zones, based on the steady-state power distribution, each of which can be controlled

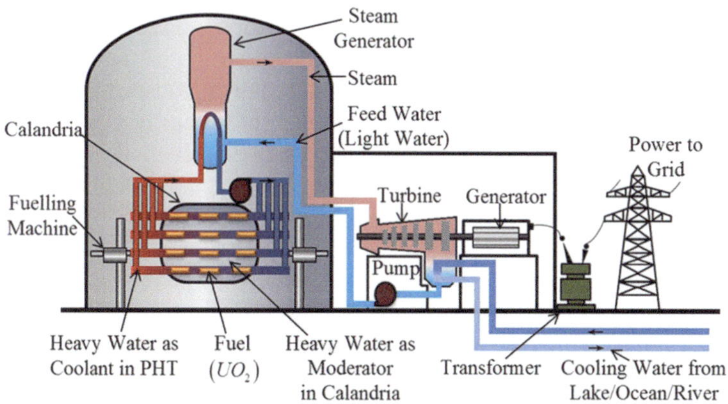

Fig. 10.2 Schematic representation of a PHWR plant [5]

by varying the height of the water column in its ZCC by a Liquid Zone Control System (LZCS). The water level in each ZCC is controlled by modulating control valves to fill or drain the compartments with light water, thus controlling the reactivity input. A comprehensive description of the LZCS is available in [22] and in this chapter the data presented in [22, 23] has been used for model presented.

In this section, it is attempted to develop an inexact transfer function model for each PHWR zone of the general form defined by Eq. 10.3. A lumped parameter point kinetic model of a PHWR zone neglecting inter zone coupling can be represented by the following set of equations.

$$\frac{\mathrm{d}P_{\mathrm{th}}}{\mathrm{d}t} = -\frac{\beta}{l}P_{\mathrm{th}} - \frac{\sigma_x X}{\Sigma_a}\frac{P_{\mathrm{th}}}{l} + \lambda C + \rho_{\mathrm{nett}}\frac{P_{\mathrm{th}}}{l} \tag{10.18}$$

$$\frac{\mathrm{d}C}{\mathrm{d}t} = \beta\frac{P_{\mathrm{th}}}{l} - \lambda C \tag{10.19}$$

$$\frac{\mathrm{d}I}{\mathrm{d}t} = (\lambda_I + \sigma_I I)P_{\mathrm{th}} + \gamma_I \Sigma_f P_{\mathrm{th}} \tag{10.20}$$

$$\frac{\mathrm{d}X}{\mathrm{d}t} - (\lambda_X + \overline{\sigma}_X P_{\mathrm{th}})X + \lambda_I I + \gamma_x \Sigma_f P_{\mathrm{th}} \tag{10.21}$$

$$\frac{\mathrm{d}T_f}{\mathrm{d}t} = \frac{1}{m_f c_{p_f}}\left\{P_{\mathrm{th}} - \Omega(T_f - T_c)\right\} \tag{10.22}$$

$$\frac{\mathrm{d}T_c}{\mathrm{d}t} = \frac{1}{m_c c_{p_c}}\left\{\left[\dot{m}c_{pc}T_{c_{\mathrm{input}}} - \dot{m}c_{pc}T_{c_{\mathrm{exit}}}\right] + \Omega(T_f - T_c)\right\} \tag{10.23}$$

where P_{th} is the zonal reactor thermal power (W). In Eq. 10.18 ρ_{nett} is the net reactivity input to the reactor and is the algebraic sum of the reactivity change due to change in fuel and coolant temperatures and the external reactivity introduced by the reactivity control devices. The term C denotes delayed neutron precursor concentration and I and X represent the iodine and xenon concentrations respectively for a particular zone (m^{-3}). Similarly, β denotes for delayed neutron fractional yield (averaged over several groups) and λ denotes the corresponding decay constant (s^{-1}). The terms σ_x and σ_I represent the microscopic thermal neutron absorption cross sections of the built up xenon and iodine respectively (m^2). Similarly, γ_X and γ_I denote xenon and iodine yield per fission, λ_X and λ_I denote corresponding decay constants (s^{-1}) and Σ_a and Σ_f represent thermal neutron absorption and fission cross sections (m^{-1}).

Eqs. 10.18–10.23 represent a coupled point kinetics and thermal-hydraulic model of a 540 MWe PHWR and in Eqs. 10.18–10.23, m_f and m_c denote mass of the fuel and mass of the coolant in the core (kg), C_{pf} and C_{pc} represent specific heat of the fuel and specific heat of the coolant respectively at constant pressure (J/kg °C) while A is the active heat transfer area (m^2) and h is the fuel to coolant heat transfer coefficient (W/m^2 °C). The symbol $\dot{m}$ denotes the mass flow rate of the coolant (kg/s), T_f (°C) denotes the average fuel temperature and T_c denotes the average coolant temperature (°C) defined as

$$T_c = \frac{T_{C_{\text{inlet}}} + T_{C_{\text{outlet}}}}{2} \tag{10.24}$$

where $T_{C_{\text{input}}}$ and $T_{C_{\text{outlet}}}$ are the inlet and outlet coolant temperatures respectively in °C. In the formulation Eqs. (10.22–10.23) $T_{C_{\text{inlet}}}$ are assumed to be constant.

With the assumptions presented above, the net reactivity input to a zone ρ_{nett} may be expressed

$$\rho_{\text{nett}} = \rho_{\text{ext}} + \alpha_f (T_f - T_{f_{\text{ss}}}) + \alpha_c (T_c - T_{c_{\text{ss}}}) \tag{10.25}$$

where ρ_{ext} denotes the reactivity introduced in the zone due to control rod movement and/or change in light water level in the corresponding ZCC, $T_{f_{\text{ss}}}$ and $T_{C_{\text{ss}}}$ represent the initial (equilibrium) values of average fuel and coolant temperatures and α_c, α_f represent the temperature coefficients of reactivity of the coolant and fuel respectively (°C^{-1}).

Using Eqs. (10.18–10.23), an incremental state-space model for the PHWR zone can be represented as

$$\dot{\tilde{X}} = A\tilde{X} + B\tilde{u} \text{ and } \tilde{y} = C\tilde{X} \tag{10.26}$$

where the set of incremental state variables are denoted as $\tilde{\mathbf{X}} = [\tilde{P}_{\text{th}}\ \tilde{C}\ \tilde{I}\ \tilde{X}\ \tilde{T}_f\ \tilde{T}_c]^T$ around an equilibrium state $[P_{\text{ss}}\ C_{\text{ss}}\ I_{\text{ss}}\ X_{\text{ss}}\ T_{\text{ss}}\ T_{\text{ss}}]^T$. Similarly, the incremental change in net reactivity denoted by $\tilde{\mathbf{u}}$ which can be represented by ρ_{nett} with, $\mathbf{C} = [1\ 0\ 0\ 0\ 0\ 0]$ being the output matrix and the output state vector a $\tilde{\mathbf{y}}$ representing incremental reactor zonal power $\tilde{P}_{\text{th}}$ calculated around P_{ss}. Substituting $\Omega = Ah$, $M_c = \dot{m}c_{p_c}$, $\mu_f = m_f c_{p_{cf}}$ and $\mu_c = m_c c_{p_c}$ in Eqs. (10.18–10.23), the relevant matrices may be computed as

$$\mathbf{A} = \begin{bmatrix} -\left(\beta + \frac{\sigma_X X_{ss}}{\Sigma_a}\right)\frac{1}{l} & \lambda & 0 & -\frac{\sigma_X X_{ss}}{\Sigma_a l} & \alpha_f \frac{P_{ss}}{l} & \alpha_c \frac{P_{ss}}{l} \\ \frac{\beta}{l} & -\lambda & 0 & 0 & 0 & 0 \\ -\left(\lambda_I + \sigma_I I_{ss} + \gamma_I \Sigma_f\right) & 0 & \sigma_I P_{ss} & 0 & 0 & 0 \\ (\gamma_X \Sigma_f - \sigma_X X_{ss}) & 0 & \lambda_I & -(\lambda_X + \sigma_X P_{ss}) & 0 & 0 \\ \frac{1}{\mu_f} & 0 & 0 & 0 & -\frac{\Omega}{\mu_f} & \frac{\Omega}{\mu_f} \\ 0 & 0 & 0 & 0 & \frac{2\Omega}{\mu_c} & -\left\{\frac{2\dot{m}c_{p_c}+\Omega}{\mu_c}\right\} \end{bmatrix} \tag{10.27}$$

$$\mathbf{B} = \left[\frac{P_{ss}}{l}\ 0\ 0\ 0\ 0\ 0\right]^T \tag{10.28}$$

Equations 10.19 and 10.20 represent a 6th order interval system with the uncertainty vector

$$\left.\begin{array}{c} \mathbf{p} = [P_{ss}, \alpha_c, \alpha_f, \Omega] \\ P_{ss} \in \left[\underline{P}_{ss}, \overline{P}_{ss}\right]; \alpha_f \in \left[\underline{\alpha}_f, \overline{\alpha}_f\right]; \alpha_c \in \left[\underline{\alpha}_c, \overline{\alpha}_c\right]; \Omega \in \left[\underline{\Omega}, \overline{\Omega}\right] \end{array}\right\} \tag{10.29}$$

Next, using the aggregation method detailed in [24], the 6th order system represented by Eqs. 10.27–10.29 can be reduced to derive an interval plant conforming to the general template Eq. 10.3 which can be represented as

$$L = L_0 + \Delta L = \frac{P_{ss}\lambda_X\,(2M_c + 2\Omega)\,k_{LZC}}{\left(2M_c P_{ss} + \beta + \frac{\sigma_X X}{\Sigma_a} - 2M_c\beta - \Omega\alpha_c P_{ss} + 2M_c + \Omega\alpha_f P_{ss}\right)} \tag{10.30}$$

$$a = a_0 + \Delta a = \frac{\frac{\Omega}{\mu_f} + \frac{2M_c+\Omega}{\mu_c} + \lambda + \frac{\beta+\frac{\sigma_X X}{\Sigma_a}}{l}}{\frac{\Omega\lambda}{\mu_f\mu_c l}\left(2M_c P_{ss} + \beta + \frac{\sigma_X X}{\Sigma_a} - 2M_c\beta - \Omega a_c P_{ss} + 2M_c + \Omega\alpha_f P_{ss}\right)} \tag{10.31}$$

$$b = 0 \tag{10.32}$$

It is further assumed that with the control rods stationary, the LZCS output can be expressed as $\rho_{ext} = k_{LZC}.\Delta P_d$, where ΔP_d denotes the power error and the gain term k_{LZC} has been assumed to be equal to 2.5×10^{-6} [22]. The parametric values in Eqs. 10.29–10.31 have been adopted from [22, 23]. Using Eqs. 10.30–10.32, the nominal system model of the form Eq. 10.4 has been developed considering the reactor operated at 80% Full Power (FP) for which the controller is designed. The data used in modelling the thermal-hydraulic characteristics of the reactor is represented in Table 10.1. In Table 10.1, P_{ss_0} and Ω_0 denote the nominal values of the variables P_{ss} and Ω respectively and Table 10.1 does not include the neutronic parameters (i.e. $\beta, \sigma_X, \sigma_I, l, \lambda, \lambda_X, \lambda_I, \gamma_X, \gamma_I, \Sigma_a$ and Σ_f) values for the sake of brevity. Readers are referred to [22] for relevant details.

Table 10.1 Thermal-hydraulic parameters for nominal system at 80% FP [17]

Zone	P_{ss0}	Ω_0	μ_f	μ_c	M_c
1, 8	108.48×10^6	3.128×10^5	1.661×10^6	5.113×10^6	5.181×10^6
2, 9	110.88×10^6	3.203×10^5	1.701×10^6	5.236×10^6	5.305×10^6
3, 10	100.66×10^6	2.907×10^5	1.54×10^6	4.753×10^6	4.816×10^6
4, 11	80.35×10^6	2.321×10^5	1.23×10^6	3.794×10^6	3.844×10^6
5, 12	100.66×10^6	2.907×10^5	1.54×10^6	4.753×10^6	4.816×10^6
6, 13	108.48×10^6	3.128×10^5	1.661×10^6	5.113×10^6	5.181×10^6
7, 14	110.88×10^6	3.203×10^5	1.701×10^6	5.236×10^6	5.305×10^6

Table 10.2 Interval parameters with ±10% uncertainties [17]

Zone	$[\underline{P}_{ss}, \overline{P}_{ss}]$	$[\underline{\Omega}, \overline{\Omega}]$	$[\underline{L}, \overline{L}]$	$[\underline{a}, \overline{a}]$
1, 8	$[97.57, 119.26] \times 10^6$	$[2.81, 3.44] \times 10^5$	[0.0055, 0.0074]	[1.3287, 1.7793]
2, 9	$[99.79, 121.96] \times 10^6$	$[2.88, 3.52] \times 10^5$	[0.0054, 0.0072]	[1.3235, 1.7725]
3, 10	$[90.58, 121.96] \times 10^6$	$[2.61, 3.19] \times 10^5$	[0.0061, 0.0082]	[1.3030, 1.7450]
4, 11	$[72.31, 88.38] \times 10^6$	$[1.98, 2.43] \times 10^5$	[0.0075, 0.0101]	[1.3247, 1.7740]
5, 12	$[90.58, 121.96] \times 10^6$	$[2.61, 3.19] \times 10^5$	[0.0061, 0.0082]	[1.3030, 1.7450]
6, 13	$[97.57, 119.26] \times 10^6$	$[2.81, 3.44] \times 10^5$	[0.0055, 0.0074]	[1.3287, 1.7793]
7, 14	$[99.79, 121.96] \times 10^6$	$[2.88, 3.52] \times 10^5$	[0.0054, 0.0072]	[1.3235, 1.7725]

The study presented in this chapter assumes an equilibrium reactor core. For the nominal system model, the reactor has been chosen to be operated at 80% FP for the purpose of handling both positive and negative changes in reactor power i.e. from 60% FP to 100% FP with a single controller.

The nominal values of α_c and α_f are taken as -5.59×10^{-4} °C^{-1} and -3.2×10^{-5} °C^{-1} respectively for all 14 zones at 80% FP, these parameter values at other power levels have been adopted from [22]. The intervals of α_c and α_f and the corresponding intervals $[\underline{L}, \overline{L}]$ and $[\underline{a}, \overline{a}]$ for each zone, computed using rule-set Eqs. 10.2 and 10.22–10.23 with a $a \pm 10\%$ of uncertainty bound for each interval variable $P_{ss}, \alpha_c, \alpha_f$ and Ω, at 80% FP have been reported in Table 10.2. The intervals associated with $\alpha_c = [-6.149, -5.031] \times 10^{-4}$ °C^{-1} and $\alpha_f = [-3.52, -2.88] \times 10^{-5}$ °C^{-1} remain same for all zones and are presented in Table 10.2. The parameter b remains zero for all zones as evident from Eqs. 10.22–10.24.

To derive the values of the variable L_0, a_0 at 80% FP, 100% FP, Eqs. 10.22–10.24 have been used and the values at 60% FP are represented in Table 10.3. The

Table 10.3 Nominal values of 'L, a' at 100% FP, 80% FP and 60% FP [17]

Zone	100% FP		80% FP		60% FP	
	L_0	a_0	L_0	a_0	L_0	a_0
1, 8	0.0051034	1.55889	0.0064626	1.55496	0.0084124	1.533248
2, 9	0.0050102	1.56477	0.0062857	1.54870	0.00825668	1.540570
3, 10	0.0052687	1.53291	0.0071854	1.524855	0.0093849	1.510010
4, 11	0.0068255	1.55388	0.0088249	1.549330	0.01149299	1.533319
5, 12	0.0052685	1.53291	0.0071854	1.524855	0.0093849	1.510017
6, 13	0.0051034	1.55889	0.0064626	1.554968	0.00841247	1.533248
7, 14	0.0050102	1.56477	0.0062856	1.54870	0.0082566	1.540575

corresponding values of $\left[\alpha_c, \alpha_f, \Omega\right]$ at 60% FP and, have been used for computing L, a as reported in [22].

The data corresponding 80% FP as reported in Table 10.2 has been used for deriving the transfer functions for all the 14 PHWR zones constituting the nominal plant, conforming to Eq. 10.4. As $b = 0$ for all zones, a study of the parameter a for each zone of the PHWR corresponding to different power levels, presented in Table 10.3, depicts that each plant is an integrator system whose real pole moves closer to the origin power P_{ss} reduces. Similar finding has been reported in [15] also.

10.3.2 Mathematical Modelling of Small a Modular PWR

In this section, an interval based approach is proposed to design a robust, load-following controller for a compact PWR which can be used as a Small Modular Reactor (SMR) in a modular NPP or as a compact unit. Such a module is assumed to be composed of a compact PWR with an integrated Helical Coil Steam Generator (HCSG) constituting a module or a unit. A schematic model of a 100 MWe PWR with its integrated HCSG is presented in Fig. 10.3 with the following assumptions:

1. The core is assumed to be neutronically small [25]. Neutron poison (Xenon and Iodine) buildup related issues are neglected [25, 26].
2. The heat transfer in the reactor core is single-phased [27].
3. It is assumed that the boiling rate is equal to the steaming rate of HCSG.
4. A pressure control mechanism is assumed to exist which keeps the pressure at the HCSG outlet constant by keeping its saturated boiling length constant using a coordinated variation of steam and feed-water flow rates [28, 29].

The following sets of equations are formulated based on neutronic and thermal hydraulics considerations presented in [13, 14, 25, 29]:

$$\frac{\mathrm{d}P_{\mathrm{th}}}{\mathrm{d}t} = \frac{\Delta\rho_{\mathrm{net}} - \beta}{l} P_{\mathrm{th}} + \lambda C \tag{10.33}$$

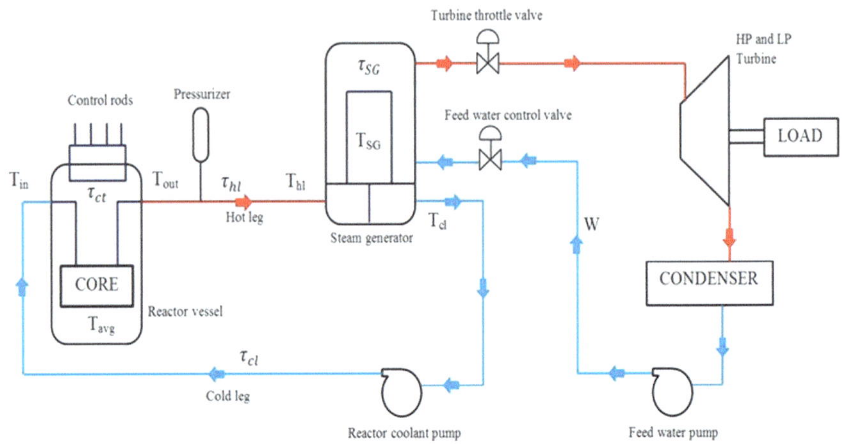

Fig. 10.3 Schematic of a PWR module

$$\frac{dC}{dt} = \frac{\beta}{l} P_{th} - \lambda C \tag{10.34}$$

As in the case of a 540 MWe PHWR discussed in the previous sections, ρ_{net} is the net reactivity input (mK) and C denotes the delayed neutron precursor concentration.

$$\frac{dT_f}{dt} = \frac{J}{\mu_f} P_{th} - \frac{\Omega}{\mu_c}(T_f - T_{avg}) \tag{10.35}$$

$$\frac{dT_{avg}}{dt} = \frac{\Omega}{\mu_c}(T_f - T_{avg}) - \frac{\omega_c}{\mu_c}(T_{out} - T_{in}) \tag{10.36}$$

The parameter T_{avg} (average coolant temperature in the primary side) is mathematically represented as

$$T_{avg} = \frac{T_{in} + T_{out}}{2} \tag{10.37}$$

The transport delay equations can be expressed as

$$\frac{dT_{in}}{dt} = \frac{2T_{sg} - T_{hl} - T_{in}}{\tau_{cl}} \tag{10.38}$$

$$\frac{dT_{hl}}{dt} = \frac{T_{out} - T_{hl}}{\tau_{hl}} \tag{10.39}$$

where T_{hl} denotes the coolant outlet temperature on the primary side of the steam generator and T_{cl} denotes the corresponding inlet temperature. With these, the average coolant temperature on the primary side of the steam generator can be expressed as

$$T_{sg} = \frac{T_{hl} + T_{cl}}{2} \tag{10.40}$$

Similarly, defining T_s as the steam temperature at the steam generator outlet, its variation can be expressed as a function of the power transferred from the steam generator primary to secondary as

$$\frac{dT_s}{dt} = \frac{U_{sg}A_{sg}}{M_{sg}C_{psg}}\left(T_{sg} - T_s\right) - \frac{W\left(h_g - h_f\right)}{M_{sg}C_{psg}} \tag{10.41}$$

with the implicit assumption that the following holds for the steam generator primary

$$2W_{hl}\left(T_{hl} - T_{sg}\right) = U_{sg}A_{sg}\left(T_{sg} - T_s\right) \tag{10.42}$$

Table 10.4 represents the significance of the symbols used in Eqs. 10.33–10.42 [27] and enumerates their representative values with units at 100% FP. In Eqs. 10.35–10.42, the parameters $\Omega = UA$, $\mu_f = m_f C_{pf}$, $\mu_c = m_c C_{pc}$, $\omega_c = \dot{m} C_{pc}$ are the thermal-hydraulic parameters approximated as piecewise linear functions of reactor power P_{th} and their values are represented in Table 10.5 [31].

With the coefficients of reactivity, α_c, α_f varying with temperature change caused by a change in reactor power, it is clear that the net reactivity change associated with change of reactor power can be expressed in a sum of integral forms represented by Eq. 10.43.

$$\Delta\rho_{\text{net}} = \Delta\rho_{\text{cont}} + \int_{T_{f0}}^{T_f} \alpha_f dT_f + \int_{T_{\text{avg0}}}^{T_{\text{avg}}} \alpha_c dT_{\text{avg}} \tag{10.43}$$

10.3.3 Modes of PWR Operation

A typical PWR can be operated in any of the following two modes:

1. *A reactor-following-turbine mode or the normal mode*: In this mode, the controller attempts to average coolant temperature (T_{avg}) constant, allowing the steam temperature at the HCSG outlet (T_s), to vary with the change in demand power. The controller action at the reactor end is initiated by a change in the steam flow to the turbine and hence the feed flow to the steam generator which tend to change the average coolant temperature (T_{avg}) on the primary side.
2. A *turbine-following-reactor mode* or the *alternate mode*: In this mode the control system attempts to keep the steam temperature at the HCSG outlet (T_s) constant while allowing the average coolant temperature (T_{avg}) to vary in accordance with a change in the demand power set-point of the reactor $\left(P_{\text{th}}^*\right)$. This causes a change in the power extracted by the primary coolant to vary and match the demand on the secondary side. The controller action is initiated at the reactor end.

Table 10.4 Reactor parameter values at 100% power [30]

Parameters	Significance	Values at 100% FP
P_{th}	Reactor thermal power	300 MWth (100 MWe)
β	Average of six group delayed neutron fraction	0.0045
λ	Average of six group delay neutron decay constant	$0.076\ s^{-1}$
l	Prompt neutron life time	0.001 s
J	Fraction of total power generated in the fuel	0.97
m_f	Mass of fuel	2555.3 kg
m_c	Mass of coolant	1760 kg
C_{pf}	Specific heat of fuel	236.77 J/kg °C
C_{pc}	Specific heat of coolant	5594.26 J/kg °C
$\dot{m}$	Mass flow-rate of coolant in the core	1410 kg/s
U	Fuel-to-coolant heat transfer coefficient	$1859\ J/s\ m^2\ °C$
A	Active heat transfer area	$1160\ m^2$
α_f	Fuel coefficients of reactivity	$-2.97 \times 10^{-5}\ °C^{-1}$
α_c	Coolant coefficients of reactivity	$-7.2 \times 10^{-4}\ °C^{-1}$
T_{in}	Coolant temperature at reactor inlet	292 °C
T_{out}	Coolant temperature at reactor outlet	328 °C
T_f	Average fuel temperature	445.3 °C
T_s	Steam temperature at HCSG outlet	298.5 °C
τ_{ct}	Transit delay in coolant movement from reactor core to the hot leg	0.66 s
τ_{cl}	Transit delay in coolant movement from steam generator outlet to reactor inlet	0.35 s
τ_{hl}	Transit delay in coolant movement from reactor to the steam generator	0.06 s
W_{hl}	Coolant flow-rate into steam generator	1272 kg/s
M_{hl}	Mass of coolant entering into steam generator	1760 kg
C_{phl}	Specific heat of coolant in the hot leg entering the steam generator	5594.3 J/kg °C
M_{sg}	Total mass of steam and water inside the steam generator	1760 kg
C_{psg}	Specific heat of water in the steam generator	5594.3 J/kg °C
U_{sg}	Effective heat transfer coefficient of steam generator	$5217.8\ J/s\ m^2\ °C$
A_{sg}	Active area of the steam generator tubes	$4641\ m^2$
W	Matched steam and feed flow-rate	150.9 kg/s
h_g	Enthalpy of the steam	2920.8×10^3 J/kg
h_f	Enthalpy of feed-water into the steam generator	1032.64×10^3 J/kg

Table 10.5 Parametric variation with varying power level [30]

Parameters	Base power levels (MWth)				
	$P_{0,1}$	$P_{0,2}$	$P_{0,3}$	$P_{0,4}$	$P_{0,5}$
	20% FP	40% FP	60% FP	80% FP	100% FP
α_f (°C^{-1})	-3.8×10^{-5}	-3.5×10^{-5}	-3.3×10^{-5}	-3.2×10^{-5}	-2.9×10^{-5}
α_c (°C^{-1})	-5.56×10^{-4}	-5.97×10^{-4}	-6.36×10^{-4}	-6.39×10^{-4}	-7.2×10^{-4}
Ω (J/s °C)	2.6553×10^{6}	2.5239×10^{6}	2.3980×10^{6}	2.2775×10^{6}	2.1571×10^{6}
μ_f (J/°C)	0.5580×10^{7}	0.57×10^{7}	0.5820×10^{7}	0.5930×10^{7}	0.6050×10^{7}
μ_c (J/°C)	0.9320×10^{7}	0.9434×10^{7}	0.9652×10^{7}	0.9705×10^{7}	0.9848×10^{7}
ω_c (J/s °C)	0.7464×10^{7}	0.7564×10^{7}	0.7653×10^{7}	0.7676×10^{7}	0.7888×10^{7}

10.4 Controller Design

In this section, interval based techniques are proposed to design controllers for a PHWR for load-following operation, followed by a modular PWR described in the previous section. It is a well-known fact that conventional reactor control approaches aim to achieve a stable reactor period around a designated reactor power level—which is usually 100% FP, with refinements like flux-tilt control and zonal power level variations within a narrow range using reactivity devices distributed across the reactor core. A bulk power controller is invoked either to raise the reactor power to a steady operational level or during a sharp reduction, known as a step-back and seldom in a demand following mode.

Since 90% of the industrial controllers are Proportional-plus-Integral-plus Derivative (PID) controllers-marked by their simple structure, excellent tracking, good disturbance rejection and reasonable noise immunity, for a PHWR for which the dynamics is well studied, it is first attempted to explore the possibility of using a PID controller for a load-following operation It is also known that PID controllers can be designed for a Single Input Single Output (SISO) plant, such as a single zone of a 540 MWe PHWR in a reduced order form, to achieve desired stability robustness in terms of specified gain and phase margins either by explicit analytical design techniques or by the Linear Quadratic Regulator (LQR) approach. However applicability of such a controller to nuclear reactor control for a load-following operation is restricted to a small power level variation due to the design based on a linearized model of a reactor. As a natural corollary, it follows that if a PID design methodology that ensures a certain level of stability robustness is augmented with add-ons that ensure parametric robustness covering a bounded range of power variation, such a controller can be used for controlling the reactor over the chosen range of varying reactor power. The controller design should also be capable of tackling uncertainties in sensor parameters. In this section, the interval model of a PHWR defined in Sect. 10.3 is used to design such an optimal PID controller for each zone of a 540 MWe PHWR which is capable of controlling a PHWR from 60% FP to 100% FP in presence of measurement uncertainties, with bounded sub-optimality.

The choice of a LQR designed PID controller with increased parametric robustness for load-following operation of a PHWR over a standard nonlinear controller raises a discussion which merits special consideration. While a nonlinear controller approach can be employed to achieve control of reactor power over the entire operating regime, such controllers tend to be complex. Stability and performance analyses with such controllers, when applied to load following reactor control, tend to be extremely complicated due to multiple rate and saturation constraints e.g. rate and quantum of reactivity addition. Therefore, for a limited load-following operation, as in the case of a PHWR, the advantages of a nonlinear controller are traded with the simplicity in controller structure, design and analysis while adopting an interval based approach.

However, for a modular PWR, the method developed for a PHWR fails as the range of power variation is much wider and it becomes necessary to explore trans-domain approaches for nonlinear plant control. With the evolution of nonlinear control methodologies as an effective means for controlling nonlinear plants, the technique of NDI [32] has gained importance, particularly in the domain of aero-space engineering. It is based on feedback linearization [33] where a nonlinear system is forced to take a *linear-looking* form in the transformed co-ordinates by carrying out a co-ordinate transformation. Thus, NDI transforms a nonlinear plant into a feedback linearized nonlinear plant with a stable linear error dynamics allowing easier synthesis of a controller with a closed-form structure which ensures effective tracking of the reference signal and a global asymptotic stability of a nonlinear system. Over the years, NDI technique have been merged with various others like static optimization to formulate *optimal dynamic inversion* [34], a constrained static optimization technique to formulate *constrained optimal nonlinear dynamic inversion* [35], model-based neuro-adaptive control design [36], H_∞ robust control [37], etc. In this chapter, the NDI methodology has been merged with interval based approach to select controller gains that allow load-following operation with the parameter variations i.e. the variation in (T_s) in the Normal mode and the variation in $\left(T_{\text{avg}}\right)$ in the alternate remaining bounded in user-specified intervals. The variations in other reactor parameters like power, reactivity, feed and steam flow, fuel, coolant and steam temperatures, rate of change of reactor power, etc. consequently lie within bounded intervals. The assumed interval bounds include parametric and/or measurement uncertainties and controller gains are static, spanning the entire operating range of the reactor.

10.4.1 LQR Based PID Controller Design for PHWR Using Interval Approach

In this section, a methodology is proposed for design of an optimal PID controller for zonal power(and hence global power) control of a PHWR described in Sect. 10.3.1. For this purpose, the original 6th order SISO representing a PHWR zone is reduced to a template of a second order interval plant in Eq. 10.44 as reported in [15]. For each and every zone of PHWR this template can be formed by using Table 10.2 and Eqs. 10.30–10.32. This results in

$$G(s, \mathbf{p}) = \frac{L}{s^2 + as + b}; \; L = \left[\underline{L}, \overline{L}\right], a = [\underline{a}, \overline{a}], b = \left[\underline{b}, \overline{b}\right] \tag{10.44}$$

where the nominal plant is

$$G_0(s) = \frac{L_0}{s^2 + a_0 s + b_0} \tag{10.45}$$

The nominal plant is obtained for each zone using data from Table 10.3. A methodology reported in [38] is used to design a PID controller using a Linear Quadratic Regulator (LQR) approach for a standard second order plant described as Eq. 10.45. It is possible to show that the gains of an optimal PID controller for the plant can be expressed as

$$\begin{bmatrix} K_i & K_p & K_d \end{bmatrix} = \mathbf{R}_0^{-1}\mathbf{B}_0^T\mathbf{P} \tag{10.46}$$

where $\mathbf{P} = \mathbf{P}^T > \mathbf{0}$ is the solution of an Algebraic Riccati Equation (ARE)

$$\boldsymbol{A_0^T P + P A_0 - P B_0 R_0^{-1} B_0^T P + Q_0} = \mathbf{0} \tag{10.47}$$

with the cost function defined as

$$J = \frac{1}{2}\int_0^{\infty} \left[\mathbf{x}^T(t)\mathbf{Q}\mathbf{x}(t) + u^T(t)\mathbf{R}u(t)\right] \mathrm{d}t \tag{10.48}$$

The system state vector for Eq. 10.45 is assumed to be augmented as

$$\mathbf{x}(t) = \left[\int y(t) \;\; y(t) \;\; \dot{y}(t)\right] \tag{10.49}$$

and accordingly the matrices $\mathbf{A}_0$, $\mathbf{B}_0$ become

$$\mathbf{A}_0 = \begin{bmatrix} 0 & 1 & 0 \\ 0 & 0 & 1 \\ 0 & -b_0 & -a_0 \end{bmatrix}; \mathbf{B}_0 = \begin{bmatrix} 0 \\ 0 \\ L_0 \end{bmatrix} \tag{10.50}$$

$\mathbf{Q}_0$ and $\mathbf{R}_0$ can be obtained as below

$$Q_0 = \begin{bmatrix} \frac{r}{L_0^2}\left[\sigma_0^2\omega_0^4\right] & 0 & 0 \\ 0 & \frac{r}{L_0^2}\left[\omega_0^4 + \left[4\xi_0^2 - 2\right]\sigma_0^2\omega_0^2 - b_0^2\right] & 0 \\ 0 & 0 & \frac{r}{L_0^2}\left[\left\{\sigma_0^2 + \left(4\xi_0^2 - 2\right)\omega_0^2\right\} - a_0^2 + 2b_0\right] \end{bmatrix} \geq 0 \text{ and } \mathbf{R}_0 = [r] \tag{10.51}$$

If the nominal-closed loop system with the optimal PID controller obtained as Eq. 10.52

$$G_{cl}(s) = \frac{L_0\left(k_p + \frac{k_i}{s} + k_d s\right)}{(s+\lambda_0)\left(s^2 + 2\zeta_0^{cl}\omega_0^{cl}s + \left(\omega_0^{cl}\right)^2\right)} : \lambda_0 \in \mathbb{R}^+ \tag{10.52}$$

The poles of the closed-loop nominal plant with the designed optimal PID controller are at $\left(-\sigma_0, -\xi_0\omega_0 \pm \omega_0\sqrt{1-\xi_0^2}\right)$. The controller Eq. (10.46) is the termed the *nominal* PID controller. It is an optimal controller because the control law $\mathbf{u}(t) = -\left[K_i\ K_p\ K_d\right]X(t)$ minimizes the cost function Eq. 10.48. Now if $G_0(s)$ is replaced by an interval plant $G(s, \mathbf{p})$ and the controller Eq. 10.46 is used, then the poles of the closed-loop system shall shift depending on the shift of reactor power from the nominal power and the nominal controller may or may not remain an optimal controller any more for a different power level. Thus, if a condition can be established, which is sufficient to ensure that the nominal controller Eq. 10.46 designed for the nominal plant in Eq. 10.45 remains an optimal PID controller for the corresponding interval system Eq. 10.44 with a varying steady-state cost function an interval plant Eq. 10.53 in closed-loop as shown below

$$G_c(s, \mathbf{q}) = \frac{L\left[K_p + \frac{K_i}{s} + K_d s\right]}{(s+\sigma)\left(s^2 + 2\zeta\omega s + \omega^2\right)} \tag{10.53}$$

then it can be said that such a controller will be able to control the PHWR for a load-following operation within the chosen interval of power variation with different degree of sub-optimality with respect to the nominal controller. This, in turn, ensures an acceptable response of the reactor under a load-following operation.

The acceptable response is specified by acceptable bounds of the interval variables $[\alpha, \zeta, \omega]$ for the closed loop plant Eq. 10.53 The implicit assumption is that the plant Eq. 10.44 remains controllable over the convex box of $[L, a, b]$.

To visualize the above statement, interval matrices for the corresponding interval plant Eq. 10.44 can be represented as
$\mathbf{A} = \mathbf{A}_0 + \Delta\mathbf{A}$ and $\mathbf{B} = \mathbf{B}_0 + \Delta\mathbf{B}$ where

$$\mathbf{A}_0 = \begin{bmatrix} 0 & 1 & 0 \\ 0 & 0 & 1 \\ 0 & -\Delta b_0 & -\Delta a \end{bmatrix}; \Delta a \in \left[\Delta\underline{a}, \Delta\overline{a}\right], \Delta b \in \left[\Delta\underline{b}, \Delta\overline{b}\right] \tag{10.54}$$

$$\Delta\mathbf{B} = \left[0\ 0\ (\gamma - 1)\ L_0\right]^T : \gamma \in \left[\underline{\gamma}, \overline{\gamma}\right] \tag{10.55}$$

Let $\mathbf{P} = \begin{bmatrix} P_{11} & P_{12} & P_{13} \\ P_{12} & P_{22} & P_{23} \\ P_{13} & P_{23} & P_{33} \end{bmatrix}$ be the solution of the ARE in Eq. 10.47, then

$$\mathbf{P} = \begin{bmatrix} P_{11} & P_{12} & \frac{rK_i}{L_0} \\ P_{12} & P_{22} & \frac{rK_p}{L_0} \\ \frac{rK_i}{L_0} & \frac{rK_p}{L_0} & \frac{rK_d}{L_0} \end{bmatrix} \tag{10.56}$$

Now if $\tilde{\mathbf{P}} = \mathbf{P} + \Delta\mathbf{P}$ be the solution of the ARE

$$\boldsymbol{A}^T \tilde{\boldsymbol{P}} + \tilde{\boldsymbol{P}}\boldsymbol{A} - \tilde{\boldsymbol{P}}\boldsymbol{B}\boldsymbol{R}^{-1}\boldsymbol{B}^T\tilde{\boldsymbol{P}} + \tilde{\boldsymbol{Q}} = \mathbf{0} \tag{10.57}$$

Then in order that the controller for $(\mathbf{A}, \mathbf{B})$ remains the same as that obtained for $(\mathbf{A}_0, \mathbf{B}_0)$,

$$\tilde{\mathbf{P}} = \begin{bmatrix} P_{11} + \Delta P_{11} & P_{12} + \Delta P_{12} & \frac{rK_i}{\gamma L_0} \\ P_{12} + \Delta P_{12} & P_{22} + \Delta P_{22} & \frac{rK_p}{\gamma L_0} \\ \frac{rK_i}{\gamma L_0} & \frac{rK_p}{\gamma L_0} & \frac{rK_d}{\gamma L_0} \end{bmatrix} \tag{10.58}$$

Expanding the ARE Eq. 10.57 as

$$\begin{aligned} &(\boldsymbol{A}_0 + \Delta\boldsymbol{A})^T (\boldsymbol{P} + \Delta\boldsymbol{P}) + (\boldsymbol{P} + \Delta\boldsymbol{P})(\boldsymbol{A}_0 + \Delta\boldsymbol{A}) \\ &- (\boldsymbol{P} + \Delta\boldsymbol{P})(\boldsymbol{B}_0 + \Delta\boldsymbol{B})\boldsymbol{R}_0^{-1}(\boldsymbol{B}_0 + \Delta\boldsymbol{B})^T(\boldsymbol{P} + \Delta\boldsymbol{P}) + \tilde{\boldsymbol{Q}} = \mathbf{0} \end{aligned} \tag{10.59}$$

yields a diagonal matrix $\tilde{\mathbf{Q}}$ and the corresponding matrix $\tilde{\mathbf{P}} = \tilde{\mathbf{P}}^T$ that satisfies Eq. 10.58 defined as

$$\left.\begin{aligned} \tilde{\mathbf{Q}} &= \begin{bmatrix} \tilde{Q}_{11} & 0 & 0 \\ 0 & \tilde{Q}_{22} & 0 \\ 0 & 0 & \tilde{Q}_{33} \end{bmatrix} \\ \tilde{Q}_{11} &= \tfrac{r}{L_0^2}\left[\sigma_0^2\omega_0^4\right] \\ \tilde{Q}_{22} &= \tfrac{r}{L_0^2}\left[\omega_0^4 + \left[4\xi_0^2 - 2\right]\sigma_0^2\omega_0^2 - b_0^2\right] + 2\left\{\tfrac{bP_{23}}{\gamma} - b_0P_{23} + a_0P_{13} - \tfrac{aP_{13}}{\gamma}\right\} \\ \tilde{Q}_{33} &= \tfrac{r}{L_0^2}\left[\left\{\sigma_0^2 + (4\xi_0^2 - 2)\,\omega_0^2\right\} - a_0^2 + 2b_0\right] + 2\left\{\tfrac{aP_{33}}{\gamma} - a_0P_{33} - P_{23}\left(\tfrac{1}{\gamma} - 1\right)\right\} \end{aligned}\right\} \tag{10.60}$$

and

$$\tilde{\mathbf{P}} = \begin{bmatrix} P_{11} + b\frac{P_{13}}{\gamma} - b_0 P_{13} & P_{12} + a\frac{P_{13}}{\gamma} - a_0 P_{13} & \frac{P_{13}}{\gamma} \\ P_{12} + a\frac{P_{13}}{\gamma} - a_0 P_{13} & P_{22} + a\frac{P_{23}}{\gamma} - a_0 P_{23} - P_{13}\left(\frac{1}{\gamma} - 1\right) + b\frac{P_{33}}{\gamma} - b_0 P_{33} & \frac{P_{23}}{\gamma} \\ \frac{P_{13}}{\gamma} & \frac{P_{23}}{\gamma} & \frac{P_{33}}{\gamma} \end{bmatrix} \tag{10.61}$$

Since the interval system $(\mathbf{A}, \mathbf{B})$ remains controllable, and $\mathbf{R}_0 = r > 0$, the condition $\tilde{\mathbf{Q}} \geq 0$ over the convex box of $[L, a, \sigma, \zeta, \omega]$ is sufficient to ensure that the ARE Eq. 10.57 shall be satisfied only by an interval matrix $\tilde{\mathbf{P}} = \tilde{\mathbf{P}}^T \geq 0$. This implies that $\left[K_i\ K_p\ K_d\right] = \mathbf{R}_0^{-1}\mathbf{B}^T\tilde{\mathbf{P}}$ yields an invariable set of controller gains $\left[K_i\ K_p\ K_d\right]$ with the substitution $\mathbf{B} = \gamma \mathbf{L}$.

Again, for a second order system $G(s, \mathbf{p})$ represented by Eq. 10.44 and controlled by a PID controller Eq. 10.46, the following set of equations hold

$$\left.\begin{aligned} K_i &= \frac{\sigma\omega^2}{L} \\ K_p &= \frac{(\omega^2 + 2\zeta\sigma\omega) - b}{L} \\ K_d &= \frac{(2\zeta\omega + \sigma) - a}{L} \end{aligned}\right\} \tag{10.62}$$

Substituting these values in the expression for $\tilde{\mathbf{Q}}$ in Eq. 10.60 produces

$$\left.\begin{aligned} \tilde{\mathbf{Q}} &= \begin{bmatrix} \tilde{Q}_{11} & 0 & 0 \\ 0 & \tilde{Q}_{22} & 0 \\ 0 & 0 & \tilde{Q}_{33} \end{bmatrix} \\ \tilde{Q}_{11} &= \frac{r}{L_0^2}\left[\sigma_0^2\omega_0^4\right] \\ \tilde{Q}_{22} &= \frac{r}{L_0^2}\left[\omega_0^4 + \left[4\xi_0^2 - 2\right]\sigma_0^2\omega_0^2 - b_0^2\right] \\ &+ \frac{2r}{\gamma L_0^2}\left\{(\omega^2 + 2\zeta\omega\sigma - b)\left(\frac{b}{\gamma} - b_0\right) + \sigma\omega^2\left(a_0 - \frac{a}{\gamma}\right)\right\} \\ \tilde{Q}_{33} &= \frac{r}{L_0^2}\left[\left\{\sigma_0^2 + \left[4\xi_0^2 - 2\right]\omega_0^2\right\} - a_0^2 + 2b_0\right] \\ &+ \frac{2r}{\gamma L_0^2}\left\{(2\zeta\omega + \sigma - a)\left(\frac{a}{\gamma} - a_0\right) - \left(\frac{1}{\gamma} - 1\right)(\omega^2 + 2\zeta\omega\sigma - b)\right\} \end{aligned}\right\} \tag{10.63}$$

It is evident that the nominal PID controller Eq. 10.46 designed for a nominal plant $G_0(s)$ in Eq. 10.45 remains an optimal controller for the interval plant $G(s, \mathbf{p})$ Eq. 10.44 and it produces a closed-loop system $G_c(s, \mathbf{q})$ defined by Eq. 10.53 with poles lying in the complex plane specified by $[\alpha, \zeta, \omega]$ only if the condition $\tilde{\mathbf{Q}} \geq 0$ is satisfied over the convex box of $[L, a, b, \sigma, \zeta, \omega]$.

The positive semi-definiteness of the matrix $\tilde{\mathbf{Q}}$ over the entire convex box of $[L, a, b, \sigma, \zeta, \omega]$ can be tested by following Eqs. 10.64–10.66 reported in [39].

$$\tilde{\mathbf{Q}}_c = \frac{1}{2}(\underline{\tilde{\mathbf{Q}}} + \overline{\tilde{\mathbf{Q}}}) \tag{10.64}$$

and

$$\tilde{\mathbf{Q}}^{\Delta} = \frac{1}{2}(\underline{\tilde{\mathbf{Q}}} + \overline{\tilde{\mathbf{Q}}}) \tag{10.65}$$

then $\tilde{\mathbf{Q}}$ is positive semi-definite for the entire interval $\left[\underline{\tilde{\mathbf{Q}}} + \overline{\tilde{\mathbf{Q}}}\right]$ if

$$\lambda_{\min}\left(\tilde{\mathbf{Q}}_c\right) \geq \rho\left(\tilde{\mathbf{Q}}^{\Delta}\right) \tag{10.66}$$

If the initial state for both nominal system $(\mathbf{A}_0, \mathbf{B}_0)$ and the interval system **(A, B)** be $\mathbf{x}(0)$, then the incremental cost of control when the nominal PID controller is used to control the interval plant is

$$\Delta J = \mathbf{X}^T(0)\Delta\mathbf{P}\mathbf{X}(0) \tag{10.67}$$

where

$$\Delta\mathbf{P} = \begin{bmatrix} rK_i\left(\frac{b}{L} - \frac{b_0}{L_0}\right) & rK_i\left(\frac{a}{L} - \frac{a_0}{L_0}\right) & rK_i\left(\frac{1}{L} - \frac{1}{L_0}\right) \\ rK_i\left(\frac{a}{L} - \frac{a_0}{L_0}\right) & rK_p\left(\frac{a}{L} - \frac{a_0}{L_0}\right) + rK_d\left(\frac{b}{L} - \frac{b_0}{L_0}\right) - rK_i\left(\frac{1}{L} - \frac{1}{L_0}\right) & rK_p\left(\frac{1}{L} - \frac{1}{L_0}\right) \\ rK_i\left(\frac{1}{L} - \frac{1}{L_0}\right) & rK_p\left(\frac{1}{L} - \frac{1}{L_0}\right) & rK_d\left(\frac{1}{L} - \frac{1}{L_0}\right) \end{bmatrix} \tag{10.68}$$

As it is seen that the incremental cost is bounded and the bound can be expressed as a function of the bound of parametric uncertainties.

Thus the steps involved in controller design are as follows:

Step1: For a given nominal plant in Eq. 10.45 obtain the augmented state system $(\mathbf{A}_0, \mathbf{B}_0)$ and choose the position of the closed-loop real pole $-\sigma_0$, closed-loop damping ζ_0 and closed-loop frequency ω_0 and value of scalar r to obtain the matrix $\mathbf{Q}_0$ using Eq. 10.60.

Step2: Design the nominal PID controller $\left[K_i \ K_p \ K_d\right]$ for $(\mathbf{A}_0, \mathbf{B}_0, \mathbf{Q}_0, \mathbf{R}_0)$ and the value of the **P** by solving the ARE in Eq. 10.47.

Step3: Select suitable intervals$\left[\underline{\sigma}, \overline{\sigma}\right]$, $\left[\underline{\zeta}, \overline{\zeta}\right]$, $\left[\underline{\omega}, \overline{\omega}\right]$ and check if Eq. 10.66 holds for the entire convex box of $[L, a, \sigma, \zeta, \omega]$. If this is satisfied, then the controller $\left[K_i \ K_p \ K_d\right]$ obtained in *Step1* is the desired controller, else modify r, σ_0, ζ_0, ω_0 and/or the intervals $\left[\underline{\sigma}, \overline{\sigma}\right]$, $\left[\underline{\zeta}, \overline{\zeta}\right]$, $\left[\underline{\omega}, \overline{\omega}\right]$ and repeat from *Step1*.

The LQR based PID controller design technique presented above and detailed in Eq. 10.15 has been extended in [18] for controlling a large 700 MWe PHWR with limited voiding consideration in coolant channel during the transient time.

10.4.2 Controller Design Methodology Using Nonlinear Dynamic Inversion and Interval Approach

A nonlinear system can be represented in a control-affine form as [34]:

$$\dot{\mathbf{X}} = \mathbf{f(X)} + \mathbf{g(X)U} \tag{10.69}$$

$$\mathbf{Y} = \mathbf{h(X)} \tag{10.70}$$

where $\mathbf{X} \in \mathbb{R}^n$, $\mathbf{Y} \in \mathbb{R}^m$ and $\mathbf{U} \in \mathbb{R}^p$ are the state-vector, output vector and the control input vector respectively, $\mathbf{f(X)} = (n \times 1)$ vector, $\mathbf{g(X)} = (n \times p)$ matrix and $\mathbf{h(X)} = (m \times 1)$ vector. If $\mathbf{Y}^*(t) \in \mathbb{R}^m$ represents the reference output vector along with the assumption that $\mathbf{Y}^*(t)$ is smooth, bounded and slowly varying, then the control objective is to make $\mathbf{Y}(t) \to \mathbf{Y}^*(t)$ as $t \to \infty$. From Eq. 10.70 it follows

$$\dot{\mathbf{Y}} = \frac{\partial h}{\partial \mathbf{X}} \dot{\mathbf{X}} \tag{10.71}$$

$$\text{where, } \frac{\partial h}{\partial \mathbf{X}} = \begin{pmatrix} \frac{\partial h_1}{\partial x_1} & \cdots & \frac{\partial h_1}{\partial x_n} \\ \vdots & \ddots & \vdots \\ \frac{\partial h_m}{\partial x_1} & \cdots & \frac{\partial h_m}{\partial x_n} \end{pmatrix} \tag{10.72}$$

Substituting Eq. 10.69 in Eq. 10.71 yields

$$\dot{\mathbf{Y}} = \mathbf{f}_Y(\mathbf{X}) + \mathbf{g}_Y(\mathbf{X})\mathbf{U} \tag{10.73}$$

where $\mathbf{f}_Y(\mathbf{X}) = \frac{\partial h}{\partial \mathbf{X}} \mathbf{f(X)}$; $\mathbf{g}_Y(\mathbf{X}) = \frac{\partial h}{\partial \mathbf{X}} \mathbf{g(X)}$.

It is next attempted to achieve

$$\dot{\mathbf{E}} + \mathbf{KE} = 0 \tag{10.74}$$

where the error $\mathbf{E} \triangleq \mathbf{Y}^* - \mathbf{Y}$ and the gain matrix $\mathbf{K}$ is a positive definite matrix. From Eq. 10.74, the following can be written:

$$\mathbf{E} = \exp^{-Kt} \mathbf{E}_0 \to 0; t \to \infty \tag{10.75}$$

Equation 10.75 implies that the instantaneous error being an exponential function of initial error $\mathbf{E}_0$, exponentially decays to 0. The rate of convergence parameter, $\mathbf{K}$, determines how fast the error tends to zero, guaranteeing asymptotic stability and leading to a closed form of the solution. The necessary condition is $\mathbf{K}$ should be always positive definite matrix. Thus from Eq. 10.74, the following can be written as

$$\left(\dot{\mathbf{Y}}^* - \dot{\mathbf{Y}}\right) + \mathbf{K}\left(\mathbf{Y}^* - \mathbf{Y}\right) = 0 \tag{10.76}$$

Substituting Eq. 10.59 in Eq. 10.62, the control output $\mathbf{U}$ can be synthesized as

$$\mathbf{U} = \frac{\dot{\mathbf{Y}}^* - \mathbf{K}\left(\mathbf{Y}^* - \mathbf{Y}\right) - \mathbf{f}_{\mathbf{Y}}(\mathbf{X})}{\mathbf{g}_Y(\mathbf{X})} \tag{10.77}$$

which yields

$$\dot{\mathbf{Y}}^* - \mathbf{KE} = \mathbf{f}_Y(\mathbf{X}) + \mathbf{g}_Y(\mathbf{X})\mathbf{U} \tag{10.78}$$

In the formulation above, a square system has been assumed, where $m = p$ (the number of inputs equal to the number of outputs) and choosing $\mathbf{K}$ as a diagonal matrix such that $\mathbf{K} = \text{diag}(k_1, \ldots k_p)$ where $k_i (i = 1, \ldots, p)$ represents the static controller gain of the ith error channel. The system dynamics therefore possesses a decoupled structure where any two error channels are independent from one another, i.e. the output of the ith error channel would only depend upon the ith input. In such case, Eq. 10.78 can be written as

$$\begin{bmatrix} f_{y_1}(\mathbf{X}) \\ \vdots \\ f_{y_p}(\mathbf{X}) \end{bmatrix} + \begin{bmatrix} g_{y_1}(\mathbf{X}) & \ldots & 0 \\ \vdots & \ddots & \vdots \\ 0 & \ldots & g_{y_p}(\mathbf{X}) \end{bmatrix} \begin{bmatrix} u_1 \\ \vdots \\ u_p \end{bmatrix} = \begin{bmatrix} \dot{y}_1^* \\ \vdots \\ \dot{y}_p^* \end{bmatrix} - \begin{bmatrix} k_1 & \ldots & 0 \\ \vdots & \ddots & \vdots \\ 0 & \ldots & k_p \end{bmatrix} \begin{bmatrix} e_1 \\ \vdots \\ e_p \end{bmatrix} \tag{10.79}$$

In Eq. 10.79, $f_y(\mathbf{X})$ represents the modified nonlinear state equation corresponding to the ith output element and $g_{y_i}(\mathbf{X})$ represents the modified nonlinear output equation associated with the ith input element u_i. The parameter $\dot{y}_i$ represents the rate of change of the ith reference input element and k_i represents the gain associated with the ith error term $e_i = \left[y_i^* - y_i\right]$. Equation 10.79 may be used to derive the control law for the system as

$$\left.\begin{aligned} u_1 &= \frac{\dot{y}_1^* - k_1\left(y_1^* - y_1\right) - f_{y_1}(\mathbf{X})}{g_{y_1}(\mathbf{X})} \\ &\vdots \\ u_p &= \frac{\dot{y}_p^* - k_p\left(y_p^* - y_p\right) - f_{y_p}(\mathbf{X})}{g_{y_p}(\mathbf{X})} \end{aligned}\right\} \tag{10.80}$$

From Eq. 10.66 it follows that if the variables u_i, y_i, $\dot{y}_i^*$ and the scalar functions $f_{y_i}(\mathbf{X})$ and $g_{y_i}(\mathbf{X})$ are assumed to be associated with intervals $\left[\underline{u}_i, \overline{u_i}\right]$, $\left[\underline{y}_i, \overline{y}_i\right]$, $\left[\underline{\dot{y}}_i^*, \overline{\dot{y}}_i^*\right]$, $\left[f_{y_i}(\underline{\mathbf{X}}), f_{y_i}(\overline{\mathbf{X}})\right]$ and $\left[g_{y_i}(\underline{\mathbf{X}}), g_{y_i}(\overline{\mathbf{X}})\right]$ with $\mathbf{X}$ varying in an interval $\left[\underline{\mathbf{X}}, \overline{\mathbf{X}}\right]$, then Eq. 10.80 can be used to derive permissible interval for each controller gain k_i such that the controller ensures that each error e_i lies within a user specified interval $\left[\underline{e}_i, \overline{e}_i\right]$ for all $i = 0, 1, 2, \ldots, p$ using the following formulation

$$\left[\underline{k}_i, \overline{k}_i\right] = \frac{\left[\underline{u}_i, \overline{u}_i\right]\left[g_{y_i}(\underline{\mathbf{X}}), g_{y_i}(\overline{\mathbf{X}})\right] + \left[f_{y_i}(\underline{\mathbf{X}}), f_{y_i}(\overline{\mathbf{X}})\right] - \left[\underline{\dot{y}}_i^*, \overline{\dot{y}}_i^*\right]}{\left[\underline{e}_i, \overline{e}_i\right]} \tag{10.81}$$

The computations involving intervals in Eq. 10.81 have been obtained using the interval mathematics. Thus, by expressing the controller parameters as algebraic functions of the reactor parameters which vary in intervals, the valid intervals of controller gains may be obtained such that the errors lie within the specified acceptable limits.

10.4.3 Controller Design for Various Modes of PWR Operation

This section describes the controller design methodology for a PWR based on NDI and interval approaches for the following modes of operation
a. Reactor-follow-turbine mode (Normal mode):
The schematic of Normal mode is presented in Fig. 10.4. The control actions in this mode follow the following sequence:

(i) In response to a change in demand on the secondary side, a change in the set-point of the turbine header pressure is initiated. The error between the changed turbine inlet pressure set-point and the existing turbine pressure generates a pressure error which is applied to a pressure controller which regulates the throttle valve position to regulate the steam flow-rate into the turbine.
(ii) Simultaneously, with the variation of steam flow-rate, the feed-water flow-rate is changed keeping the HCSG pressure constant by keeping the length of the saturated steam portion of the HCSG constant.
(iii) With reactor power remaining unchanged, an increased feed flow into the steam generator causes a change in the degree of superheat and hence the steam temperature T_s at the outlet of the HCSG.
(iv) The power transfer from the steam generator primary to its secondary is altered by the changed T_s keeping T_{avg} constant as evident from Eq. 10.41 which requires the reactor power to change according to the demand.
(v) From (iv) it is clear that a reduction in T_s will cause an increase in load and vice versa. Now, since the pressure remains constant, it is obvious that the interval

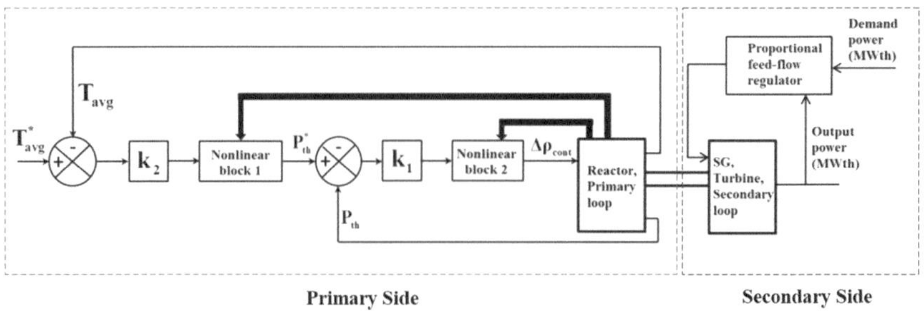

Fig. 10.4 The control scheme for a PWR in reactor-follow-turbine mode

$\left[\underline{T}_s, \overline{T}_s\right]$ should be chosen in such a manner that there is no carry-over of water droplets to the turbine or a dry-out.

The states, inputs, outputs and references of the system are defined as

$$\mathbf{X} = \left[P_{\text{th}}\ C\ T_f\ T_{\text{avg}}\ T_{\text{in}}\ T_{hl}\ T_s\right]^T \tag{10.82}$$

$$\mathbf{U} = \left[\Delta\rho_{\text{net}}\ P^*_{\text{th}}\right]^T \tag{10.83}$$

$$\mathbf{Y} = \left[P_{\text{th}}\ T_{\text{avg}}\right]^T \tag{10.84}$$

$$\mathbf{Y}^* = \left[P^*_{\text{th}}\ T^*_{\text{avg}}\right]^T \tag{10.85}$$

which yield

$$f(\mathbf{X}) = \left[\frac{-\beta}{l}P_{\text{th}} + \lambda C, \frac{\beta}{l}P_{\text{th}} - \lambda C, \frac{J}{\mu_f}P_{\text{th}} - \frac{\Omega}{\mu_f}(T_f - T_{\text{avg}}), -\frac{\omega_c}{\mu_c}(T_{\text{out}} - T_{\text{in}}), \frac{2T_{sg} - T_{hl} - T_{\text{in}}}{\tau_{cl}}, \frac{T_{\text{out}} - T_{hl}}{\tau_{hl}}, \frac{U_{sg}A_{sg}}{M_{sg}C_{psg}}(T_{sg} - T_s) - \frac{W(h_g - h_f)}{M_{sg}C_{psg}}\right]^T \tag{10.86}$$

$$g(\mathbf{X}) = \begin{bmatrix} \frac{P_{\text{th}}}{l}\ 0\ 0\ 0\ 0\ 0\ 0 \\ 0\ 0\ 0\ \frac{J}{\mu_c}\ 0\ 0\ 0 \end{bmatrix}^T \tag{10.87}$$

and

$$\dot{\mathbf{Y}} = \begin{bmatrix} -\frac{\beta}{l}P_{\text{th}} + \lambda C \\ -\frac{\omega_c}{\mu_c}(T_{\text{out}} - T_{\text{in}}) \end{bmatrix} + \begin{bmatrix} \frac{P_{\text{th}}}{l} & 0 \\ 0 & \frac{J}{\mu_c} \end{bmatrix} \begin{bmatrix} \Delta\rho_{\text{net}} \\ P^*_{\text{th}} \end{bmatrix} \tag{10.88}$$

Considering the parametric variation presented in Eq. 10.67, the final control law can be derived as

$$\left[\underline{k}_1, \overline{k}_1\right] = \frac{-\left[\frac{\mathrm{d}}{\mathrm{d}t}\underline{P}^*_{\text{th}}, \frac{\mathrm{d}}{\mathrm{d}t}\overline{P}^*_{\text{th}}\right] + \left\{-\frac{\beta}{l}\left[\underline{P}_{\text{th}}, \overline{P}_{\text{th}}\right] + \lambda\left[\underline{C}, \overline{C}\right]\right\}}{\left[\underline{e}_p, \overline{e}_p\right]} + \frac{\left\{\left[\Delta\underline{\rho}_{\text{cont}}, \Delta\overline{\rho}_{\text{cont}}\right] + \left[\Delta\underline{\rho}_{\text{fuel}}, \Delta\overline{\rho}_{\text{fuel}}\right] + \left[\Delta\underline{\rho}_{\text{coolant}}, \Delta\overline{\rho}_{\text{coolant}}\right]\right\}\left[\underline{P}_{\text{th}}, \overline{P}_{\text{th}}\right]\frac{1}{l}}{\left[\underline{e}_p, \overline{e}_p\right]} \tag{10.89}$$

$$\left[\underline{k}_2, \overline{k}_2\right] = \frac{\left[\underline{P}^*_{\mathrm{th}}, \overline{P}^*_{\mathrm{th}}\right]\frac{J}{\mu_c} - \frac{\mathrm{d}}{\mathrm{d}t}T^*_{\mathrm{avg}}}{\left[\underline{e}_{T_{\mathrm{avg}}}, \overline{e}_{T_{\mathrm{avg}}}\right]} - \frac{\frac{\omega_c}{\mu_c}\left[\left[\underline{T}_{\mathrm{out}}, \overline{T}_{\mathrm{out}}\right] - \frac{C_1}{C_2}\left[\underline{T}_{hl}, \overline{T}_{hl}\right] - \frac{C_1 - C_2}{C_2}\left[\underline{T}_{cl}, \overline{T}_{cl}\right] - \frac{C_3}{C_2}\left[\underline{T}_s, \overline{T}_s\right]\right]}{\left[\underline{e}_{T_{\mathrm{avg}}}, \overline{e}_{T_{\mathrm{avg}}}\right]} \tag{10.90}$$

b. Turbine-follow-reactor mode (Alternate mode):
The schematic of Alternate mode is presented in Fig. 10.5. The following sequence of actions occurs for PWR in the turbine-follow-reactor mode:

(i) The demand power set-point is varied by the power control program to change the reactor power and the power error ΔP_{th} is used to move the control rods to change the power produced by the reactor.
(ii) The variation in reactor power causes a change in T_{avg} and hence T_{sg} with an altered power flow from the primary to the secondary of the steam generator. This will tend to generate an error ΔT_s which can be nullified by altering the feed-water flow rate into the steam generator. Since the steaming rate equals the boiling rate in HCSG, this means a matching change in the steam flow out of the HCSG is also required.
(iii) Since the feed-control and power-control loops are interacting, the need of turning the pressurizer on to avoid fuel damage can be eliminated, only if the permissible intervals of variation of T_{avg} and T_f be known and accordingly, designing a control scheme that ensures the feed and reactivity constraints are met. The reactivity requirements are much higher than normal mode since both the coolant and the fuel temperatures change.

To derive a control law for this mode, the states remain same as defined by Eq. 10.82 while the inputs, outputs and reference are chosen as

$$\mathbf{U} = [\Delta\rho_{\mathrm{net}} \quad W]^T \tag{10.91}$$

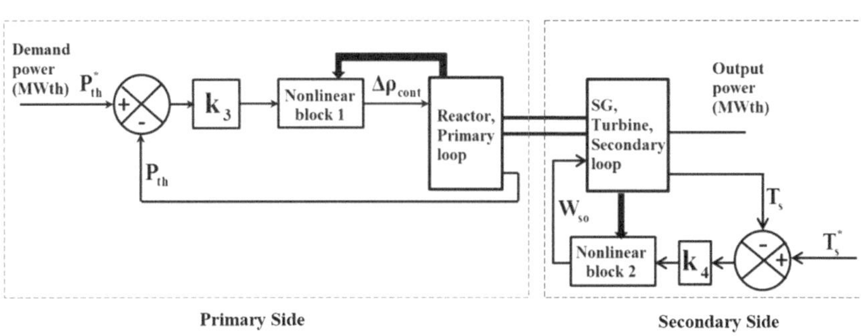

Fig. 10.5 The control scheme for a PWR in a turbine-follow-reactor mode

$$\mathbf{Y} = [P_{\text{th}} \quad T_s]^T \tag{10.92}$$

$$\mathbf{Y}^* = \left[P_{\text{th}}^* \quad T_s^*\right]^T \tag{10.93}$$

which yields

$$\begin{aligned} f(\mathbf{X}) = \Bigg[& \frac{-\beta}{l} P_{\text{th}} + \lambda C, \frac{\beta}{l} P_{\text{th}} - \lambda C, \frac{J}{\mu_f} P_{\text{th}} - \frac{\Omega}{\mu_f}(T_f - T_{\text{avg}}), \\ & \frac{\Omega}{\mu_c}(T_f - T_{\text{avg}}) - \frac{\omega_c}{\mu_c}(T_{\text{out}} - T_{\text{in}}), \\ & \frac{2T_{sg} - T_{hl} - T_{\text{in}}}{\tau_{cl}}, \frac{T_{\text{out}} - T_{hl}}{\tau_{hl}}, \frac{U_{sg} A_{sg}}{M_{sg} C_{psg}}(T_{sg} - T_s) \Bigg]^T \end{aligned} \tag{10.94}$$

$$g(\mathbf{X}) = \begin{bmatrix} \frac{P_{\text{th}}}{l} \; 0\,0\,0\,0\,0 & 0 \\ 0 \;\; 0\,0\,0\,0\,0 & \frac{-(h_g - h_f)}{M_{sg} C_{psg}} \end{bmatrix}^T \tag{10.95}$$

and

$$\dot{\mathbf{Y}} = \begin{bmatrix} -\frac{\beta}{l} P_{\text{th}} + \lambda C \\ \frac{U_{sg} A_{sg}}{M_{sg} C_{psg}}(T_{sg} - T_s) \end{bmatrix} + \begin{bmatrix} \frac{P_{\text{th}}}{l} & 0 \\ 0 & \frac{-(h_g - h_f)}{M_{sg} C_{psg}} \end{bmatrix} \begin{bmatrix} \Delta\rho_{\text{net}} \\ W \end{bmatrix} \tag{10.96}$$

The final control law can be derived as

$$\begin{aligned} \left[\underline{k}_3, \overline{k}_3\right] = & \frac{-\left[\frac{\mathrm{d}}{\mathrm{d}t} \underline{P}_{\text{th}}^*, \frac{\mathrm{d}}{\mathrm{d}t} \overline{P}_{\text{th}}^*\right] + \left\{-\frac{\beta}{l}\left[\underline{P}_{\text{th}}, \overline{P}_{\text{th}}\right] + \lambda\left[\underline{C}, \overline{C}\right]\right\}}{\left[\underline{e}_p, \overline{e}_p\right]} \\ & + \frac{\left\{\left[\Delta\underline{\rho}_{\text{cont}}, \Delta\overline{\rho}_{\text{cont}}\right] + \left[\Delta\underline{\rho}_{\text{fuel}}, \Delta\overline{\rho}_{\text{fuel}}\right] + \left[\Delta\underline{\rho}_{\text{coolant}}, \Delta\overline{\rho}_{\text{coolant}}\right]\right\}\left[\underline{P}_{\text{th}}, \overline{P}_{\text{th}}\right]\frac{1}{l}}{\left[\underline{e}_p, \overline{e}_p\right]} \end{aligned} \tag{10.97}$$

$$\begin{aligned} \left[\underline{k}_4, \overline{k}_4\right] = & \frac{-\frac{\mathrm{d}}{\mathrm{d}t} T_s^* + \frac{U_{sg} A_{sg}}{M_{sg} C_{psg}}\left[\underline{\left(T_{sg} - T_s\right)}, \overline{\left(T_{sg} - T_s\right)}\right]}{\left[\underline{e}_{T_s}, \overline{e}_{T_s}\right]} \\ & - \frac{\frac{\left[\underline{W}, \overline{W}\right]\left[\underline{\left(h_g - h_f\right)}, \overline{\left(h_g - h_f\right)}\right]}{M_{sg} C_{psg}}}{\left[\underline{e}_{T_s}, \overline{e}_{T_s}\right]} \end{aligned} \tag{10.98}$$

Table 10.6 Nominal plants and interval parameters for controller design [17]

Zone	$G_0(s)$	$\underline{L}, \overline{L}$	$\underline{a}, \overline{a}$
1, 8	$\frac{0.0064626}{S^2+1.55496s}$	[0.0051034, 0.0084124]	1.3287, 1.7793
2, 9	$\frac{0.0062857}{S^2+1.54870s}$	[0.0050102, 0.0062857]	[1.3235, 1.7725]
3, 10	$\frac{0.0071854}{S^2+1.524855s}$	[0.0052687, 0.0071854]	[1.3030, 1.7450]
4, 11	$\frac{0.0088249876}{S^2+1.54933s}$	[0.0068255, 0.0114929]	[1.3247, 1.7740]
5, 12	$\frac{0.0071854}{S^2+1.524855s}$	[0.0052687, 0.0071854]	[1.3030, 1.7450]
6, 13	$\frac{0.0064626}{S^2+1.55496s}$	[0.0051034, 0.0084124]	[1.3287, 1.7793]
7, 14	$\frac{0.0062857}{S^2+1.54870s}$	[0.0050102, 0.0062857]	[1.3235, 1.7725]

In [40], the NDI based controller design methodology has been extended for the control of a small PWR operated in a novel coordinated control mode for accomplishing very rapid power manoeuvring without defying the system parametric constraints. The NDI technique has also been used in [14], for power level control of a large PWR with consideration of Xenon induced oscillations.

10.5 Simulation and Results

This section presents the simulation results demonstrating the efficacy of the controller design presented in the Sect. 10.4.

10.5.1 Results Related to PID Controller Designed for PHWR

The methodology presented in the controller design section is used to design robust PID controllers for each zone of the ZCC of a PHWR to achieve the following:

i. Producing satisfactory time response with a variation of $\Delta P_{\text{ss}} = \pm 20\%\ FP$ in the zonal reactor power where ΔP_{ss} is defined as $\Delta P_{\text{ss}} = P_{\text{ss}} - P_{\text{ss}_0}$. This ensures the load-following ability of the reactor within the specified interval band of the reactor power.
ii. Producing satisfactory time response with uncertainty consideration of $\pm 5\%$ in each element of the set of parameters **p** defined by Eq. 10.29 representing zonal reactor power P_{ss}, reactivity coefficients α_c and α_f, the parameter Ω. This tests the robustness of the controller under uncertainties in the measurement of power, reactivity of coolant and fuel and in estimation of heat transfer parameters.

For meeting the specifications enumerated in (i) and (ii) with a robust PID controller, the intervals $\left[\underline{L}, \overline{L}\right]$ and $\left[\underline{a}, \overline{a}\right]$ computed using Eqs. 10.30–10.31 and the corresponding values are represented in Table 10.6.

Table 10.7 Nominal PID controller parameters [17]

Zone	K_p	K_i	K_d
1, 8	44.7188	2.3984	27.0851
2, 9	45.9774	2.8432	28.8432
3, 10	40.2204	2.1572	28.5503
4, 11	32.7482	1.7564	20.4728
5, 12	40.2204	2.1572	28.5503
6, 13	44.7188	2.3984	27.0851
7, 14	45.9774	2.8432	28.8432

The nominal values of closed-loop system parameters are specified as $\alpha_0 = 1.55$, $\zeta_0 = 0.9$, $\omega = 0.1$ rad/s, for nominal controller design. An acceptable time-response is specified as one which produces a $\pm 10\%$ uncertainty bound in $[\alpha, \zeta, \omega]$ around $[\alpha_0, \zeta_0, \omega_0]$ with these specifications. Thus,

$$\alpha \in [1.39, 1.70]\,;\ \zeta \in [0.81, 0.99]\,;\ \omega \in [0.09, 0.11]\,; \tag{10.99}$$

With these considerations following the methodology presented in Section 10.4, Table 10.7 presents the nominal PID controller parameters for all the 14 zones of a PHWR.

Using MATLAB/SIMULINK, it is next attempted to study the robustness of the closed-loop plants. Figure 10.6 depicts the general schematic for MATLAB simulation. For ZCC models, the Simulink blocks have been adopted from [41] as presented in Fig. 10.6 and the PID controller blocks listed in Table 10.6 are included in the LZC Controller block. The zonal power demand set-points and zonal power errors for each zone of the PHWR are generated by The Zonal Demand Power block shown in Fig. 10.6.

In order to evaluate the performance of the designed controller in a load-following mode, the following simulation study is conducted first without parametric uncertainties and then with parametric uncertainties.

For studying the simulated response, the demand power set-point is ramped at $\pm 0.5\%$ FP/s for 20 s with the reactor operating at 80% FP (nominal plant) to set the power at 100% FP (i.e. $\Delta P_{ss} = +20\%$ FP) and then at 60% FP (i.e. $\Delta P_{ss} = -20\%$ FP)

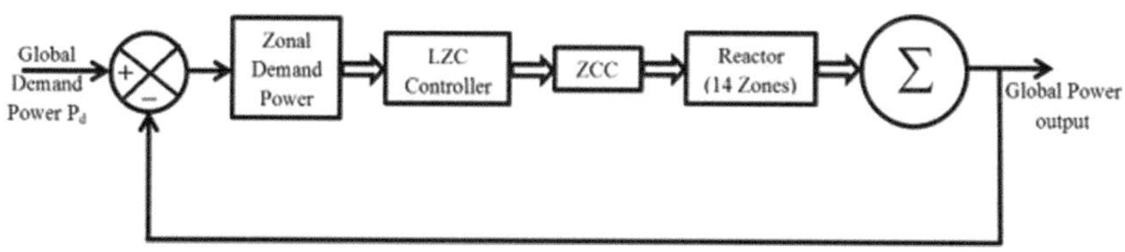

Fig. 10.6 Schematic diagram of zonal power control loop of a PHWR

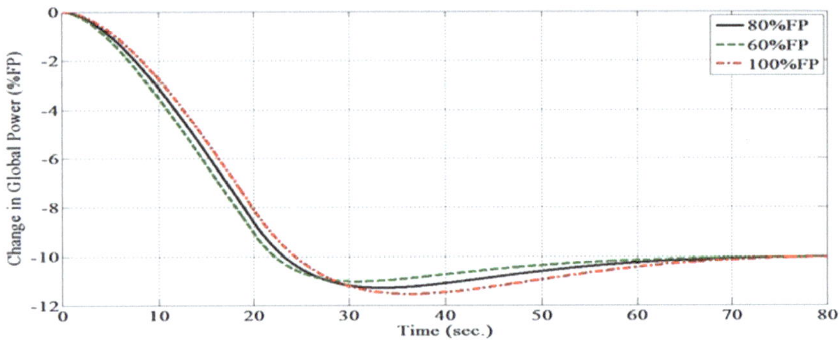

Fig. 10.7 Global power variation due to demand power set-point ramping at 100% FP, 80% FP and 60% FP with nominal PID controller

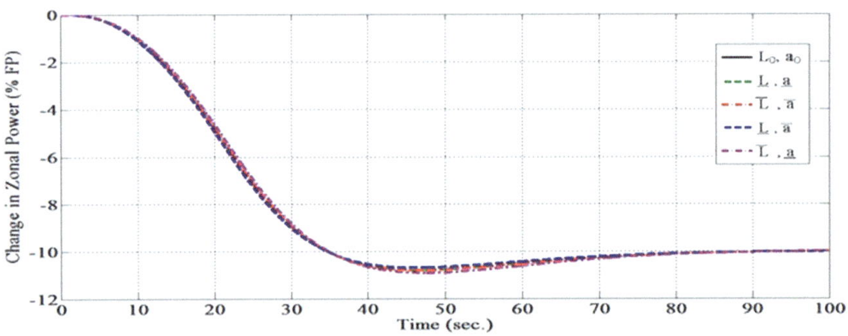

Fig. 10.8 Representative zonal reactor power variation due to demand power set-point ramping for 4 different cases arising out of parametric uncertainties

with the nominal PID controllers enumerated in Table 10.7 in loop. The simulated variations of the global reactor power (sum of individual zonal powers) with time in each case in each of these three power levels are presented in Fig. 10.7 with a corresponding representative zonal power variation in Fig. 10.8

As evident from Figs. 10.7 and 10.8 which are reproduced from [17], the nominal PID controllers are capable of accomplishing effective load-following within a ±20% FP band around the nominal power level. For each case, depicted in Figs. 10.7 and 10.8, the peak overshoot is a little more than the value corresponding to $\zeta = 0.9$. Also the bulk power settles within 2% of its steady-state value within 60 s. with marginal decrease when power is reduced to 60% FP and marginal increase at 100% FP. The settling time remains constant, the variation in closed-loop damping $|\Delta\zeta|$ is less than 10% as envisaged.

Next, for evaluating the Case (ii), ±5% uncertainty bound is considered in the parameters $\left[P_{ss}, \alpha_c, \alpha_f, \Omega\right]$ and the corresponding zonal transfer functions has been obtained for cases corresponding to the 4 vertices of the resultant convex box of $[L, a]$ computed using Eqs. 10.30–10.31 and rule-set Eq. 10.2. The controllers described

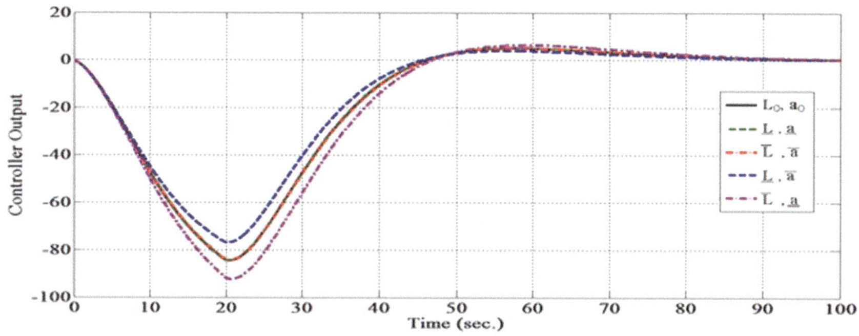

Fig. 10.9 Representative controller output due to demand power set-point ramping for 4 different cases arising out of parametric uncertainties

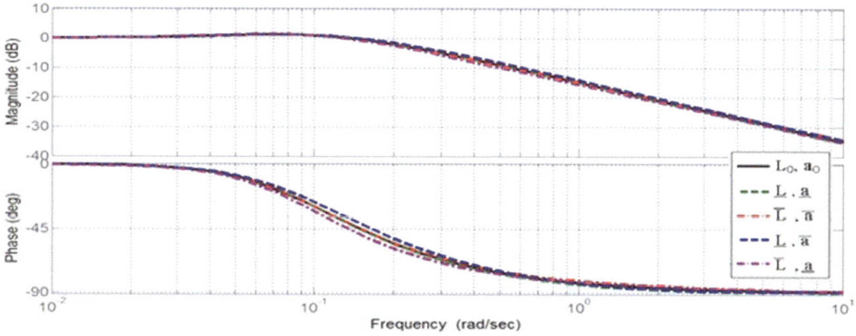

Fig. 10.10 Representative frequency response for a zone 4 with nominal PID controller for 4 different cases arising out of parametric uncertainties

by Table 10.7 are used to control these plants. The global demand power set-point is again ramped by -0.5% FP/s, which initiates proportionate variation in zonal powers. A representative variation of zonal powers (Zone 4) under this condition, for 4 cases each corresponding to a vertex of the convex box of $[L, a]$, is depicted in Figs. 10.9 and 10.10. Figure 10.9 represents the corresponding controller outputs for the same zone and Fig. 10.10 presents the frequency response for each case.

Figures 10.9 and 10.10 depict that, with the nominal PID controller in loop, the time response and frequency response for the considered Zone 4, remain almost unaltered with $\pm 5\%$ of uncertainty bound in the parameters $\left[P_{ss}, \alpha_c, \alpha_f, \Omega\right]$. For the sake of brevity, the time response of the other zones as well as the simulation result for demand power set-point ramp-down for the same zone have not been presented. Similar variations in controller outputs and frequency responses for the other zones, have also been observed in simulation studies.

Further, Fig. 10.9 reveals that for the plant in Eq. 10.18 representing the reactor zone considered, controller output is minimum when the gain term L is minimum and the pole at $s = -a$ is farthest from the origin. Similarly, at the instance when the term

L is maximum and the pole at $s = -a$ is nearest to the origin, the controller output is maximum. The shifting of L and a in the same direction, which is caused by the uncertainties, keep the controller output almost unaltered with respect to the nominal system. Eqs. 10.30–10.31 and the representative values of the parameters depicted in Table 10.1, support the above statement. The parameters L, a, for a particular power level, are mostly sensitive to changes in P_{ss} and Ω. Similar findings have also been reported in [22] where uncertainties have been injected in P_{ss} and Ω and robustness in presence of bounded parametric uncertainties have been examined. A study of Eqs. 10.30–10.31 depicts that L, a reduces in presence of positive uncertainties and increases in presence of negative uncertainties in P_{ss}, Ω. However, opposite signs uncertainties in P_{ss}, Ω keep L, a constant. The value of the ΔP matrix defined by Eq. 10.68 is negative which implies that for the condition $(L < L_0)$ the cost of control decreases and vice-versa which goes in accordance to the variation in controller output depicted in Fig. 10.9. The controller output is converted to a 4–20 mA signal which is used for the purpose of control valve opening and closure, in order to alter the level of light water in a ZCC. The steady-state value is 12 mA.

The constraints associated with a 540 MWe PHWR imposes a ramping restriction of $\pm 0.5\%$ FP for demand power and this goes at par with the existing design of the present LZCS for Indian PHWRs. Studies associated with uncertainty injection in heat transfer related parameters for large PHWR are unavailable. An uncertainty bound of 1% in P_{ss} and 2.1% in Ω for a 2MW reactor is reported in [39].

10.5.2 Results Related to NDI Based Controller Design for PWR

The performance of the NDI based controller, modified using Interval Approach, has been validated on an OpalRT based real-time Hardware-in-Loop platform and the corresponding outputs are fed directly to a storage oscilloscope to generate the traces. For evaluating controller performance, following two cases have been considered:

Case 1: Power ramp-up for 60 s from 20% FP to 100% FP using the normal mode.

Case 2: Power ramp-down for 60 s from 100% FP to 20% FP using the alternate mode.

The simulation results are presented in Figs. 10.11 and 10.12.

A detailed study of Figs. 10.11 and 10.12 reveals that the chosen controllers are not only capable of effectively achieving the desired load variations but also satisfy all the assumed constraints in terms of permissible parametric variations. It is further seen that for all presented power manoeuvres, the steam temperature T_s variation at the HCSG outlet in the normal mode and the average coolant temperature T_{avg} variation in the alternate mode remain within their respective intervals, with changing load.

It is also evident from Figs. 10.11 and 10.12 that the reactor power P_{th} tracks its set-point P_{th}^* for each case and hence the oscilloscope traces presented in these figures, depicts overlapping curves. The variation of the steam flow-rate and hence

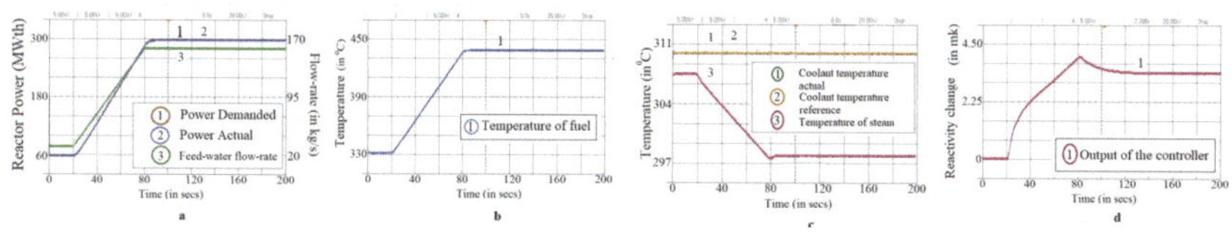

Fig. 10.11 Plant performance with proposed controller for 20–100% FP variation at normal mode **a** demand power and feed-water flow-rate **b** fuel temperature variation **c** reference and actual average coolant temperature and steam temperature variation **d** change in controller output in terms of reactivity

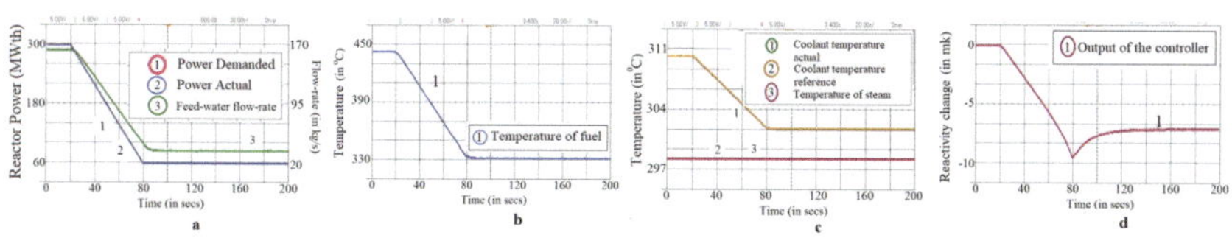

Fig. 10.12 Plant performance with proposed controller for 100–20% FP variation at alternate mode **a** tracking performance of actual and reference reactor power and feed-water flow-rate **b** fuel temperature variation **c** actual and reference steam temperature and average coolant temperature variation **d** change in controller output in terms of reactivity

the feed flow-rate are as expected. Further examination of these figures reveals that the demand power set-point variation is initiated by the coolant temperature loop after the variation of feed flow, for all power manoeuvres in normal mode. Similarly, the variation in reference reactor power and actual power occurs before the variation in feed-flow, for all power manoeuvring in alternate mode.

10.6 Conclusion

This chapter puts forward an interval approach based controller design for both PHWR and PWR class of nuclear reactors. A methodology to achieve robust control of a large PHWR under changes in demand power set-point and uncertainties in measurement of reactor power, reactivity coefficients and heat transfer related parameters has been presented, which uses an interval approach to tune conventional PID controllers to achieve robust control of PHWR. The controller obtained by this methodology is an optimal controller and the acceptable bounds in closed-loop time response parameters can also be directly specified. The robustness of the controllers has been established using credible simulation of a PHWR with representative data. The controller design methodology has been extended towards the control of a 700 MWe PHWR considering the voiding in the coolant transfer loop in transient time. Similarly, an NDI based controller design methodology has been considered for controlling the load-following operation of a small PWR which also takes into account the variation of thermal hydraulic parameters with changing reactor power.

The NDI methodology modified with Interval approach is capable of accomplishing effective trajectory tracking alongside satisfying system constraints. The NDI based controller design technique has been extended towards control of a PWR operating in Co-ordinated mode for accomplishing very rapid power manoeuvring. Also the NDI technique has been extended towards the control of a large PWR with the consideration of Xenon induced oscillations in the power level.

References

1. S. Saha, S. Das, R. Ghosh, B. Goswami, R. Balasubramanian, A.K. Chandra, S. Das, A. Gupta, Design of a fractional order phase shaper for iso-damped control of a PHWR under step-back condition. IEEE Trans. Nucl. Sci. **57**(3), 1602–1612 (2010)
2. B. Bhushan, *ABC of ABT: A Primer on Availability Tariff* (2010) [Online], http://www.nldc.in/docs/abc-abt.pdf
3. A. Lokhov, *Technical and Economic Aspects of Load-Following with Nuclear Power Plants* (NEA, OECD, Paris, France, 2011)
4. Y.J. Lee, M.G. Na, Robust controller design of nuclear power reactor by parametric method. J. KNS **34**(5), 436–444 (2002)
5. K. Suzuki, J. Shimazaki, Y. Shinohara, Application of H-infinity control-theory to power-control of a nonlinear reactor model. Nucl. Sci. Eng. **115**(2), 142–151 (1993)
6. X. Jin, A. Ray, R.M. Edwards, Integrated robust and resilient control of nuclear power plants for operational safety and high performance. IEEE Trans. Nucl. Sci. **57**(2), 807–817 (1993)
7. R.M. Edwards, K.Y. Lee, A. Ray, Robust optimal control of nuclear reactors and power plants. Nucl. Technol. **98**, 137–148 (1992)
8. D.B. Talange, B. Bandyopadhyay, A.P. Tiwari, Spatial control of a large PHWR by decentralized periodic output feedback and model reduction techniques. IEEE Trans. Nucl. Sci. **53**(4), 2308–2317 (2006)
9. T.U. Bhatt, K.C. Madala, S.R. Shimjith, A.P. Tiwari, Application of fuzzy logic control system for regulation of differential pressure in liquid zone control system. Ann. Nucl. Energy **36**, 1412–1423 (2009)
10. S. Saha, S. Das, R. Ghosh, B. Goswami, R. Balasubramanian, A.K. Chandra, S. Das, A. Gupta, Design of a fractional order phase shaper for iso-damped control of a PHWR under step-back condition. IEEE Trans. Nucl. Sci. **57**(3), 1602–1612 (2010)
11. S. Dasgupta, A. Routh, S. Banerjee, K. Agilageswari, R. Balasubramanian, S.G. Bhandarkar, S. Chattopadhyay, M. Kumar, A. Gupta, Networked control of a large pressurized heavy water reactor (PHWR) with discrete proportional-integral-derivative (PID) controllers. IEEE Trans. Nucl. Sci. **60**(5), 3879–3888 (2013)
12. G.R. Ansarifar, S. Saadatzi, Sliding mode control for pressurized-water nuclear reactors in load following operations with bounded xenon oscillations. Ann. Nucl. Energy **76**, 209–217
13. D. Bose, S. Banerjee, M. Kumar, P.P. Marathe, S. Mukhopadhyay, A. Gupta, An interval approach to nonlinear controller design for load-following operation of a small modular pressurized water reactor. IEEE Trans. Nucl. Sci. **64**(9), 2474–2488 (2017). https://doi.org/10.1109/TNS.2017.2728187
14. D.K. Yadav, A. Gupta, P. Munshi, Non linear dynamic inversion based controller design for load following operations in pressurized water reactors with bounded xenon oscillations. Nucl. Eng. Des. **328**, 241–254 (2018)
15. Y.J. Lee, H Robust controller design of reactor power control system. J. KNS **29**(4), 280–290 (1997)
16. S.S. Shyu, R.M. Edwards, A robust multivariable feedforward/feedback controller design for integrated power control of boilingwater reactor power plants. Nucl. Technol. **140**(2), 129–140 (2002)

17. S. Banerjee, K. Halder, S. Dasgupta, S. Mukhopadhyay, K. Ghosh, A. Gupta, An interval approach for robust control of a large PHWR with PID controllers. IEEE Trans. Nucl. Sci. **62**(1), 281–292 (2015). https://doi.org/10.1109/TNS.2014.2376942
18. S. Banerjee, D. Bose, A. Hazra, S. Chattopadhyay, K. Ghosh, A. Gupta, Controller design for operation of a 700 MWe PHWR with limited voiding. Nucl. Eng. Des. **357**, 110370 (2020)
19. S.M. Rump, Fast interval matrix multiplication. Numer. Algor. **61**(1), 1–34 (2012)
20. S.P. Bhattacharyya, L.H. Keel, A. Datta, *Linear Control Theory: Structure, Robustness, and Optimization* (CRC Press, 2009)
21. A.P. Tiwari, B. Banyopadhyay, G. Govindarajan, Spatial control of a large pressurized heavy water reactor. IEEE Trans. Nucl. Sci. **43**(4), 2440–2453 (1996)
22. S. Das, I. Pan, B. Majumder, S. Das, A. Gupta, Control of nuclear reactor power with thermal-hydraulic effects via fuzzy PID controllers, in *Proceedings of the International Conference on Communication and Industrial Application (ICCIA)*, vol. 1, no. 5 (2011), pp. 26–28
23. B. Bandyopadhyay, H. Unbenhauen, B.M. Patre, A new algorithm for compensator design for higher-order system via reduced model. Automatica **34** (1998)
24. M. Das, R. Ghosh, B. Goswami, A. Gupta, A.P. Tiwari, R. Balasubramanian, A.K. Chandra, Networked control system applied to a large pressurized heavy water reactor. IEEE Trans. Nucl. Sci **53**(5), 2948–2956 (2006)
25. S.R.P. Perillo, B.R. Upadhyaya, F. Li, Control and instrumentation strategies for multi-modular integral nuclear reactor systems. IEEE Trans. Nucl. Sci. **58**(5), 2442–2451 (2011)
26. S.R.P. Perillo, *Multi-modular Integral Pressurized Water Reactor Control and Operational Reconfiguration for a Flow Control Loop*. Ph.D. dissertation (Univ. Tennessee, Knoxville, 2010)
27. K. Zhao, *An Integrated Approach to Performance Monitoring and Fault Diagnosis of Nuclear Power Systems*. Ph.D. dissertation (Univ. Tennessee, Knoxville, 2005)
28. L.N.F. Guimaraes, *Non-linear Dynamics of a Once-Through Steam*. Ph.D. dissertation (Univ. Tennessee, Knoxville, 1992)
29. M.A. Schultz, *Control of Nuclear Reactors and Power Plants*, 2nd edn. (McGraw-Hill, New York, 1961)
30. D. Bose, A. Hazra, S. Mukhopadhyay, A. Gupta, A co-ordinated control methodology for rapid load-following operation of a pressurized water reactor based small modular reactor. Nucl. Eng. Des. **367**, 110748 (2020)
31. A. Piazzi, A. Visioli, Optimal dynamic-inversion-based control of overhead crane. IEE Proc. Control Theory Appl. **149**(5), 405–411 (2002)
32. J.J.E. Slotine, W. Li, Appl. Nonlinear Control **199**(1) (1991)
33. R. Padhi, S.N. Balakrishnan, Optimal dynamic inversion control design for a class of nonlinear distributed parameter systems with continuous and discrete actuators. IET Control Theory Appl. **1**(6), 1662–167 (2007)
34. C.M. Sakode, R. Padhi, S. Kapoor, V.P.S. Rallabandi, P.K. Roy, Multimodal therapy for complete regression of malignant melanoma using constrained nonlinear optimal dynamic inversion. Biomed. Signal Process. Control **13**, 198–211 (2014)
35. R. Padhi, M. Kothari, An optimal dynamic inversion-based neuro-adaptive approach for treatment of chronic myelogenous leukemia. Comput. Methods Prog. Biomed. **87**(3), 208–224 (2007)
36. A.D. Ngo, W.C. Reigelsperger, S.S. Banda, J.S. Bessolo, Multivariable control law design for a tailless airplanes, in *Proceedings of the AIAA Conference on Guidance, Navigation and Control*, July 29–31 (1996)
37. J.B. He, Q.-G. Wang, T.H. Lee, PI/PID controller tuning via LQR approach. Chem. Eng. Sci **55**(13), 2429–2439 (2000)
38. H. Milan, D. Daney, E. Tsigaridas, Bounds on real eigenvalues and singular values of interval matrices. SIAM J. Matrix Anal. Appl. **31**(4), 2116–2129 (2010)
39. H. Kazeminejad, Uncertainty and sensitivity analyses for steady-state thermal-hydraulics of research reactors. Prog. Nucl. Energy **49**, 313–322 (2007)

40. D.K. Kumar, S.K. Nagar, J.P. Tiwari, Model order reduction of interval systems using Mihailov criterion and factor division method. Int. J. Comput. Appl. (IJCA) **28**(11), 4–8 (2011)
41. M.G. Na, S.H. Shin, W.C. Kim, A model predictive controller for nuclear reactor power. Jour.-Kor. Nucl. Soc. **35**(5), 399–411 (2003)

Chapter 11
On Robotics and Control in the Nuclear Servicing Sector

Erwin Jose Lopez Pulgarin, Kaiqiang Zhang, Robert Skilton, and Guido Herrmann

11.1 Introduction

The field of robotics has been advancing rapidly in the last decade, improving the performance and reliability of classic industrial robotics and pushing for adoption in application fields outside of manufacturing. A modern robotic deployment includes a robotic device, its control unit and potentially many additional subsystems (e.g. sensors, communication network, remote control interfaces), creating a system in itself during deployment (i.e. robotic system). Civil nuclear applications such as nuclear energy systems can benefit from the use of robotic systems, as these systems would replace humans dealing with dangerous environments and substances dangerous to manipulate [1]; mobile robots and robotic arms alike have been thought to be used in such applications. In the field of fusion reactors, robotic manipulators [2] and other automated tools are key enablers for this technology to work and be maintained [3].

Nuclear energy systems bring particular difficulties for the design and deployment of robotic systems, such as shelf life of the devices, task definitions and particularly safety requirements [4]. The safety and task requirements in any nuclear set-up (e.g.

E. J. Lopez Pulgarin · G. Herrmann (✉)
Department of Electrical and Electronic Engineering (EEE), University of Manchester, Manchester, UK
e-mail: guido.herrmann@manchester.ac.uk

E. J. Lopez Pulgarin
e-mail: erwin.lopezpulgarin@manchester.ac.uk

K. Zhang
Remote Applications in Challenging Environments (RACE), United Kingdom Atomic Energy Authority, Abingdon, UK
e-mail: kaiqiang.zhang@ukaea.uk

R. Skilton
Robotics, Repurposing & Decommissioning Directorate (RRDD), United Kingdom Atomic Energy Authority, Abingdon, UK
e-mail: robert.skilton@ukaea.uk

U. Mangalanathan et al. (eds.), *Nuclear Instrumentation and Control*,
https://doi.org/10.1007/978-981-97-1283-0_11

decommissioning, maintenance) define the difficulty of deploying a robotic system, particularly due to the high precision and reliability required for the tasks being performed; these task performance needs to show minimal degradation in time and a set of counter-measurements to deal with failures. Many deployed robotic systems need to deal with harsh environments that restrict them to being single-use devices, effectively reducing their shelf life to a couple of hours [5]. All these need to fit under the current regulations and technologies used both for current and future builds, which fall into the category of Control and Instrumentation systems. Control and Instrumentation (C&I) functions in nuclear energy systems are required to operate at the highest levels of safety and reliability. Safety for these systems can often be categorised into (a) conventional safety, which relates to safety risks not associated with nuclear materials, (b) nuclear safety, which relates to radiation hazards, and (c) investment protection, which relates to the protection of the facility as an asset. All are highly important for the successful operation of a nuclear power plant. These categories and the processes behind building, maintaining and decommissioning any such system are highly regulated and follow strict standards.

The inclusion of robots supporting any function within a nuclear power plant needs to be well planned and assessed by the relevant regulatory bodies (e.g. Office for Nuclear Regulation (ONR) in the UK [6]). A meticulous assessment process [4] is performed to validate any claims around the safety of the implemented devices performing said functionality. These functions can be classified based on their importance and required safety level considering a case where they fail. For the UK, the standards EN 61226:2010 [7] and EN 61508-1:2010 [8] regulate these considerations. The main concept driving each category is the design basis event, which includes both accidents for which a facility is designed for (i.e. design basis accident) and the normal deviation expected from an operational process to occur at least once (i.e. anticipated operational occurrence). These categories are:

- Category A:
  - Functions that play a principal role in the achievement or maintenance of nuclear power plant safety to prevent a design basis event leading to unacceptable consequences.
  - Functions whose failure could directly lead to accident conditions which may cause unacceptable consequences if not mitigated by other Category A functions.
- Category B:
  - Functions that play a complementary role to the Category A functions in the achievement or maintenance of nuclear power plant safety, especially the functions required to operate after the non-hazardous stable state has been achieved, prevent design basis event from leading to unacceptable consequences or mitigate the consequences of design basis event.
  - Functions whose failure could initiate a design basis event or worsen the severity of a design basis event.

- Category C:
  - Functions that play an auxiliary or indirect role in the achievement or maintenance of nuclear power plant safety. Category C includes functions that have some safety significance, but are not Category A or B.

Current robotics deployments have been done around Category C functions, such as decommissioning or other supporting applications. Safety claims are easier to create for lower categories, considering that Category A and B tasks are inherently more complex and difficult to perform. Even for Category C functions, any robotics development needs to perform complex high-precision tasks in harsh environments during long periods of time. Technical standards are available to guide new robotics deployments across different application areas. For the application of industrial robots (e.g. robotic arm manipulators) the ISO 10218-1:2011 [9] standard describes safety requirements and considerations useful for any application (e.g. speed control, axis limit, singularity protection). Relevant technical specifications, for example, include BS 8611:2016 [10] about ethical considerations for robotic systems and ISO/TS 15066:2016 [11] about collaborative robotics designs. These norms inform robotics deployments from a bottom-up approach (i.e. from the robot perspective) and serve as blueprints for future deployments. However, there is no complete guiding document for the design, creation and deployment of robotic systems in a nuclear power plant [12] to this day.

Considering the requirements for task performance and the regulatory framework where robotic systems need to work, current research is pushing both technical and regulatory limits towards wider adoption in all possible categories.

11.2 Robotic systems for Nuclear Decommissioning and Servicing

Current work, for example, the remote maintenance and decommissioning system used at JET, largely relies on bespoke nuclear telemanipulator systems which are highly expensive and tend to make use of outdated technologies. The demand for reduced cost and improved reliability is steering the field more towards the use of Commercial Off-The-Shelf (COTS) industrial manipulators and collaborative robots, which have not been designed from the outset for teleoperation.

For future nuclear fusion power plants, regular in-reactor maintenance is necessary due to material changes in the components. As the maintenance requires shutting down of the energy generation asset, maintenance duration affects commercial viability of the technology. Similarly, in decommissioning scenarios, the cost of maintaining the plant in a safe state rapidly accumulates, driving the need for increased productivity. These productivity demands are driving the exploration of the use of more conventional automation systems and industrial robots.

11.2.1 MASCOT

The MASCOT (Manipolatore Servo Controllato Transistorizzato) system at the UKAEA is a representative robotic system that is specifically designed for teleoperation applications in a nuclear environment [13]. Back in the 1980s, the Joint European Torus (JET) was built as an experimental plant for fusion power generation solutions. It was essential to deploy a dexterous teleoperated system to undertake various kinds of remote operations, such as maintenance and reconfiguration of tokamak devices in JET for experimental research, inside the contaminated fusion reactor. Thus, at that time, the MASCOT system was developed from the Elsag Bailey's telemanipulator design [14]. Continuous development has been undertaken by the UKAEA, motivated by the need for maximising the efficiency of remote operations [3]. Over the last decades, the MASCOT system has been developed and deployed in remote reconfiguration of the Joint European Torus (JET) tokamak for thousands of hours. Recently, a programme of work has been undertaken to produce a new version of MASCOT, which leverages modern robotic techniques and control systems technologies [3]. The modern techniques, applied to the yet deployed next-version MASCOT, will significantly improve the control performance and operation efficiency. In the following, this section focuses on the common design concepts of MASCOT's control systems that are specifically developed for teleoperation applications.

The application of teleoperated robotic systems is a fundamental requirement for the development of fusion energy [13, 15, 16] . The fusion reaction for power generation results in various radiation, including significant neutron flux inside the reactor vessel. The contamination level of the reactor vessels increases as the fusion experiments undertake. Therefore, it is necessary to deploy robotic systems in the contaminated active environments to help achieve ALARA (As Low As Reasonably Achievable) for minimising the hazards to humans. Also, the fusion generation technologies are still at an experimental stage, which has a significant need for changes in the reactor design or deployment of different inspection instruments. Thus, it is essential to use teleoperated robotic systems, which are flexible human-in-the-loop systems replicating dexterous operations of human operators, for the possibility to undertake various inspection, maintenance and experiments. In practice, such a highly flexible robotic system needs to be deployed into the reactor vessels that are designed to retain a high vacuum environment. A teleoperated system, which can be commonly designed in a foldable manner, ensures the minimal access requirement for the fusion reactor rather than the access for human beings or typical machines.

Remarkably, the MASCOT system at the UKAEA is facilitated using an electromechanical design, specifically for enabling complex telemanipulations, within the dimensional constraints in JET. In operations, a main operator in a control room operates a local human interface of MASCOT comprising two robotic manipulators as shown in Fig. 11.1a. The remote teleoperated set-up in JET (Fig. 11.1b), therefore, is driven to follow the motions of the manoeuvred local human interface. The basic teleoperation scheme, therein, leverages the identical kinematics between the local human interface and the associated remote manipulator. It is able to achieve

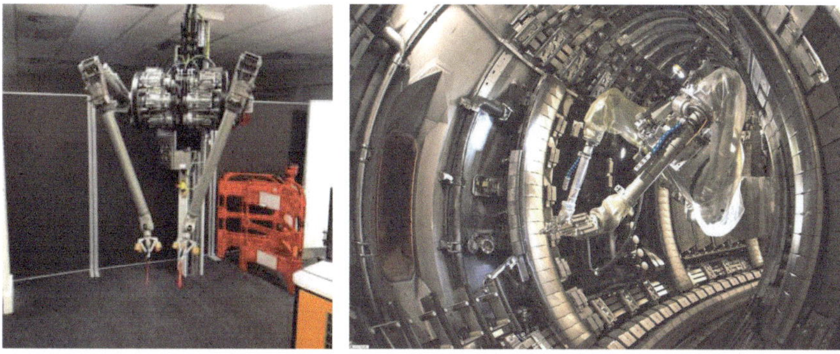

(a) Local manipulators of MASCOT at a control room

(b) Remote manipulators of MASCOT inside JET

Fig. 11.1 The MASCOT system at the UKAEA

high levels of fidelity and transparency using a simple bilateral joint position control scheme, requiring no forward or inverse kinematics calculations. Such a design concept, resulting in the ease of control system design, helps the human operators easily understand the kinematics of the remote manipulators inside the space-constrained reactor vessel, thereby, ensuring high usability and good operation safety.

The control systems and the algorithms, therein, are specifically designed and tuned to achieve high-fidelity remote control with haptic feedback. There are a few fundamental functions necessary to be implemented in the specific control system for undertaking telemanipulations [13, 17]. As aforementioned, the servomotion control design, which realises the teleoperation function with force feedback feature, employs bilateral control algorithms at each joint on both the local and remote sides. In practice, the bilateral torque control algorithm takes into account the position errors between the two sides, joint friction effects and the torque that is caused by the remote load, the local manipulation. These measurements are computed via a feedback control algorithm that utilises a proportional gain combining with a stiffness coefficient and a damping coefficient. Because the operator needs to cope with the weight loaded on the remote manipulators as a result of the bilateral control scheme, it is important to reduce the load that is transmitted from the remote side to the local side to avoid tiring the operator easily. Thus, a gravity compensation function is developed in the MASCOT system. Another key function is the ability to undertake teleoperated motions with constraints. For example, only rotating and translating motions along the bolting direction are allowed in a bolting operation. Therefore, this results in a high operation efficiency and assists avoiding collisions by constraining the telemanipulations with suitable Cartesian space limitations. Furthermore, the control system needs to allow for deploying the remote robotic manipulators automatically at specific joint positions, i.e. in a status where the remote manipulators can be easily manoeuvred to achieve the operation objectives, considering the manipulator kinematics. Specifically, in this function, the control system disables the bilateral control algorithms and drives the remote manipulators in the joint space or the Cartesian space, when the local manipulators are controlled to follow the remote manipulators' motions. Then, the bilateral control is switched on enabling the teleoperatation function, after the remote manipulators are well positioned.

There are additional features of MASCOT, necessarily extending the operation abilities to assist telemanipulations, ensuring the capability to undertake complex operations in the confined vacuum vessel. In detail, a winch with a 100 kg lifting capacity is installed at the top of the MASCOT's remote set-up, so as to allow vertical lifting of heavy loads. Power and data connectivity is designed at the remote side for the use of remotely controlled tooling. In operations, the vision information is provided by two fixed cameras at the wrists of the two remote manipulators and two 3-dof (degree-of-freedom) camera arms, which can point at various views, optically zoom the image, adjust focus and lighten the environment. Note that gaiters and the associated features are installed at the remote side of the MASCOT for contamination control. The primary method for radiation hardening in the MASCOT system is the separation of the sensitive components from the radiation environment. The manipulators (at either the remote or local side) utilise passive components for sensing and actuation (such as, resolvers, potentiometers, tachometers), all of which are connected to a control system 125 m outside the tokamak by long cables.

The local control set-up of MASCOT also includes a low-latency camera-based viewing system, a virtual reality-based viewing system, as well as a software-based graphical user interface. The MASCOT system is operated by a two-person team in its control room. Specifically, the main operator is responsible for direct manipulation in the remote environment using the local human-interface arms. The second person takes the responsibility for controlling camera views, auxiliary functions and redundant gripper release locks designed to prevent accidental dropping of components.

11.2.2 Remote Control of Industrial Robots

In the nuclear industry, there are various robotic systems that are not naturally invented for teleoperation applications, in contrast to high-fidelity systems similar to MASCOT. Typically, such robotic systems are developed for undertaking specific remote operations, such as facility decommissioning, waste storage management or accidents response in the nuclear industry (see examples in the reviews [18–20]). The development of such a robotic system, which is typically designed for completing one specific mission, requires rigorous development procedure considering the detailed operation and environmental requirements. Due to the significant improvement of the modern industrial robots [21], there is a growing interest in applying industrial robots, which are not natively designed for remote operations, in various nuclear applications. For instance, since 2015, the UK National Nuclear Laboratory (NNL) has initiated several research and development (R&D) projects motivated to remotely control industrial robots undertaking nuclear waste handling operations [22]. Moreover, in response to the national calls of the UK, two national research hubs, RAIN and NCNR [23, 24] and one major research program [25] grant have been established to exploit advanced robotic techniques in different nuclear applications. As a result, various prototype systems employing industrial robots have been developed for different remote applications, such as nuclear waste storage inspection or decommissioning. For instance, Fig. 11.2a shows a high-payload industrial robot designed to undertake remote handling operations, and Fig. 11.2b represents a dual-arm teleoperated robotic system developed for lightweight operations in nuclear gloveboxes.

In practice, there is still a wide range of operations undertaken by human operators in the nuclear industry. These kinds of operations, such as many manufacturing operations inside gloveboxes [26] or waste storage inspection operations [27], share the following common characteristics. (a) There is a significant workload. For instance, it is necessary to keep monitoring the conditions of a large amount of nuclear waste at temporary storage facilities to avoid accidents due to the material degradation of the storage containers. (b) Operation environments may vary over time. The conditions of a waste-drum storage may change due to the corrosion or deformation of the drums. (c) Special expertise is needed in complex operation sequences. The operations to repackage plutonium canisters, for example, consist of a series of delicate operations using different tools inside gloveboxes. (d) The environment typically has a low requirement for radiation tolerance, allowing for radiation-metered human operations.

Considering the aforementioned common facts in these operations, advanced robotic systems, which employ industrial robots to facilitate remote/teleoperated applications, are regarded as the future solutions for an improved safety and efficiency subject to a reduced cost [22, 27]. Specifically, experienced operators can remotely control the industrial robots staying away from the radiation hazards. This allows for easily employing the special expertise of the operators to tackle the varying operation conditions. Therefore, it improves the operation safety and reduces the mental stress by keeping human operators away from the radiation hazards. Industrial robots have been created for undertaking heavy work tasks, and they are comparably radiation hardened in relation to human operators. Thus, the remotely controlled industrial robots can undertake designated operations without performance degradation for a long period, i.e. resulting in significant efficiency improvement.

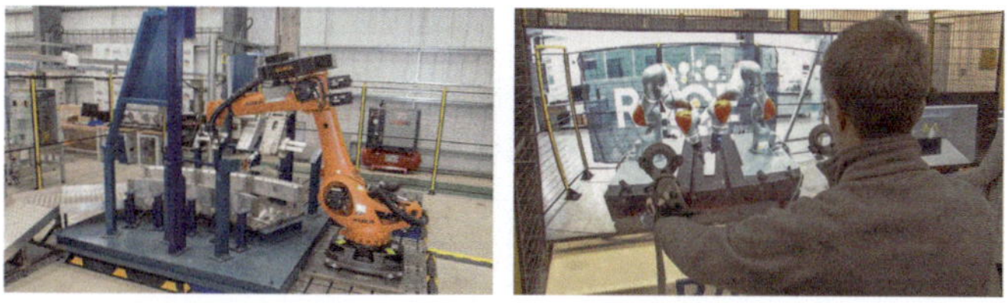

(a) A remote control system to operate a high-payload industrial robot

(b) A dual-manipulator system developed for glovebox operations

Fig. 11.2 Remotely controlled industrial robots for nuclear applications at the UKAEA

Importantly, the R&D of advanced control approaches is fundamental to realise remote control of industrial robots for such nuclear applications, e.g. glovebox operations [23], storage inspection [28] and waste handling [29]. The control solutions for such a robotic system typically comprises four elementary parts: *(a) remote manipulation device, (b) local control device, (c) supportive feedback system and (d) control system*. For instance, Fig. 11.3 illustrates a real-time control system is developed to allow human operators for remotely controlling an industrial robot manipulator [28]. A modular sensor unit, including a radiation meter and a distance sensor [30], is attached to the robot's flange, thereby, enabling remote radiation inspection operations.

(a) The *remote manipulation device* is the core actuation unit to be deployed intentionally in specific radiation environments (e.g. the industrial robot with its controller in Fig. 11.3). The key design concept is the suitability and capability of accomplishing the operations, which the remote manipulation device is designed for. Especially in nuclear applications, the adopted manipulators need to be sufficiently radiation tolerant, i.e. guaranteeing a sufficient service life, considering the radiation strength in its deployed environments. Thus, the radiation tolerance of an industrial robot is suggested to be evaluated via systematical approaches (see [5] as an example).
(b) The main purpose of a *local control device* is to create commands sent to the manipulation device. The commands, which typically include the demanded position, orientation or interaction with the environment/workpiece, can be generated by joy sticks [22], a remote controller [28] (in Fig. 11.3), hand gestures captured by optical devices [31] or even verbal inputs [32]. Note that a local control device sometimes needs to provide the operator a pseudosense (e.g. a pseudosense of force is created by the local controller in [33]), so suitable commands can be determined according.
(c) As the manipulation device is operated in a remote manner, *supportive feedback* is needed for providing the operators with sufficient information to determine the operation commands. Vision information is a basic supportive feedback. A widely applied solution providing vision data is to set up a series of cameras, maybe with pan–tilt–zoom functions, capturing live views of the operations

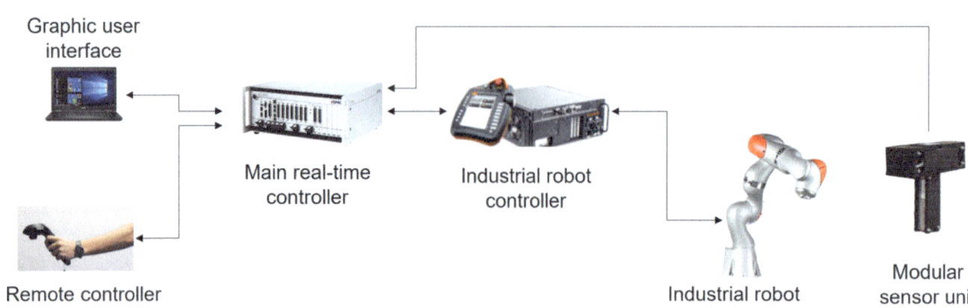

Fig. 11.3 Control system design of a remote inspection system as an example

displayed on screens [22]. Motivated to provide 3D information, virtual reality (VR) techniques can be used to create a virtual representation using the 3D measurement (from an RGB-D camera [34] for instance) of the environment. Besides the essential vision information, some remote applications also have to collect other information, such as radiation dose or temperature, which are important data for sample inspection or waste management [28, 30] (see the modular sensor unit in Fig. 11.3 as an example).

(d) The *control system* integrates all the aforementioned functional parts as a whole system. Typically, the control system design considers the operation safety and control performance according to specific operation requirements. As the safety performance is defined in the associated standards (see examples in [35]), the major investment commonly applies to the R&D activities that facilitate the demanded control functionalities. The infrastructure of a control system, such as the choice of controller and communication buses, determines the maximum capability of the robotic system. For instance, in [28], a real-time controller running at 1 kHz is required for controlling a lightweight industrial robot that is designed to achieve delicate torque-based motion control at high precision. Nevertheless, motivated to ensure high operation efficiency, it is important to develop suitable control algorithms for a robotic system deployed in designated applications. For example, in nuclear waste classification applications, an adaptive control algorithm is developed to assist verifying the dimensional information about a robot's grasped payload, which is essential for post-processing yet typically unknown a priori [36]. Also, in waste classification systems, the grasping algorithm is a key technology, impacting a system's efficiency and safety (see the automated optimal grasping algorithm in [37] for instance).

11.3 Distributed Infrastructure for Control and Robotics

Whilst the development of advanced robotic systems capable of supporting functions inside a civil nuclear plant is pursued, ancillary and supporting infrastructure is necessary for its appropriate functioning. As robotic systems add both functionality and complexity to a system, proper management of its operational cycle is necessary; this includes task planning, programming, task verification and visualisation of the robot's operation. In addition, networking infrastructure is crucial for a robotic system's functioning, connecting the control processing unit with the robot's moving parts; as mixed wiring solutions for harsh environments are complex and expensive, solutions to this problem such as wireless connectivity need to be further developed. These two aspects will be explored in this section.

11.3.1 Robotics Control Platform—CorteX

Reduction of through-life cost of robotic systems in nuclear facilities is of key importance. This can be achieved through the use of standardised architectures and interfaces which promote improved modularity and ease of integration, as well as reducing training costs through standardisation of human interfaces in line with industrial standards and accessibility guidelines.

Traditional Object-Oriented methods go part way towards addressing these challenges through providing strong modularity by encapsulation and hiding the inner-workings of subsystems. Abstraction and inheritance allow some level of interoperability by allowing assumptions to be placed on the behaviour of a novel child module, by making use of its inherited interfaces (polymorphism).

CorteX [38] is a software and data framework that attempts to solve the main problems associated with interoperable, plug-and-play, distributed robot systems-of-systems, at least from a data/communications perspective. Cortex can be considered as:

- A standardised graphical data representation for robotic systems (Fig. 11.5a)
- A method for communicating this representation (Fig. 11.4)
- A software framework that implements the above
- Additional software tools to add further useful functionality (Fig. 11.5b).

The standardised representation includes definitions of the data and commands available in modular units of the system, described using a name-value pair method that allows a generic description to be used for all modules. In software, these can be represented as, for example, C++ classes, and they can be communicated via standard network protocols, allowing for distribution of systems.

Improved interoperability can be gained through enhancing this by standardising type definitions using an ontological type system which includes structural and morphological rules defining expected structure and behaviour. In CorteX this is achieved whilst still allowing for a wide range of functionality using the standardised data representation. In contrast to other robotics middlewares such as ROS, CorteX facilitates all functionality without allowing bespoke or custom messages which reduce interoperability.

The standard CorteX data representation can be transferred across communications networks, forming a communication protocol, thus CorteX can be considered as middleware agnostic or as middleware-over-middleware. The architecture is compatible with a number of underlying communications architectures including publish-subscribe and service-oriented architectures. The current implementation makes use of DDS with options for using shared memory locally, and there are plans to extend to other protocols.

One useful by-product of self-description and the standardised data representation is the ability for a system to dynamically adjust to the configuration of a remote system. A Graphical User Interface (GUI) can be generated at runtime based on pre-defined rules for interpreting the remote system. This allows for standardisation

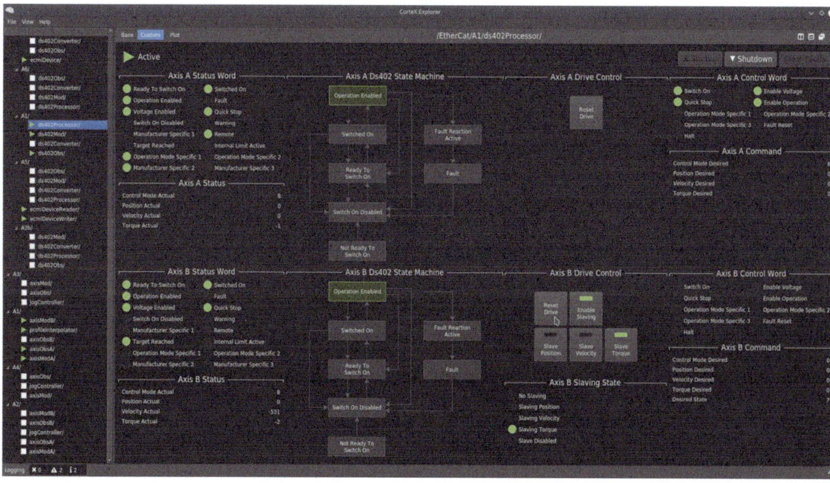

Fig. 11.4 CorteX explorer interface displaying the operation of a robot and a task description as a state machine diagram

of human factors, alignment with accessibility standards and achieving a consistent look-and-feel across the whole facility, reducing training needs.

The system has been developed for generic operations including in fusion energy remote maintenance and nuclear decommissioning operations. A range of deployments have been made in lab-based trials, including the control of the extremely large TARM (Telescopic Articulated Remote Mast) manipulator [39] at Culham. Further development and active deployments are planned for future work.

11.3.2 Wireless Nodes for Control—ICON

Reliable infrastructure is necessary to power and communicate the different devices comprising a nuclear energy system. Such infrastructure has traditionally included specialised shielded wires designed to withstand harsh environments (i.e. high temperature and radiation levels). Wired solutions have proven feasible to both power and communicate any physical asset, with high costs and complexity related to their deployment and maintenance. The use of wireless communication within the civil nuclear industry can bring many benefits over wired solutions, such as reducing lifecycle costs and enabling new applications in asset and process management (e.g. improve asset integrity, enable intelligent monitoring). If self-powered devices are considered together with wireless devices, the benefits mentioned before are extended to assets and devices in places hard to access and maintain. These benefits could apply to new builds, existing plants and decommissioning setups alike.

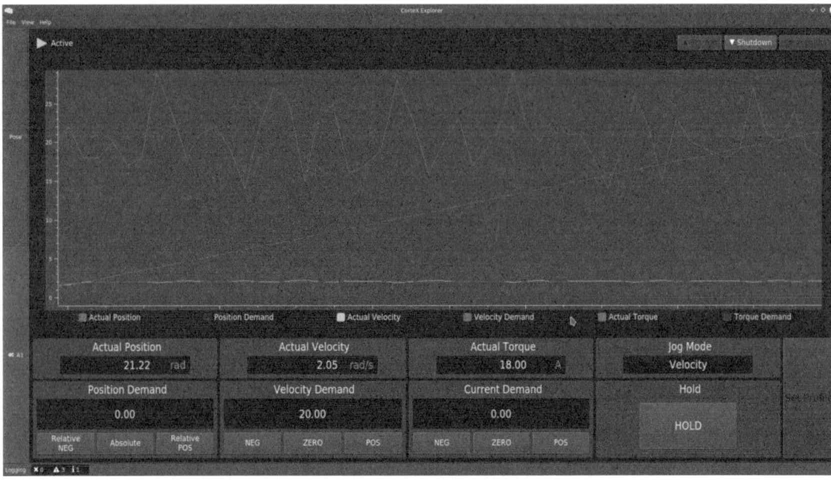

(**a**) Generic tools such as plotting data are available to all systems.

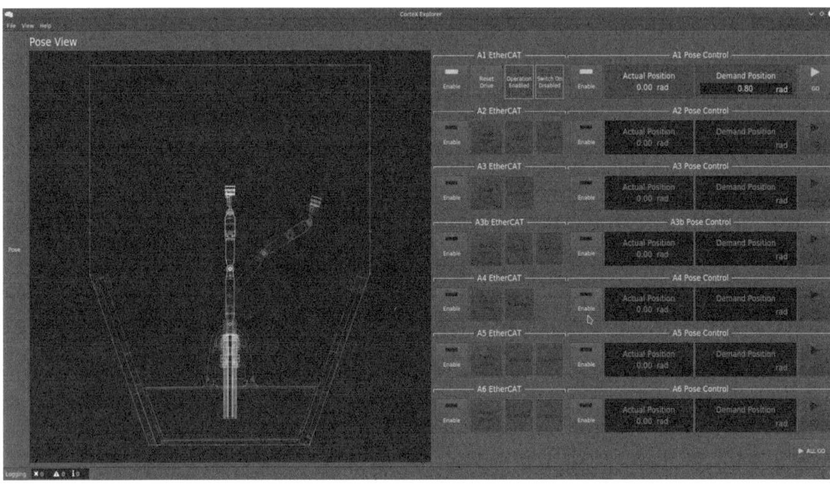

(**b**) Using custom device-specific interfaces as plugins requires minimal effort and re-validation.

Fig. 11.5 CorteX allows for the automated, procedural generation of standardised GUI elements and layouts allowing for a consistent look-and-feel and ease of use

Creating distributed wireless nodes to enable robust and safe remote control of, generally lower risk, nuclear asset operations and robots is the aim of the 'Intelligent Control for Nuclear—ICON' [40] between two industrial stakeholders (Altran Ltd (now Capgemini), Moltex Ltd) and an academic institution (University of Bristol, previous affiliation of Lopez & Herrmann); its follow-on project between brings together an even larger consortium of industrial and academic stakeholders. Using an assortment of wireless transceivers connected to each other, resilient mesh networks can be created to ensure data transmission. This mesh would communicate wirelessly to a gateway device that allows a user to both inspect and control the robotic device (Fig. 11.6). The wireless nodes would receive control signals, commands or control set-points that would drive the robot itself, whilst being real time monitored by the user in the control room or an automated algorithmic system. The development of such a technological solution comes with different challenges currently being faced, which detail is expanded down below.

Most wireless applications within the nuclear power industry so far have been placed outside the reactor, within environments where temperature and radiation are less harsh compared to within the reactor. When describing a wireless application, they involve the transmission of data using transceivers that both sense and generate electromagnetic waves. These applications include sensors providing background radiation and personal communication devices used by human operators. As described in the guidelines available in the technical report IEC/TR 62918:2014 [41], wireless communications are recommended to be used in non-safety-critical and C&I systems; following the standards mentioned during the introduction [7, 8], wireless communications shall not be used in systems supporting Category A and B functions, and their inclusion for Category C functions shall be done only after proper suitability analyses (e.g. feasibility tested in [40]).

A guiding principle for wireless system design is understanding the deployed environment, including any environmental considerations and its physical layout. Regardless of its robustness and ease of use, wireless technology performance is sensitive to environmental considerations including, but not limited to temperature, humidity and physical layout (e.g. walls and objects around and between transceivers, their location, material and geometry). These effects have been studied and understood (e.g. temperature and humidity in wireless propagation [42], electromagnetic environment [43]) hence its importance. This highlights the role of careful design and simulation [44] of wireless solutions, which require strong knowledge of the deployed environment to understand and plan for its optimal performance.

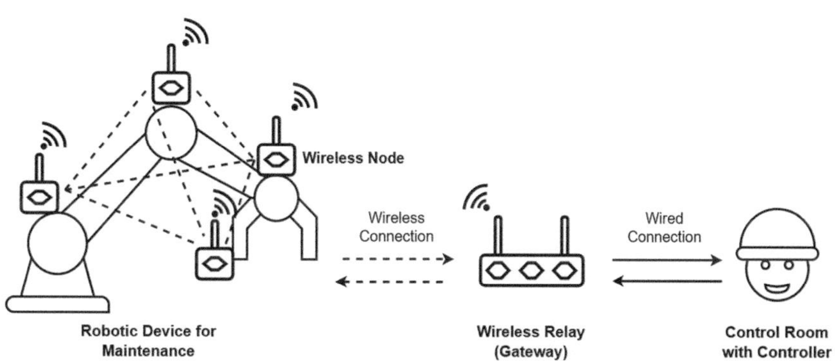

Fig. 11.6 Proposed communication architecture for distributed robotics control using wireless nodes

Following the importance of environmental considerations, a strong safety concern regarding wireless system's deployment is them producing electromagnetic interference and radio-frequency interference (EMI/RFI) that leads to unexpected failure of C&I equipment. Documented cases of false alarms and faulty performance related to RFI do exist [45] as many legacy devices has not been designed or even tested for EMI/RFI. However, careful design and testing of equipment can avoid these problems, together with exclusion zones from wireless deployments for safety-critical legacy equipment [45].

The inclusion of wireless technologies in safety-critical systems [46] is an active R&D topic, which brings many additional challenges such as extra failure modes, cybersecurity concerns and regulatory challenges. To bring wireless technologies to more Category C functions [40] and to create a case for inclusion in Categories A and B in nuclear power plants, a set of complex requirements need to be met. Deterministic behaviour of the timings and performance of the network (i.e. latency constrained) is a significant main challenge for these networks. A high link availability or dedicated link with deterministic timings is needed, meaning connection and reconnection happening in deterministic time; low delay is more important than high throughput, considering that payload for control messages can be less than 1 kilobyte long. Inside such a network, various fixed sampling periods between signal inputs and outputs are suited and desired to control multiple-input multiple-output plant systems or various plants at the same time. Delays should be bounded to a fraction of the periodic sampling period, as large delay values can be manageable as long as they follow a bounded deterministic behaviour.

Most current wireless sensor protocols [47] for wireless sensor networks [48] do not include a mechanism to implement tightly coupled, deterministic networks. A popular option for this is the IEEE 802.15.4 standard [49], which revisions starting from 802.15.4e onwards include changes in the MAC modes protocol that improve the overall reliability, robustness and latency of the network. Time-slotted Channel Hopping (TCH) and Deterministic and Synchronous Multi-channel Extension (DSME) MAC modes provide dedicated and reliable communication between devices by using techniques such as multi-channel frequency hopping and time slotted scheduled communication. These modes improve on the IEEE 802.15.4 to produce deterministic low-latency networks (also suggested for [40]). The 802.15.4e standard, initially published in 2012 [50], is still under continuous development, specifying the implementations of a variety of further mechanisms inside the MAC modes, which provides room for further improvement and customizations [51, 52].

11.3.3 Open Challenges

Although much interest has been given to the use of robots in nuclear civil applications, many challenges around its development and deployment remain. The most common one is survivability of the robotic system, as harsh environments typical in nuclear applications can reduce the robot's shelf life to just a tenth of its nominal value. Environmental factors reducing the life of typical off-the-shelf devices include temperature, humidity and neutron radiation, with the most significant impacting factor in both fusion and fission use cases is the exposure to gamma radiation. Gamma radiation can have various effects on semiconductor electronics, optical glasses, polymers and lubricants. In high gamma environments, it results in reduced survivability of typical devices such as encoders, cameras, LIDAR and microprocessors as the power electronics inside these are particularly vulnerable to dosage [53]. The effects of such exposure manifests in faulty or intermittent operation, with various effects including single interactions (i.e. sudden damage to components) and accumulated dose (i.e. reduced performance of components leading to failure). Testing methodologies for COTS systems have been developed [5], but due to the technical complexity of modern robotic systems, both finding the failing module and repairing it is many times unfeasible. The use of radiation-hardened components and shielding for traditional components to increase their radiation tolerance are still the most popular solutions to increase shelf life of electronics in civil nuclear and aerospace industries. However, cost, weight and geometrical constraints limit the feasibility of these methods, as components are costly, electronic redesign is expensive and shielding increase the overall size and weight of the system. Significant work is required in the development of new materials and new designs of components to enable modern robotics to be conveniently deployed in extreme environments.

Besides the challenges related to a robotic system's physical deployment (e.g. installation, powering, maintenance and refurbishment), further development is needed in the interfaces available to control and inspect the robot. The needs and well-being of the human user have to be considered whilst controlling the robot [54], regardless of the role the human is performing (e.g. operator, inspector, plant manager). Aspects of Human Robot Interaction (HRI) are needed, which go beyond user interface design and verification. The overall experience of using a robotic device must be designed, verified and validated around task performance and compliance with both performance and safety criteria. These aspects should complement any training and certification processes performed for new users and maintainers of the system.

Robots need to be deployed together with a connectivity infrastructure, either wired or wireless [46]. Such connectivity introduces additional failure modes related to data transmission (e.g. delayed and lost data, loss of connectivity, critical failure for non-resilient controllers). Connecting civil nuclear plants to public networks such as the internet enables many useful capabilities, such as remote supervision and intra-plants control. However, major security concerns still exist as it enables malicious access to critical infrastructure, leading to potential safety threats to larger populations if the plant's operation is remotely manipulated.

Within fusion energy and other applications dealing with extreme environments, reliability, safety and hence verification and assurance are highly important. Where human access is permissible, the design of robot cells must consider conventional safety [55]. Where nuclear hazards are present, safety processes are usually validated by a regulating institution [56]. Additionally, the robustness and reliability of any control system is critical to ensuring both the protection of investments such as the robots and the facilities in which they operate, and the productivity and operational efficiency of the plant [18]. Traditional verification techniques that make use of probabilistic or deterministic arguments do not provide the tools required to guarantee the performance of modern data-driven, learning-based and probabilistic robotic control and perception algorithms yet [57]. If the benefits of state-of-the-art robotics and AI techniques are to be realised in these facilities, which is highly desirable, then new ways to assure and build trust in these methods must be developed.

Nevertheless, it is worth noting the importance of leveraging merging technologies, such as modern sensing and actuating devices, in both nuclear fission and fusion applications. For instance, in the continuous development work on MASCOT, the performance and payload of the existing system can be significantly improved by replacing the legacy motors with modern motors [3]. Also, new robotic systems can be designed with a high radiation tolerance, due to the development of radiation hardening techniques (e.g. [58]) and the improved understanding of radiation effects (e.g. [59]). Moreover, driven by the current industrial demands, the modern sensing and actuating devices are designed in a modular manner, which allows implementing systems in compact size at ease of maintenance [60, 61]. Prospectively, in nuclear applications, it is significantly beneficial for future robotic systems to embrace the emerging technologies, thereby, permitting to improve legacy systems and develop superior new systems.

11.4 Summary

Future nuclear power systems are to be enabled by new operational systems that increase efficiency, improve safety to human service staff, lower downtime and enable later decommissioning. This has to be achieved through novel control infrastructure and robotic systems, initially tested and limited to plant functions of auxiliary nature, impacting safety the least, i.e. Safety Category C. In this sense, this chapter has revised one of the oldest robotic service systems, the teleoperated MASCOT-robot for servicing JET, i.e. in a nuclear fusion environment. MASCOT has served as an inspiration and initial seed for other robotic and control systems in the nuclear sector, leading also to the technical elevation of COTS-based systems to Nuclear Devices through focused Research & Development, specifically for robotic systems (e.g. RAIN) and control infrastructure (e.g. CORTEX & ICON). In this sense, technical challenges within the domain of radiation tolerance/hardening, security, safe HRI-based teleoperation, sensing and actuation remain.

Acknowledgements This work was supported by the EPSRC's Robotics and Artificial Intelligence for Nuclear project—(RAIN) [EP/R026084/1], the Innovate UK ICON project [103908] (see also [40]) and the SPARC collaboration fund [1051].

References

1. J. Iqbal, A.M. Tahir, R. Ul Islam, Riaz-Un-Nabi, Robotics for nuclear power plants-challenges and future perspectives. In *2012 2nd International Conference on Applied Robotics for the Power Industry, CARPI 2012* (IEEE Computer Society, 2012), pp. 151–156
2. R. Buckingham, A. Loving, Remote-handling challenges in fusion research and beyond. Nat. Phys. **12**(5), 5 (2016)
3. R. Skilton, N. Hamilton, R. Howell, C. Lamb, J. Rodriguez, MASCOT 6: achieving high dexterity tele-manipulation with a modern architectural design for fusion remote maintenance (2018)
4. ONR, ONR GUIDE safety systems document type: nuclear safety technical assessment guide. Technical report, Office for Nuclear Regulation (2018)
5. K. Zhang, C. Hutson, J. Knighton, G. Herrmann, T. Scott, Radiation tolerance testing methodology of robotic manipulator prior to nuclear waste handling. Front. Robot. AI **7**, 6 (2020)
6. ONR, ONR GUIDE Categorisation of safety functions and classification of structures, systems and components. Technical report, Office for Nuclear Regulation (2019)
7. BSI Standards Publication, Nuclear power plants—Instrumentation and control important to safety—Selection and use of industrial digital devices of limited functionality (2010)
8. BSI Standards Publication, Functional safety of electrical/electronic/programmable electronic safety related systems Part 7: overview of techniques and measures (2010)
9. BSI Standards Publication, Robots and robotic devices—Safety requirements for industrial robots, part 1: Robots (ISO 10218-1:2011) (2014)
10. BSI Standards Publication, Robots and robotic devices guide to the ethical design and application of robots and robotic systems (2016)
11. BSI Standards Publication, PD ISO/TS 15066:2016 Robots and robotic devices—Collaborative robots (2016)
12. R. Smith, E. Cucco, C. Fairbairn, Robotic development for the nuclear environment: Challenges and strategy. Robotics **9**(4), 94 (2020)
13. L. Galbiati, T. Raimondi, P. Garetti, G. Costi, Control and operational aspects of the Mascot 4 force feedback servomanipulator of JET, in *IEEE/NPSS Symposium on Fusion Engineering* (1991)
14. cyberneticzoo.com, 1958—MASCOT Remote Servo-manipulator—Carlo mancini et al. (Italian) (2014)
15. M. Keilhacker, A. Gibson, C. Gormezano, P.H. Rebut, The scientific success of jet. Nucl. Fusion **41**(12), 1925 (2001)
16. R. Buckingham, Towards a remote handling toolkit for fusion: lessons learnt and future challenges-17360, in *Towards a Remote Handling Toolkit for Fusion: lessons Learnt and Future Challenges-17360* (2017)
17. D. Hamilton, G. Preece, Development of the MASCOT telemanipulator control system (European Fusion Development Agreement, Project, 2001)
18. F.E. Gelhaus, H.T. Roman, Robot applications in nuclear power plants. Progr. Nucl. Energy **23**(1), 1–33 (1990)
19. Y. Yokokohji, The use of robots to respond to nuclear accidents: applying the lessons of the past to the fukushima daiichi nuclear power station. Ann. Rev. Control Robot. Auton. Syst. **4** (2020)
20. J. Iqbal, A.M. Tahir, R. ul Islam, et al., Robotics for nuclear power plants—Challenges and future perspectives, in *2012 2nd International Conference on Applied Robotics for the Power Industry (CARPI)* (IEEE, 2012), pp. 151–156
21. J.F. Engelberger, *Robotics in Practice: management and Applications of Industrial Robots* (Springer Science & Business Media, 2012)
22. National Nuclear Laboratory, National nuclear laboratory news (2015). https://www.nnl.co.uk/news/. Accessed 13 Oct 2019
23. Robotics and AI in Nuclear, Demonstration & testing (2018), https://rainhub.org.uk/demonstration-and-testing/. Accessed 01 Mar 2021

24. National Centre for Nuclear Robotics, National centre for nuclear robotics: home (2018), https://www.ncnr.org.uk/. Accessed 01 Mar 2021
25. Robotics for Nuclear Environments, Robotics for nuclear environment (2018)
26. M. Rentetzi, Determining nuclear fingerprints: glove boxes, radiation protection, and the International Atomic Energy Agency. Endeavour (2017)
27. Nuclear Decommissioning Authority. Cleaning up our nuclear past: faster, safer and sooner (2021), https://nda.blog.gov.uk/. Accessed 15 Oct 2019
28. C. Hutson, A. Banos, D. Connor, K. Zhang, S. Kaluvan, S. White, Y. Verbelen, M.R. Tucker, J. Knighton, K. Wood et al., Development of robotic inspection systems for in-situ characterisation prior to decommissioning-20306, in *WM 2020 Conference. Waste Management 2020 Conference* (2020)
29. N. Marturi, M. Kopicki, A. Rastegarpanah, V. Rajasekaran, M. Adjigble, R. Stolkin, A. Leonardis, Y. Bekiroglu, Dynamic grasp and trajectory planning for moving objects. Auton. Robots **43**(5), 1241–1256 (2019)
30. S.R. White, D.A. Megson-Smith, K. Zhang, D.T. Connor, P.G. Martin, C. Hutson, G. Herrmann, J. Dilworth, T.B. Scott, Radiation mapping and laser profiling using a robotic manipulator. Front. Robot. AI **7** (2020)
31. I. Jang, J. Carrasco, A. Weightman, B. Lennox, Intuitive bare-hand teleoperation of a robotic manipulator using virtual reality and leap motion, in *Annual Conference Towards Autonomous Robotic Systems* (Springer, 2019), pp. 283–294
32. A. Ghosh, D.A.P. Soto, S.M. Veres, J.A. Rossiter, Human robot interaction for future remote manipulations in industry 4.0, in *IFAC-PapersOnLine. International Federation of Automatic Control (IFAC)* (2020)
33. F. Abi-Farraj, C. Pacchierotti, O. Arenz, G. Neumann, P.R. Giordano, A haptic shared-control architecture for guided multi-target robotic grasping. IEEE Trans. Haptics **13**(2), 270–285 (2019)
34. J.M. Aitken, S.M. Veres, A. Shaukat, Y. Gao, E. Cucco, L.A. Dennis, M. Fisher, J.A. Kuo, T. Robinson, P.E. Mort, Autonomous nuclear waste management. IEEE Intelligent Syst. **33**(6), 47–55 (2018)
35. G. Virk, S. Cameron. ISO-IEC standardization efforts in robotics, in *Proceedings of the Workshop on Standardized Knowledge Representation and Ontologies for Robotics and Automation*, pp. 5–6 (2014)
36. S. Wang, K. Zhang, G. Herrmann, An adaptive controller for robotic manipulators with unknown kinematics and dynamics. IFAC-PapersOnLine **53**(2), 8796–8801 (2020)
37. A. Altobelli, O. Tokatli, G. Burroughes, R. Skilton, Optimal grasping pose synthesis in a constrained environment. Robotics **10**, 1–22 (2021)
38. I. Caliskanelli, M. Goodliffe, C. Whiffin, M. Xymitoulias, E. Whittaker, S. Verma, C. Hickman, R. Skilton, CorteX: a software framework for interoperable, plug-and-play, distributed robotic systems-of-systems, in *Software Engineering for Robotics*, eds. by A. Cavalcanti, J. Woodcock (Springer, 2020)
39. T. Raimondi, The jet experience with remote handling equipment and future prospects (1989)
40. ICON, Gtr-Intellligent Control (fOr efficient Nuclear applications) (2018)
41. BSI Standards Publication. Nuclear power plants—Instrumentation and control important to safety—Use and selection of wireless devices to be integrated in systems important to safety (2014)
42. J. Luomala, I. Hakala, Effects of temperature and humidity on radio signal strength in outdoor wireless sensor networks, in *2015 Federated Conference on Computer Science and Information Systems (FedCSIS)* (2015), pp. 1247–1255
43. Z. Changlin, Z. Zhan, Q. Xuebing, Y. Hongtao, Z. Weidong, Research on the electromagnetic environment effect on wireless communication systems, in *2008 8th International Symposium on Antennas, Propagation and EM Theory* (2008), pp. 1478–1481
44. M. Cappelli, V. Lopresto, R. Cecchi, G. Marrocco, Application of wireless technologies in a nuclear plant: evaluation of electromagnetic propagation with different computational techniques. J. Nucl. Eng. Radiat. Sci. **6**(2), 021112 (2020)

45. D.Y. Ko, S.I. Lee, Applicable approach of the wireless technology for Korean nuclear power plants. Nucl. Eng. Des. **265**, 519–525 (2013)
46. A.B. Johnston, W. Schiffers, A.H. Kharaz, Towards wireless technology for safety critical systems. J. Phys.: Conf. Ser. **1065**, 72032 (Institute of Physics Publishing, 2018)
47. J. Song, S. Han, A. Mok, D. Chen, M. Lucas, M. Nixon, W. Pratt, WirelessHART: applying wireless technology in real-time industrial process control, in *2008 IEEE Real-Time and Embedded Technology and Applications Symposium* (IEEE, 2008), pp. 377–386
48. B. Ataul, J. Jin, A review of wireless sensor networks for nuclear power plant applications. Nucl. Saf. Simul. **6**(1) (2015)
49. IEEE, IEEE Std 802.15.4-2015 (Revision of IEEE Std 802.15.4-2011): IEEE Standard for Low-Rate Wireless Networks (IEEE, 2016)
50. IEEE Standard for Local and metropolitan area networks–Part 15.4: low-rate wireless personal area networks (LR-WPANs) Amendment 1: MAC sublayer (2012)
51. IEEE Stand. Low-Rate Wirel. Netw. (2020)
52. N. Choudhury, R. Matam, M. Mukherjee, J. Lloret, A performance-to-cost analysis of IEEE 802.15.4 MAC with 802.15.4e MAC modes. IEEE Access **8**, 41936–41950 (2020)
53. B. Todd, S. Uznanski, Radiation risks and mitigation in electronic systems, in *CERN Accelerator School: Power Converters, CAS 2014—Proceedings* (CERN, 2018), pp. 245–263
54. M. Talha, Human factors issues in telerobotic decommissioning of legacy nuclear facilities. Ph.D. thesis (University of Birmingham, 2019)
55. P. Chemweno, L. Pintelon, W. Decre, Orienting safety assurance with outcomes of hazard analysis and risk assessment: a review of the ISO 15066 standard for collaborative robot systems. Saf. Sci. **129**, 104832 (2020)
56. Wood, Research supporting regulatory guidance for new technologies and new materials (2020), https://www.onr.org.uk/documents/2020/onr-rrr-076.pdf. Accessed 05 July 2021
57. A. Gannous, A. Andrews, B. Gallina, Toward a systematic and safety evidence productive verification approach for safety-critical systems, in *2018 IEEE International Symposium on Software Reliability Engineering Workshops (ISSREW)* (IEEE, 2018), pp. 329–336
58. P. Leroux, W. Van Koeckhoven, J. Verbeeck, M. Van Uffelen, S. Esqué, R. Ranz, C. Damiani, D. Hamilton, Design of a MGY radiation tolerant resolver-to-digital convertor IC for remotely operated maintenance in harsh environments. Fusion Eng. Des. **89**(9–10), 2314–2319 (2014)
59. J.T. Howard, E.J. Barth, R.D. Schrimpf, R.A. Reed, L.C. Adams, D. Vibbert, A.F. Witulski, Methodology for identifying radiation effects in robotic systems with mechanical and control performance variations. IEEE Trans. Nucl. Sci. **66**(1), 184–189 (2018)
60. M. Grebenstein, A. Albu-Schäffer, T. Bahls, M. Chalon, O. Eiberger, W. Friedl, R. Gruber, S. Haddadin, U. Hagn, R. Haslinger et al., The DLR hand arm system, in *2011 IEEE International Conference on Robotics and Automation* (IEEE, 2011), pp. 3175–3182
61. A. Albu-Schäffer, S. Haddadin, C. Ott, A. Stemmer, T. Wimböck, G. Hirzinger, The DLR lightweight robot: design and control concepts for robots in human environments. Ind. Robot: Int. J. (2007)

Author Index

U. Mangalanathan et al. (eds.), *Nuclear Instrumentation and Control*,
https://doi.org/10.1007/978-981-97-1283-0

Index

U. Mangalanathan et al. (eds.), *Nuclear Instrumentation and Control*,
https://doi.org/10.1007/978-981-97-1283-0

P

R

S

T

W

X

MIX
Papier aus verantwortungsvollen Quellen
Paper from responsible sources
FSC® C105338

If you have any concerns about our products,
you can contact us on
ProductSafety@springernature.com

In case Publisher is established outside the EU,
the EU authorized representative is:
Springer Nature Customer Service Center GmbH
Europaplatz 3, 69115 Heidelberg, Germany

Printed by Libri Plureos GmbH
in Hamburg, Germany